T0177303

Mathematics and Logic in History and in Contemporary Thought

Mathematics and Logic in History and in Contemporary Thought

Ettore Carruccio
Translated by Isabel Quigly

Routledge
Taylor & Francis Group

LONDON AND NEW YORK

First published 2006 by Transaction Publishers

Published 2017 by Routledge
2 Park Square, Milton Park, Abingdon, Oxon OX14 4RN
711 Third Avenue, New York, NY 10017, USA

Routledge is an imprint of the Taylor & Francis Group, an informa business

Copyright © 1964 by Faber and Faber Limited.

All rights reserved. No part of this book may be reprinted or reproduced or utilised in any form or by any electronic, mechanical, or other means, now known or hereafter invented, including photocopying and recording, or in any information storage or retrieval system, without permission in writing from the publishers.

Notice:
Product or corporate names may be trademarks or registered trademarks, and are used only for identification and explanation without intent to infringe.

Library of Congress Catalog Number: 2006042890

Library of Congress Cataloging-in-Publication Data

Carruccio, Ettore.
　　[Matematica e logica nella storia e nel pensiero contemporaneo. English]
　　Mathematics and logic in history and in contemporary thought / Ettore Carruccio ; translated by Isabel Quigly.—1st pbk. printing.
　　　　p. cm.
　　Includes bibliographical references and index.
　　ISBN 0-202-30850-2 (alk. paper)
　　1. Mathematics—History. 2. Mathematics—Philosophy. 3. Logic—History. I. Title.

QA21.C2483　2006
510.9—dc22　　　　　　　　　　　　　　　　2006042890

ISBN 13: 978-0-202-30850-0 (pbk)

Contents

Contents

CHAPTER I

The Meaning, Purpose and Methods of the History of Mathematics and of Logic

§1. *The meaning of the history of science, and the object of studying the history of mathematics*

The history of science, of which the history of mathematics is a branch, is still often considered a collection of the results of scientific enquiry through the ages, one which establishes the relative importance of discoveries and inventions and the outward facts in the life and work of scientists: in fact, a museum of documents and scientific curiosities. But this is not really the definition that applies in the case of this book: here, the history of science is identified with science itself considered in its development, in the thought of those who constructed it, and in its relation with the various aspects of culture and of human life (1). Scientific problems in general and mathematical problems in particular show their full meaning when they are considered in their own history (2). The need for a historical view arises mainly in times of crisis, when scientific thought is being renewed. As Enriques (3) writes: 'The historical view of science appears above all as a dynamic view, which can measure the future by the past; and it is therefore the answer to the idealism that, consciously or unconsciously, inspires the efforts of researchers, who never pause in a rigid present but are ever striving to break out of it.'

In teaching, a knowledge of the history of mathematics not only enables the teacher to give his pupils cultural analogies and to enliven the theoretical side of the work, but provides a

(1) See Enriques, **13** [such references as this are to the bibliography, at the end of this book].
(2) See ch. XII, §7, of this book.
(3) See Enriques, **14**, p. 4.

9

greater understanding of the way in which thought builds up its mathematical world. Of course there is not enough time to give the historical background in all its detail, nor would young people be capable of profiting from it; but the various phases in the development of mathematics, from the rules of empirical origin to the gradual conquest of rigorous logic, can be brought alive in teaching, from primary school to university, not literally perhaps but in spirit (4), if Vico's relationship between the development of civilization and the development of the human personality is borne in mind.

A broad historical understanding of the mathematical problems undertaken and of what led to the construction of rational theories is of great value to the student's intellectual development: it takes him to the heart of mathematical questions, and a study of the development of science may lead him to contribute to the progress of science itself.

§2. *Methods. Subject of this book*

The history of mathematics, understood in this sense, should as far as possible be studied directly from documents and from originals; what is required is not so much erudition as an effort to enliven the texts by interpretation and by relating them to their own times. The field which the historian of mathematics should cover in tracing the development, ancient and modern, of the subject, and its relationship with other aspects of life and culture, is broad enough to daunt the student; but while obviously no one can cover the whole of it in detail, each student must choose what best suits his own intellectual interests.

The subjects treated in this book have been chosen mainly to show the development of the most important fundamental concepts of mathematics, and particularly the contribution made by mathematical thought to the evolution of logic. This will allow us to cover the entire history of mathematics from ancient times to our own day, observing the changes that have taken place in the conception of the structure of a rational theory, until we reach the delicate, and often lively and disconcerting, problems of contemporary logic. Though narrowed down like this, our field of enquiry is still too vast to be treated

(4) See Frajese, 5.

all in one piece. Yet, as our object is essentially formative rather than informative, we prefer to leave occasional gaps, in order to emphasize subjects which seem fundamental in the training of those who will do research in the history and philosophy of mathematics.

At this point the reader may wish for a precise definition of what we mean by the terms 'mathematics' and 'logic'. But the meaning of the terms themselves, as we shall see, keeps changing in the course of the history of thought. To understand adequately what thinkers have meant by these terms we need, therefore, a broad idea of the historical development of mathematics and logic. To understand clearly what is meant by such terms today we need too, I think, a knowledge of the evolution of the ideas in question, and it is this evolution that we shall follow in this book.

For the moment, in so far as the concept of mathematics is concerned, which historically and today is linked with that of logic, we shall confine ourselves to recording the classic definition of mathematics, valid for centuries and still accepted by Leibniz (5), who gave hints for further developments. According to this concept mathematics was understood as a science of quantity. Descartes, more precisely, in his *Discours de la Méthode* and in the *Regulae ad directionem ingenii*, says that only speculations in which order and measurement are examined apply in mathematics, apart from the objects ordered, or on which the measurements are made (6).

But this old and respectable concept no longer corresponds entirely to the object of the science which today we call mathematics: topology and theory of groups, for example, do not refer to quantity and its measurement.

From the start the concept of mathematics has undergone a profound evolution through the centuries, especially through the critique of Euclid's fifth postulate, the rise of non-Euclidean geometry, the introduction of theories of infinite sets, the construction of abstract geometry and the development of symbolic logic. The development of mathematical thought has, as we shall see, powerfully influenced the evolution of logic,

(5) 'Mathesis universalis est scientia de quantitate in universum . . .' (Leibniz, **1**; III–53, referred to by Enriques, **12**, p. 140).
(6) Descartes, vol. VI, pp. 19–20, vol. X, pp. 377–8.

with which, in a sense, mathematics was destined to be fused (7).

(7) For the development of logic in reference to mathematics, up to the beginning of the twentieth century, see Enriques, 2. On the history of formal logic, see Bocheński, 4, 5. For a general survey of the bibliography of history of mathematics Enriques, 12, and Geymonat, 2, may be consulted.

CHAPTER II
Pre-Hellenic Mathematics

§1. *Primitive mathematics*

The origins of mathematics, considered in its simplest aspects —numeration, elementary properties of figures, etc.—presumably arose in the first dawn of primitive man's intelligence. But, while traces have remained of the cavemen's art, there is nothing which allows us to reconstruct the mathematical thought of the first human beings who lived in the forests and caves. Efforts have been made to reconstruct primitive mathematics indirectly, for instance by Brunschvicg (1), who tried to get round the difficulties by substituting for research— which would be impossible in this subject—the observations which can be made today on existing primitive tribes; in fact, substituting ethnography for history. It is not part of our plan to discuss the legitimacy of this practice, or to judge the results obtained from it, but anyone who wishes to look further into the matter can study the work of Brunschvicg and the relevant bibliographical material.

§2. *Mesopotamia*

The most ancient scientific documents to reach us come from the civilizations which flourished in Mesopotamia, and consist of tens of thousands of clay bricks covered in cuneiform characters, taken from the surface-soil and preserved in various European and American museums (2).

The most interesting of these texts come from the Sumerian civilization, the development of which, starting from very ancient imprecise origins, lasts to about the end of 2100 B.C. The period more correctly called Babylonian starts with the reign of Hammurabi about 1800 B.C. and ends with the conquest by the Assyrians. Their capital Nineveh fell in 606 B.C., and

(1) Brunschvicg, ch. 1, pp. 3 ff.
(2) Bortolotti, **13**; Cipolla, pp. 11 ff.

Babylonian power arose again through Nebuchadnezzar. In 528 B.C. Mesopotamia became a province of the Persian Empire, and in its turn was conquered by Alexander the Great (330 B.C.). After this conquest the oriental and Greek civilizations each influenced one another (3).

It is said that we owe to the Sumerians the first system of rationally constituted numeration, the sexagesimal system, later adopted by the Babylonians. In this system we find a principle of position, but with the disadvantage that the magnitude of the various units is not clearly indicated: the digit corresponding to the unit, for instance, may indicate any positive or negative or zero power of 60: the order of magnitude must appear from the context.

As far as arithmetical operations were concerned, for addition and subtraction the Babylonians seem to have used some mechanical means like the abacus. For multiplication they had to know the products in pairs of the 59 unitary digits, and for this they had appropriate tables; other tables were used to find the reciprocals of given numbers, for division. But in Hammurabi's time the Babylonians could not effect division in cases where the divisor contained prime factors other than 2, 5, 3 (those contained in 60). Because of the complexity of the sexagesimal system Sumerio-Babylonian mathematics was used exclusively by a small circle of initiates who cultivated it chiefly for astronomical and astrological purposes (4). With regard to the calculation of the diagonal of a rectangle, on a brick of Hammurabi's time we find an attempt to calculate the square root of a number which is not a perfect square: but it is only an approximate process.

There has been talk of Babylonian algebra, but the term must be used with great caution. Can an equation be called algebraic when the root is known in advance? Can a problem exist in which there is nothing unknown to determine? In the extant Babylonian problems there is nothing unknown to the setters. It was a question of finding the process of working with figures in such a way that solutions which were settled beforehand should be attained. For instance, 650 systems of Babylonian

(3) Bortolotti, **20**, p. 553.
(4) Bortolotti, **20**, pp. 553–5. On the Assyrian astronomy, see Schiaparelli: '*I primordi dell'astronomia presso i Babilonesi*', pp. 41–89; '*I progressi dell'astronomia presso i Babilonesi*', pp. 91–123.

equations have been found, all of which have the solution $x = 30$, $y = 20$ (5).

'It can be seen,' Bortolotti observed, 'that the processes indicated by the solution are conceptually valid, independently of the numerical value of the figures but practically usable, in Sumerian mathematics, only for certain values, which were fixed beforehand, in such a way as to avoid operations which did not fit into the narrow numerical system of those times.

'This shows that Sumerian mathematics was not used for the solution of problems in practical life, but only for enjoyment or for exultation of the spirit' (6).

Until the fourth millennium B.C., the Sumerians regularly solved the numerical problems which we can reduce to first degree equations, and had rules for solving problems which we call of the second degree, without, however, using the algebraic method as we understand it and the theory of equations. But the Sumerian rules for second degree problems could be used only for problems that had been specially prepared (7). Neugebauer recognizes in the Babylonian problems relating to a rectangular parallelepiped the solution of cubic equations. The problems considered would lead to an equation, which, in our symbolism, could be reduced to the formula $x^3 + x^2 = a$. The Babylonian process for the solution is of empirical character and it develops by means of trial methods and direct verifications. Neugebauer has also compared the solution with a tablet on which are calculated the sums of the square and cube of the first 30 numbers.

It has even been said that trigonometrical and exponential equations have been found on Assyrio-Babylonian tablets. Trigonometrical equations, however, appear only through a verbal quibble, in so far as the ratio between two sides of a right-angled triangle is interpreted as the cotangent of an angle, while Babylonian mathematics makes no mention of angles or cotangents.

The solution of exponential equations is also thought to have been found on Assyrio-Babylonian tablets.

(5) Bortolotti, **13**.
(6) Bortolotti, **20**, pp. 558–9: against Cipolla, p. 23, who claimed that both Assyrian and Egyptian mathematics aimed exclusively at practical ends. I consider Bortolotti is right.
(7) Bortolotti, **18**.

'Take the capital sum of one mina, which, at an interest of 20%, will be doubled in five years; if the capital thus obtained is used fruitfully, after another five years it will be doubled again, and so on. After how many years will a given sum be accumulated?'

If the problem is set out in general form and a indicates the initial capital, k the accumulated capital, and n the time passed, expressed in periods of five years, we get the equation $k = 2^n a$ (as Neugebauer says), the solution of which is obtained on the basis of the theory of logarithms to base 2. But, as Bortolotti observed, on examining the text we see that Babylonian mathematicians counted on their fingers, saying:

After 1 period of five years there will be 2 minae.

After 2 periods of five years there will be 4 minae.

. . . .

until the sum established was reached (8).

'Sumerian mathematics,' Bortolotti concludes, and we conclude with him, 'cultivated by the narrow circle of the exclusive priestly castes, lost contact with civilized life and, when the Sumerian civilization was overcome and partly absorbed by the Babylonians, it was degenerating into cabal and into the mysticism of numbers. This happened at the beginning of the second millennium B.C. As far as mathematics is concerned, the Assyrio-Babylonian period is one of progressive decadence' (9).

§3. *Egypt*

Apart from the indirect references of a number of Greek writers, the most important documents of Egyptian mathematics are two papyri (10). One of these is the Rhind papyrus (11), written by the scribe Ahmes about 1650 B.C., on the basis of a document which appeared in 1800 B.C. The papyrus in question is entitled *Ways of investigating Nature and knowing all that exists, every mystery . . . every secret*. As it turns out, the elementary rules of arithmetic and geometry found in the Rhind papyrus would disappoint anyone expecting

(8) Bortolotti, **20**, pp. 559–60.
(9) Bortolotti, **18**.
(10) For a survey, and a bibliography, of Egyptian mathematics, see Bortolotti, **20**, pp. 549–52, and Cipolla, pp. 15–24.
(11) Peet.

the contents to correspond to the title; and yet the title presumably expresses the author's hope of finding a mathematical interpretation of the universe, a hope we shall meet in more advanced forms at various times in the history of thought.

The other document is the Moscow papyrus, which appeared in 1850 B.C.

For the integers the Egyptians used a numeration on a decimal basis, and fractions with the numerator 1, called unitary; besides this they had a special symbol to indicate $\frac{2}{3}$. But fractions of a different kind, for instance $\frac{5}{12}$, did not exist for them as a numerical entity, but as a statement of the problem: divide 5 into 12 equal parts; the result would be written in the form $\frac{1}{3} + \frac{1}{12}$. In general the quotients were in the form of sums of unitary fractions. With this in mind they drew up tables for dealing with the fractions that appeared most often in their calculations. We should note that the decomposition of a quotient into unitary fractions is not unique.

The Egyptians solved first degree problems by processes which some have called algebraic; but, according to Bortolotti, the processes used may be considered the result of primitive and universal forms of reasoning common to all men of good sense (12).

According to several Greek authors (Herodotus, Hero, Strabo, Proclus), Egyptian geometry, as the etymological meaning of the word shows, had its origins in the measurement of land (13).

In the Egyptian papyri we find exact rules for determining the areas of rectangles and volumes of rectangular parallelepipeds. For the area of the circle the Rhind papyrus gives us approximate rules that consist in taking $\frac{8}{9}$ of $\frac{8}{9}$ of the area of the square circumscribed by the circle; that is equivalent to giving π the value of 3·16......

For the area of the triangle and the trapezium the Egyptian papyri give us rules whose interpretation is dubious.

In the Moscow papyrus we find the calculation of the volume of a particular truncated pyramid; yet it is not likely that the Egyptians knew how to apply this in a general way. There, too, we find a rough rule for calculating the area of a surface

(12) Bortolotti, **13**.
(13) See Enriques, **8**, pp. 4–5; Frajese, **4**, pp. 3–5.

which can be interpreted as a hemisphere; but we are very far from the rational perfection of Archimedes in his work on the surfaces of spheres.

Undoubtedly Egyptian geometry developed through the needs of technology, not in agriculture alone, but in building, in the placing of buildings, in hydraulics, etc.; but we have Aristotle's word for it that Egyptian science was not merely practical in character: 'Now that practical skills have developed enough to provide adequately for material needs, one of those sciences which are not devoted to utilitarian ends has been able to arise in Egypt, the priestly caste there being in possession of the *otium* necessary for disinterested research' (14).

§4. *Phoenicia, China, India*

Among the most advanced eastern civilizations, we must mention the Phoenician, the Chinese, and the Indian.

To the Phoenicians Proclus attributes the origins of arithmetic, developed for the practical needs of commerce and business (15).

About ancient Chinese mathematics, we know little for certain. It does not appear that the Chinese possessed a theoretical arithmetic and a geometric science in ancient times; but they had a very highly developed practical arithmetic which was applied to commerce, administration and statistics (16). Besides, in their books of divination they had a system of numeration to base 2; in which, that is, they used only two digits: zero and one. In Europe this idea appeared later, through Napier (the inventor of logarithms), and was taken up again by the philosopher and mathematician G. W. Leibniz (17). In our own day this system finds its application to electronic computers. As Peano has pointed out, in relation to the Chinese ideography Leibniz observed: 'Je ne sçais s'il y a jamais eu dans l'écriture chinoise un avantage approchant de celui qui doit être nécessairement dans une charactéristique que je projette . . .' (18).

(14) Aristotle, 11, I–981b, a passage cited by Bortolotti, 20, pp. 547–8.
(15) Proclus, Prologus II. In Frajese, 4, p. 20, there is a translation of this passage.
(16) Vacca, 4, p. 171.
(17) Vacca, 12, p. 17.
(18) Peano, 1.

It is doubtful whether any of the Indian mathematical writings that have reached us go back to the time before Alexander's expedition (19). We shall have occasion to return to Indian mathematics in dealing with the origins of algebra.

To sum up the impressions gained from what has been said about mathematics in the pre-Hellenic oriental civilizations: mathematics appears mostly, but not exclusively, linked to the needs of technology. Rational elements appear, but fragmentarily, as in the solution of particular problems; we do not yet find the systemization of theories that is most characteristic of the Greek genius.

(19) Bortolotti, **20**, p. 548.

CHAPTER III

Greek Mathematics before Euclid

§1. *The sources of pre-Euclidean mathematics*

The most ancient Greek mathematical work which has reached us complete is Euclid's *Elements*. But the perfection of this work, the refinement of the critical standards according to which it was written, indicates, even if there were no other documents to do so, that Euclid's writings are not the beginning, but rather the conclusive result, of a long period of evolution of mathematical thought.

The period in which pre-Euclidean Greek mathematics was elaborated cannot be reconstructed on the basis of complete works, but by patching together fragments of works which are not merely mathematical. Philosophical writings are particularly important in this reconstruction, as at that time the sciences were not yet specialized, and mathematics, according to those who cultivated it, was closely linked with philosophy: the writings of Plato (429–348 B.C.) and of Aristotle (384–322 B.C.) are therefore especially important for our purpose.

Other references reach us from a commentator of Aristotle called Simplicius (sixth century A.D.) who transmits to us a passage from Hippocrates of Chios (about 450 B.C.) on the squaring of the lunes; from Archimedes (287–212 B.C.), from Pappus (who lived in the time of Diocletian), from Proclus (A.D. 412–485), and from later collectors who, quite devoid of scientific spirit, brought together scattered fragments taken from the period of greatest illumination of the classical thought (1).

§2. *Ionic school*

A brief historical survey of pre-Euclidean geometry comes to

(1) For the sources of pre-Euclidean geometry, see Enriques, 8, pp. 2–4; Enriques–De Santillana, 1, pp. 17–20. The fragments of the pre-Socratics have been collected in the master-work of Diels.

us from Proclus (2); here is his account of the relationship
between Egyptian and Greek science and the contribution of
Thales (624–548 B.C.):

'Since we must consider the beginnings of the arts and
sciences of the present period (3), we must say that many tell us
that geometry was first found among the Egyptians, with its
origin in the measurement of land. In fact they needed it,
because when the Nile overflowed it obliterated all the
boundaries. And so it is not surprising that this science and
others arose from practical necessity, since everything proceeds
from the imperfect towards the perfect: and so it is likewise
true that men will pass from sensation to reason, and from this
to [pure] intelligence. And thus, as the Phoenicians acquired
an exact knowledge of numbers for reasons of commerce and
business, the Egyptians studied geometry for the reasons I
have mentioned.

'Thales, having gone to Egypt, was the first to bring this
science [geometry] back to Greece, and he himself discovered
many things, and indicated many principles to those who came
after him, dealing with some in an abstract way and with
others in a more practical manner.'

The first mathematical relationships between the ancient
pre-Hellenic civilizations and the Greek civilization are therefore, according to Proclus, those connected with the journey to
Egypt of Thales, who, according to the testimony of Flavius
Josephus, must have also been a disciple of the Chaldeans (4).

'But,' observes Enriques (5), 'when the Greeks were just
approaching civilization, oriental science seems to have fallen
into a period of decadence, and all that has been preserved of it
is what was used for practical purposes: rules of calculation and
of measurement, which were now adopted without question.'

Among the Greeks, on the contrary, the wish to construct

(2) Proclus: pp. 64 ff. See Frajese, 4, pp. 20 ff.; see also pp. 19–25 of the
same work, dealing with the sources and the reliability of the résumé of Proclus's
work, made perhaps by Eudemus, a disciple of Aristotle.

(3) A few lines above, Proclus said: '...now, we have to deal with the
origin of geometry in our own times. Aristotle, inspired by his genius, said
that the same opinions prevailed many times in the history of man, following
a periodical movement; it is not the first time that the sciences have assumed
this position. There have been preceding cosmic periods (we don't know
how many they were), when those sciences have risen and have fallen.'

(4) Diels, p. 8.

(5) Enriques, **12**, p. 8.

rational theories, so characteristic of their genius, already appears in Thales, either in a search for the origin of all things (identified by him with water), or else in formulating principles from which geometrical theorems could be deduced, as well as in generalizing known results.

Thales' contributions to the progress of geometry, according to ancient writers, were the following (6): according to Proclus, he 'proved' that a circle is bisected by its diameter (but it is not clear in what sense this is to be understood), he was the first to assert that the angles at the base of an isosceles triangle are equal, and he was the first to find that vertically opposite angles are equal; while Eudemus, as Proclus himself said, attributes to Thales the second criterion for congruence of triangles, used to determine a ship's distance from the shore.

According to Diogenes Laertius (at the beginning of the third century), Pamphilus (who lived at the time of Nero) says that Thales was the first to inscribe a right-angled triangle in a semi-circle, while Apollodorus the logician attributes this to Pythagoras.

Diogenes Laertius and Pliny attribute to him a method of determining the height of a pyramid, by measuring the shadow at the moment in which the body and its shadow are equal; while Plutarch attributes to him a generalization of this method for any inclination of the sun's rays, setting down these observations as if they were addressed to Thales:

'Above all [the Egyptian King Amasi] esteems you for the way you measure the pyramids without trouble and without instruments, merely by laying your stick at the end of the shadow which a pyramid casts; from the two triangles which are made by meeting the sun's rays, you show that one shadow has the same ratio to the other as the pyramid has to the stick.'

We shall not linger over the learned and ingenious researches of mathematical historians in their efforts to reconstruct Thales' geometry, in which appears the wish, though imperfectly and fragmentarily acted upon, to make a rational system out of what geometrical knowledge there was.

(6) For a critical analysis of the ancient traditions concerning Thales' contribution to geometry, see: Frajese, 4, pp. 3–18. The following passages are referred to : Proclus, pp. 157, 250, 299, 352; Diogenes Laertius, I ch. I, n. 3; I ch. I, n. 6; Plutarch.

To the Ionic school founded by Thales belongs the philosopher Anaximandrus (611–545 B.C.), whose thought concerns us above all because we owe to him the introduction of the concept of infinity as the origin of all things (7). Anaximandrus's conception has been interpreted in two diverse, but not irreconcilable, ways: primitive substance is indefinite, primitive substance is infinite. The second conception seems preferable for several reasons.

Thus the concept of the infinite, which is of the highest importance in the development of mathematics, appears in Greek philosophy (8).

A profound Ionic thinker, Heraclitus of Ephesus (530–470 B.C.) (9) seems, among the above-mentioned philosophers, the most remote from the mathematical spirit, but he is no less interesting for that. He saw fire as the primitive substance and the origin of all things in contrast, and is the philosopher of the eternal becoming, who liked to make statements on the identity of contraries: 'The road going up and the road going down are one and the same thing' (10). 'Good and evil are one and the same thing. Doctors who cut and burn ask for a fee for doing so' (11). 'In the same river we flow down or don't flow down, we are and we are not' (12).

His mental position seems to precede that of idealistic logic (for instance, Croce's), of which we shall be speaking later (Chapter XX, §1). But for Heraclitus there existed the supreme Reason, the *Logos*: 'It is wise to listen not to me but to reason and to confess that all things are one' (13).

§3. *School of Pythagoras*

In the school of Pythagoras of Samos (580–504 B.C.), founded in Southern Italy, we find for the first time the term *mathematics* in the sense of a rational science (14).

For reasons of space we will omit the many interesting but

(7) Enriques–De Santillana, 1, pp. 54–5.
(8) Concerning infinity in Greek thought, from many standpoints, see Mondolfo, and ch. VI of this book.
(9) For further information about Heraclitus, see Enriques–De Santillana, 1, pp. 64–9.
(10) Diels, fragm. 60, p. 70, translated in: Enriques–De Santillana, 1, p. 66.
(11) Diels, fragm. 58, p. 70, translated in: Enriques–De Santillana, 1, p. 66.
(12) Diels, fragm. 49, a, p. 69.
(13) Diels, fragm. 50, p. 69.
(14) Enriques, 12, p. 9

mostly legendary stories handed down by tradition about Pythagoras, and the many fruitful philosophical, religious, and political aspects of his school (15), and will confine ourselves to those most closely linked with the development of mathematics and logic.

It is impossible to distinguish Pythagoras's contributions from those of his followers with any certainty, since he left no written works, and the work of the school was done in common. According to some scholars (Burnet, E. Frank, I. Levy) Pythagoras's school had in its early days an exclusively religious and political character, and only in the time of Archytas of Tarentum, around 400 B.C., undertook mathematical studies. Again for reasons of space we cannot go far into this, but will note with Frajese that a mathematical bent appears in the most ancient accounts we have of the school, and that the complex and refined solution by Architas of the problem of Delos cannot be found in the early stages of Greek mathematics. It would seem, therefore, that the discoveries of the school of Pythagoras began at the time of Pythagoras himself (16).

Aristotle gives an important account of Pythagorean thought (17):

'The so-called Pythagoreans, having begun to do mathematical research and having made great progress in it, were led by these studies to assume that the principles used in mathematics apply to all existing things. And as the first which are met with are, by their nature, numbers, they felt they had found in these many more analogies with what exists and happens in the world, than can be found in fire, earth and water. . . . Having then discovered that the properties and the relations of musical harmony correspond to numerical relationships and that in other natural phenomena analogies corresponding to numbers are found, they were more than ever disposed to say that the elements of all existing things are found in numbers, and that all heaven is proportion and harmony.'

This passage and others like it make us ask what was meant by saying that numbers are the elements of all things. A statement of the kind would obviously be meaningless if

(15) Enriques–De Santillana, 1, pp. 71–96.
(16) Frajese, 4, pp. 30–9.
(17) Aristotle, 11; I., cited by Enriques–De Santillana, 1, p. 80.

'number' were understood in its abstract modern sense. What did the Pythagoreans mean by 'number'? A passage from Aristotle is helpful (18): the Pythagoreans, he says, 'compose all heaven of numbers, not of numbers in the purely arithmetical sense, though, but assuming that monads have size.'

Thus, like P. Tannery, and following Enriques, we are led to interpret Pythagorean 'number' as an aggregate of monads, which in their turn are meant as corpuscles with very small, but not infinitesimal or non-existent, dimensions. This Pythagorean concept is equivalent, therefore, to considering that all bodies are formed of monads arranged in a particular geometrical order; an interpretation confirmed by Pythagorean research on figured numbers: triangular, square, cube, pyramidal, etc.

The Pythagorean theory of monads led to a quantitative explanation of the universe, and to a rough idea of mathematical physics which, applied to geometry, led to a synthesis of geometry and arithmetic, while the theory of measure became extremely simple: if two segments contained respectively m and n monads, their ratio would have to be m/n. Therefore incommensurable magnitudes could not exist. As we shall shortly see, this conclusion is irreconcilable with the theorem of Pythagoras, as the Pythagoreans themselves discovered.

Zeuthen suggests (19)—and he is probably right—that the first nucleus of geometrical propositions rationally deduced from a few obvious principles emerged from the need to demonstrate the theorem attributed to Pythagoras, empirical verifications and particular cases of which (20) were known. We do not possess Pythagoras's demonstration of his theorem, which Plutarch, Athenius, and Diogenes Laertius attributed to him (21). In Proclus's summary, quoted elsewhere, we read: '... Pythagoras transformed this study [geometry] into a kind of liberal teaching, investigating its principles from above, and working out the theorems abstractly and intellectually'.

(18) Aristotle, 8 III, I 300, 16, reported by Brunschvicg, p. 34. See also Maddalena, p. 264.
(19) See Zeuthen, 2.
(20) We know a very ancient particular case of Pythagoras's theorem, from a Chinese manuscript, probably from the second millennium B.C. The same case is also given in Indian MSS. See Enriques–De Santillana, 1, p. 91.
(21) Enriques, 8, p. 8.

Other geometrical achievements attributed to Pythagoras by Proclus (22) are the following: the theorem according to which the sum of the angles of a triangle is equal to two right angles; the application of areas, that is the solution of problems on rectangles which we can reduce to first or second degree equations; the construction of regular polyhedra (or at least, according to other writers, of the cube, the tetrahedron and the dodecahedron).

But the Pythagorean school's most important achievement, for its influence on the evolution of the logical structure of mathematical theory, was the discovery of incommensurable segments.

The reasoning which led the Pythagoreans to this discovery, or more precisely to that of the incommensurability of the diagonal of a square with one of its sides, is presumably the reasoning we find on the subject in Aristotle (23), and in a scholium of Book X of the *Elements* of Euclid (24), which can be expressed in the following way: given a square *ABCD* (see fig. 1), let us reason *ad absurdum* and suppose that the

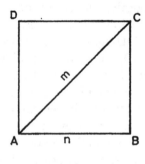

Fig. 1.

diagonal *AC* is commensurable with the side *AB*. On such a hypothesis the ratio *AC/AB* can be expressed by the fraction which we can consider reduced to its smallest terms m/n, where m and n are integers prime to each other, which indicate how many times the highest common sub-multiple of *AC* and *AB* goes respectively into *AC* and *AB*. Applying the theorem

(22) Proclus, pp. 397, 419, 420, 65.
(23) Aristotle, 4, I, 23.
(24) Euclid, t. III, p. 408.

of Pythagoras to the triangle ABC and considering the number of squares having the highest common sub-multiple considered as one side we have:

$$m^2 = 2n^2$$

where, from what we have said, m and n cannot both be even. But from the preceding equality it results that m^2 is even, and therefore that m is even and n is odd. We can therefore say that $m = 2m_1$, from which:

$$4m_1^2 = 2n^2$$
$$2m_1^2 = n^2$$

Therefore n^2 is even and so is n, contrary to the preceding result: the hypothesis of the commensurability of the side of a square and of its diagonal is therefore absurd (25).

This discovery must have made a strong impression in Pythagorean circles and must presumably have caused profound bewilderment too, since the existence of incommensurable magnitudes irreparably wrecked the early monadic Pythagorean theory. The echo of the impression made reaches us very eloquently in a scholium of Book X of Euclid, attributed to Proclus (26): 'It is well known that the man who first made public the theory of irrationals perished in a shipwreck, in order that the inexpressible and unimaginable should ever remain veiled. And so the guilty man, who fortuitously touched on and revealed this aspect of living things, was taken to the place where he began and there is for ever beaten by the waves.'

This is of interest because it shows that the ancients had realised that there were *inexpressible* elements in the continuum. The man who wrote it had perhaps caught a glimpse of a discovery of Cantor's on which Richard and E. Borel throw further light, and to which we shall return: '*In the geometrical continuum there exist* (if it is not an abuse of the term to say "exist") *elements that cannot be defined*' (27).

Another remarkable thing about the discovery of incom-

---

(25) For further knowledge concerning the development of the theory of irrational numbers, see Zariski, 1.

(26) Euclid, t. V, p. 417. The reader may find this passage in Colerus.

(27) Geymonat, 2, pp. 292–4.

mensurable magnitudes is the fact that it goes beyond the sensory. There is no sense in wondering, from the point of view of physical measurement, if the edges of two physically given rulers are or are not commensurable, because inevitably one cannot decide between the two alternatives: there is a sub-multiple that goes into the two lengths to be measured; and what is left is too small to be observed physically. In fact from the most refined physical point of view one could say outright that there is no sense in talking of the exact measurement of magnitudes, and therefore even less of incommensurable magnitudes.

Perhaps just because of this, Plato found the discovery of incommensurable magnitudes very important. In his *Theaetetus* he writes at length about irrationals, citing the results obtained by the mathematician Theodore of Cyrene, and by Theaetetus himself, who treats the subject in a more general way. And elsewhere he scorns the ignorance of those who believe that all magnitudes are commensurable (28).

The discovery of incommensurable magnitudes made classical mathematics develop in a particular direction.

The primitive theory of measure, which, as we have seen, was presumably held by the early Pythagoreans, had to be abandoned, and, in the absence of a purely arithmetical theory of irrationals, the primacy of the concept of number in mathematics and a systematic arithmetical interpretation of geometrical properties were renounced as well. A geometrical theory of proportion, wonderfully rigorous and logically most subtle, was arrived at; a theory that is attributed to Eudoxus of Cnidus (a contemporary of Plato) and given in Book V of Euclid's *Elements*. Thus in classical times geometry was held to be a more general doctrine than the science of numbers, and all mathematics was generally considered geometrically.

Mathematical interest shifted from geometry to the theory of numbers, as we shall see, towards the end of the classical era among the neo-Pythagoreans and the neo-Platonists, with Diophantus who flourished about A.D. 300, and with St. Augustine (354–430).

The primacy of numerical analysis over geometry was to be

(28) Plato, 9, VII, 819, 820, chiefly 820 B

successfully affirmed by R. Descartes (1596–1650), but a rigorous and purely numerical theory of irrational and, in general, of real numbers was to be propounded only with R. Dedekind (1831–1916) and other mathematicians of his time.

Today questions relating to the continuum still have their suggestive and disconcerting aspects, which we will deal with in Chapter XVII, §1.

§4. *Parmenides*

The fundamental concepts of geometry, which now, as we have seen, could no longer be based on the monads (with dimensions) of the early Pythagoreans, were still criticized by the school of Elea (29), dominated by Parmenides (born about 500 or else about 515). He knew Xenophanes of Colophon (580–488?), whose conception of a single and immutable God perhaps influenced the vision of the single and immutable being of the school of Elea (30). But a Pythagorean, Ameinia of Croton, taught Parmenides, whose thought grows clearer if it is considered as a critical revision of the early Pythagorean position, which the discovery of irrationals made untenable.

Parmenides, who, besides thinking profoundly on questions of thought and of being, left lasting laws to his city, expressed the results of his meditations in a poem, large parts of which have been preserved. The poem, the allegorical meaning of which is pretty clear, opens in this way (31):

'The horses which carry me have taken me far, as far as my heart could desire; because they have taken me and put me down on the famous way of the Goddess, who alone shows the man of knowledge the way through everything. There was I led; the swift horses bore me and girls showed me the way. And the axles, burning in the hub—that was hemmed in on all

---

(29) About the Eleatic school, see Enriques–De Santillana, 1, pp. 97–120. We have drawn from this book the preceding translations from Xenophanes, Parmenides, and Plato.

(30) Xenophanes not only raises objections to the anthropomorphic conceptions of gods, but elevates his mind to an 'only God, the greatest above gods and men, which is not like men, neither in its shape, nor in its thought . . . ; it sees everything, thinks and understands, and rules calmly everything with the strength of its spirit . . . , and lives always in the same place, as it is not convenient for it to wander.' (Diels: Xenophanes, fragm. 23, 24, 25, 26, p. 50.)

(31) Diels: Parmenides, fragm. 1, pp. 114–15.

sides by whirling wheels—gave out a great noise, when the daughters of the sun, wishing to lead me to the light, took the veils from their faces and left their abode in the Night.

'There are the portals from which go the ways of Day and Night, with an architrave on high and below a threshold of stone. The gates are high with two great doors, and vengeful Justice holds the keys that lock and unlock them.

'The girls said sweet words to her and cleverly persuaded her to unbolt the gates. And when they were flung wide, they showed a great void; and their doors of bronze, embossed and studded, swung back on their hinges. Straight through the doors the girls urged the horses and their chariot: the Goddess greeted me with friendliness, took my right hand between hers and spoke:

'"Welcome, young man," she said, "for coming to my dwelling in a chariot drawn by immortal steeds! This is no misfortune; justice and righteousness have set you on this road out of the way trodden by man! But you must know everything, both the heart fearless of the complete truth, and the vain opinions of mistaken mortals, far from the true faith."'

At the beginning of his poem Parmenides establishes at once a definite antithesis between the Truth (of rational character) and opinion (that originates in the senses).

One of the Eleatic school's main points on the subject of truth, understood in this sense, is the unity, that is the indivisibility, of being; referring to which Parmenides says (32): 'It is not divisible, because everything is similar to itself. There is nowhere a *more existing* that can break its connection or a *less existing*; but all is full of the existing. Thus everything is perfectly compact: the existing confined with the existing.'

It may, of course, be possible to interpret the passage we have examined in subtler and more general terms, but the simplest explanation, and the one that best seems to fit into the framework of the pre-Socratean philosophers' efforts to find a rational explanation of the universe, is this: according to Enriques's interpretation, the entity Parmenides deals with is extended matter, solidified space, which is said to be impenetrable, while emptiness is said not to exist. According to this interpretation, in the universe of Parmenides only geometrical properties are

(32) Diels: Parmenides, fragm. 8, vv. 22–5, p. 120.

possible, while another passage aims to make precise the extra-sensory character of geometrical beings and the concept of surfaces without thickness (33). 'What does not touch your senses [literally: absent things], consider it firmly present to your reason. You will not separate the existent from the connection with the existent, either by removing it from all the parts at all regularly [as in the case of a closed surface circum-scribing a solid] or by joining it [as the surface common to two contiguous solids does].'

Parmenides' work in making more exact the fundamental concepts of geometry is confirmed by a comment of Proclus on the first of the terms of Euclid: 'A point is that which has no parts.' At this point Proclus observes that, accord-ing to Parmenides, the negative definitions belong to the principles (34).

So, in the school of Elea, fundamental concepts of rational geometry (point without dimensions ... surface without thick-ness), clearly appeared which, in spite of ancient and modern anti-mathematical criticism (empirical, idealist, etc.), have not shown any intrinsic contradictions, and still remain essential elements of our geometrical theory.

The second important element in Eleatic thought is the immutability of being, and it is a result of the concept of a single undifferentiated being, in which, by the principle of sufficient reason, nothing can happen. In considering the problem of the origin of becoming, Parmenides wonders (35):

'What origins do you search for? As, for instance, from where [the existent can be] grown? You cannot say or think that it is from the non-existent: in fact it is not sayable or intelligible, because it is not. In fact what need is there for it to have begun early or late, to come from nothing? So it must either exist for ever or not exist at all.

'Nor does the force of persuasion allow us to believe that from nothing can be generated something which is not in itself nothing. So justice holds birth and death in its strong bonds. The whole problem therefore lies in this: to exist or not to exist ... what is once born does not exist, nor does it exist

(33) Diels: Parmenides, fragm. 2, p. 116.
(34) Proclus, p. 94.
(35) Diels: Parmenides, fragm. 8, p. 119.

either if it is to be born in the future.   Therefore birth vanishes and death is inconceivable.'

In the face of this paradoxical view that led to the denial of movement—the world 'fixed in its strong bonds' (36)—the historians of science have tried to give a rational interpretation to the fragments that have reached us on the subject from the school of Elea.

Enriques has rightly supported the hypothesis that the movement Parmenides denied is absolute movement.   Speaking of the world, Parmenides says (37): 'The same and remaining in the same, is quiet in relation to itself, and in such a guise it is also [absolutely] motionless.'

In Zeno we shall see more clearly the conception that movement is only relative.

The world of Parmenides is limited: 'if there were no limits there would be nothing' (38).

Plato was to express his admiration for Parmenides, the thinker who had sacrificed sensible appearances to the demands of reason, and called him, in Homer's words: 'august and terrible in his greatness' (39).

§5. *Zeno*

Parmenides' disciple Zeno, twenty-five years his junior, tried to uphold the positions taken up by the Eleatic school, using suggestive arguments, and trying to reduce the ideas that opposed it to absurdity.

Let us examine some of the paradoxes of Zeno, starting with the one in which the relativistic view is clearly seen.   Let us consider three rows *a*, *b*, *c*, of material points set out, on each row at an equal distance from one another (40), on parallel straight lines.

$$\ldots\ldots\ldots a\ldots\ldots\ldots$$
$$\ldots\ldots\ldots b\ldots\ldots\ldots$$
$$\ldots\ldots\ldots c\ldots\ldots\ldots$$

(36) Diels: Parmenides, fragm. 8, p. 120.
(37) Diels: Parmenides, fragm. 8, p. 120.
(38) Diels: Parmenides, fragm. 8, p. 120.
(39) Plato, **I**, 183 e.   Parmenides' rationalism is expressed synthetically by the words: 'Truly, thought and being are the same . . .' (Diels: Parmenides, fragm. 5, p. 117).
(40) Diels: Zenon, fragm. 28, p. 132.

with *a* motionless while *b* and *c* move in opposite directions with equal velocity *v*. The velocity of a point on the line *c* is *v* in relation to a point of *a* and 2*v* in relation to a point on *b*. The reasoning, as Enriques says, tries to reduce to absurdity the idea that velocity is a character of motion in itself. And thus is reaffirmed the idea of the relativity of motion, which we have already met in Parmenides.

The paradoxes of dichotomy (41), and of Achilles and the tortoise (42), have other meanings. Here is the latter:

Swift Achilles and the slow tortoise agree to run a race. The paradox consists in proving that if Achilles gives the tortoise a start he will not be able to catch up with him. Let Achilles (see fig. 2) be at *A* and the tortoise at *T*. The race is run along the straight line *AT*.

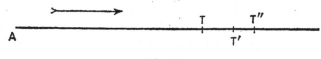

Fig. 2.

When he is given the signal at the beginning of the race, Achilles dashes from *A* to *T* faster than it takes to say it, but when he arrives there the tortoise is no longer at *T*, but a little further ahead, at *T'*. Achilles reaches *T'*, but the tortoise is no longer at *T'*; he is a little further ahead at *T''*, and so on, to the infinite. To catch up with the tortoise Achilles must cover the infinite segments *AT*, *TT'*, *T'T''* ...; to run infinite segments would take an infinite time, and therefore he will never be able to catch up with the tortoise.

Some neo-Kantian philosophers have tried to explain the paradox by saying that in fact Achilles would catch up with the tortoise because it is impossible to divide matter into an infinite number of parts. If this explanation were correct, Enriques acutely observed, the paradox of Achilles and the tortoise would be 'a theorem of rational mechanics'!

Mathematicians have another answer: it is true that the segments *AT*, *TT'*, *T'T''* ... are infinite in number, but their

(41) Diels: Zenon, pp. 130–1.
(42) Diels: Zenon, fragm. 26, p. 131.

sum is a finite segment that can be run in a finite time and calculated as the sum of an infinite geometrical progression of common ratio less than 1. If AT is equal to 100 times the unity of measure, and if the speed of Achilles is ten times the speed of the tortoise, the segment Achilles must cover to catch up with the tortoise can be measured by:

$$S = 100 + 10 + 1 + \frac{1}{10} + \dots = \frac{100}{1 - \frac{1}{10}} = \frac{1000}{9} = 111\cdot11\dots$$

But what was this paradox aiming at? According to the traditional explanation, the negation of movement, and in this case it was wrongly argued. But according to Tannery the paradox was leading to the confutation of the primitive Pythagorean monadic theory: if the segment AT, TT', $T'T''$, $\dots$ were always greater than or equal to the extended monad of small but infinite dimension ε, the segment that Achilles would have to cover to catch up with the tortoise would be infinite because made up of infinite segments larger than or equal to ε.

§6. *Movements of thought originating from the Eleatic school*

In Zeno's arguments we find the process of reduction to absurdity, which, to be conducted correctly, needs special logical caution, since it is hard to apply intuition in this form and there is no way of checking it through the experience of the senses. So it is understandable that Diogenes Laertius considered Zeno the founder of dialectics, that is, of logic.

In the school of Elea the development of logic, with special emphasis on reasoning through the absurd, is linked to an interest in questions of infinitesimal character. Here originated the method of exhaustion with which the ancients made their infinitesimal reasoning exact; this method will be dealt with in Chapter VI, §2 and §3.

The Greek philosophers opposed Eleatic thought, which is highly suitable for the development of mathematics and logic, but not for explaining phenomena that could be observed in the world of becoming, in three ways. The pluralists (Empedocles and Anaxagoras) denied the qualitative unity of being. The sophists (Protagoras, Gorgias) undervalued rational truth in the name of sensible experience. The atomists (Leucippus,

Democritus) affirmed the void, in which atoms move. Plato left us his synthesis of the eternal becoming of Heraclitus and the immutable being of Parmenides.

§7. *The pluralism of Anaxagoras and of Empedocles* (43)

Anaxagoras of Clazomenae (500–428 B.C.) attributes an objective existence to the qualities of matter, not considered as pure appearances. In every fragment of matter there exist infinite qualitatively different dimensionless elements (the homeomeries); in a fragment of iron, for instance, there are prevalently the homeomeries of iron, although there are others as well; and the same thing is true of the other substances.

This view of the physical world suggested geometrical ideas of infinitesimal character to Anaxagoras (44). 'With regard to the small there is no smallest, but always an even smaller, because the existent cannot be annihilated [by division]. Thus, with regard to the large, there is always a larger, and this larger is like to the small in plurality, and in itself everything considered as the sum of infinite infinitesimal parts is at the same time large and small.'

In this passage Anaxagoras presents the continuum to us as structurally irreducible to points, according to a conception which in a way foreshadows the theory of the modern intuitionists, to which we shall return (45).

It is interesting to observe that in the fragment we have considered Anaxagoras shows us two sets made up of infinite elements, as is done systematically in the modern theory. According to an hypothesis formulated by Bortolotti (46), Anaxagoras would have been able to state that two segments of different lengths have the same number of points insofar as, through a geometrical construction, a one-to-one correspondence can be established between the points of the segments themselves. This hypothesis becomes more likely when one

(43) Further knowledge on the pluralists may be obtained, e.g. from Enriques–De Santillana, 1, pp. 121–40; see, at p. 123, a translation of Anaxagoras's fragment, referred to in this paragraph.
(44) Diels, fragm. 3, p. 314.
(45) Geymonat, 2, p. 272. On the intuitionists, see ch. XIX, §3, and ch. XX, §2 of this book.
(46) Bortolotti, 14, pp. 153–4.

recalls that, according to Vitruvius, Anaxagoras studied perspective (47).

'When Aeschylus was staging one of his tragedies in Athens, Agatharcus first of all made a stage set and left a short treatise on it. Inspired by this, Democritus and Anaxagoras wrote on the same subject, showing how necessary it was that, fixing a certain point as the centre, the lines [that surrounded the object] should naturally correspond with the point of vision [of the spectator] and with the distance; so that actual images of an illusory reality will give, in painted scenery, the appearance of buildings, and objects drawn on straight and uniform surfaces would seem to be placed at varying distances.'

In Proclus's summary, which we have quoted several times, we learn that 'Anaxagoras of Clazomenae studied geometrical questions'.

Empedocles of Agrigentum (490–430 B.C.), however, who held the theory of the four elements (earth, water, air, fire), lacked the geometrical outlook, and was a physician and poet.

§8. *The sophists*

The sophists had, and still have, a rather bad reputation, mainly because we know them best through the works of their great adversaries Plato and Aristotle (48). But we must admit that they contributed to the development of logic and the analysis of language, if only because they wanted to confound their enemies and arouse interest in the quarrel.

Enriques supposes that Protagoras questioned the validity of the Parmenidean principle of non-contradiction. With his relativistic views, Protagoras must have observed that opposite qualities can be attributed to the same object when this object is considered from varying points of view. Aristotle admitted this limitation of the principle.

The sophists did not all hold the same beliefs, but in general their movement had an empirical character, which in modern parlance we should call positivistic; they upheld the value of the experience of facts against the metaphysical rationalism

(47) Vitruvius, VII, praef. II (see also Diels: Anaxagoras, A 39, p. 300, a passage translated by Enriques–Mazziotti, p. 209).

(48) For further information about the sophists, see, e.g., Enriques–De Santillana, **1**, pp. 159–2, whence we have drawn the interpretation given in this paragraph. See also Timpanaro Cardini.

of the school of Elea, and this led to an attitude which presents some analogies with illuminism. Besides this, the sophists enjoyed the analysis of language and logical subtleties so that they can, in a sense, be considered distant precursors of the neo-positivists (see Chapter XX, §3 of this book). They favoured, too, the diffusion of culture and removed the main interest of thinkers from the nature of man.

In the *Theaetetus* (49) of Plato, when Socrates asks the young mathematician Theaetetus for a definition of science, he first of all answers: 'Science is sensation.' Socrates replies: 'You give an opinion that cannot be despised, since it was Protagoras's. Yet he expressed it in another way, by saying that man was the measure of all things.' This means that whereas Parmenides sought for the nature of things in themselves, looking at what existed for itself, Protagoras of Abdera (485–410 B.C.) maintained that existence is not a property of the object in itself, but is related to the sentient man.

The paradoxical title of Protagoras's work, *Writings demolishing truth*, is interpreted according to the meaning which the term 'truth' had assumed in Parmenides' system; therefore the meaning of what Protagoras wrote becomes: *Writings demolishing metaphysical truth*. The title of another work attributed to the sophist Gorgia of Lentini in Sicily (483–375 B.C.) can be interpreted in the same way: *Of Nature or what does not exist*. Here nature is interpreted as the essence of things, an essence that for the empiricist does not exist.

This mental attitude was to lead the sophists into an anti-mathematical position, since they opposed the consideration of the extra-sensory geometrical entities studied by the school of Elea: point without dimension, line without width. . . .

And in fact we find Aristotle (50) writing on the subject, taking up Protagoras's thesis.

'Indeed the sensible lines are not what the geometers say, because nothing sensible is thus [rigorously] straight or curved. In fact the circumference does not touch the line [tangent] at a point, but, as Protagoras said on the subject of geometry, along an element of a certain length α.'

The sophists held the same sort of views on the squaring of

(49) Plato, 1 (152).
(50) Aristotle, 11, II, 2 (20).

Democritus's conception of the movement of atoms precedes the principle of inertia, as is shown in a passage of Cicero (57): 'Ille [Democritus] atomos quas appellat, id est corpora individua propter soliditatem, censet in infinito inani, in quo nihil nec summum nec infimum nec medium nec ultimum nec extremum sit, ita ferri, ut concursionibus inter se cohaerescant, ex quo efficiantur ea, quae sint quaeque cernantur, omnia; eumque motum atomorum nullo a principio, sed ex aeterno tempore intellegi convenire.'

Democritus's position can also be interpreted as a reaction against the empiricism of the sophists. In fact, according to Sextus Empiricus, Democritus and Plato believed in 'the existence of the intelligible', contradicting Protagoras. With regard to this, Enriques and Santillana (58) make a precise distinction between 'intelligible' and 'sensible', in order to apply the principles of logic. It is generally thought that these principles can be applied to sensible data, but we notice that two sensible objects (sensibly) equal to a third, may not be (sensibly) equal to one another. The question is illustrated by the neo-Eleatian philosophers of the School of Megara, who observed that a heap of grain which was diminished by one grain remained a heap of grain sensibly equal to the given heap, but if you removed all the grains successively from the heap nothing would be left.

Thus it is concluded that the principles of logic do not apply to sensible objects, but to objects of thought.

While Parmenides thought it impossible to base a rational science on the sensible, and the sophists, on the contrary, valued only empirical data, Democritus stressed the need to make the thinkable agree with the sensible: 'to save phenomena'. Enriques thinks that we owe to Democritus the definition of science, discussed in Plato's *Theaetetus*, as 'true opinion accompanied by reasoning': this would be a matter of experimental rationalism, of Galileo's type.

Democritus's rationalism urged him to find in reality what is possible in thought: 'worlds with more suns and moons', 'atoms as big as a world' (59).

(57) Cicero, 1, reported by Diels, p. 362.
(58) Enriques–De Santillana, 1, pp. 194–6.
(59) Enriques–De Santillana, 1, p. 198.

It seems that the term logic appears for the first time in Democritus (60). His work on the subject has been lost, and so we do not know exactly what meaning he attributed to it. In a fragment of the same work preserved by Sextus Empiricus we read: 'There are two forms of knowledge: a pure or legitimate one and a shadowy or spurious one. To this latter form belong sight, hearing, taste, smell and touch. But pure knowledge is completely different.'

We know the titles of various mathematical works of Democritus, unfortunately lost, but Proclus does not mention them: this is explained (even if it is not justified) by the difference of philosophical opinion between Plato's late follower and his great adversary.

According to Archimedes (61), Democritus discovered the volumes of the pyramid and the cone: '. . . With regard to the theorems with which Eudoxus first discovered the demonstration—that is, that the volume of a cone is one-third of the volume of a cylinder, the volume of a pyramid one-third of the volume of the prism with an equal base and an equal altitude—a good deal of the merit should go to Democritus, who first, without demonstration, gave the propositions relating to these figures.'

We do not know the process used by Democritus to obtain these results. Enriques suggests that he may have obtained them by calculating the sum of an infinite geometrical progression. And again (with Mazziotti) Enriques (62) suggests that he may have used a principle like that of Cavalieri's indivisibles in order to prove that pyramids having the same bases and same altitudes are equivalent, and thus would have obtained his result by dividing a triangular prism into three pyramids equivalent to one another.

Certainly, according to Plutarch, Democritus was not unaware of ideas on the infinitesimal and the early difficulties which are found reflected in the subject (63): 'See how [Crisippus] argues precisely and scientifically with Democritus, who suggests the following doubt: if a cone is cut by planes

(60) Diels, B., fragm. 10, b and 11, p. 389. The interpretation we report is due to Enriques–De Santillana, 1, p. 200.
(61) Enriques, 8, pp. 21–2; Rufini, 2.
(62) Enriques–Mazziotti, pp. 199–200.
(63) Diels, B., fragm. 155, pp. 412–13.

parallel to its base, what must we think of the surfaces of the sections? Are they equal or unequal? If they were unequal the cone would be irregular, with many very rough scratches that looked like steps; and if on the other hand the surfaces were equal, the sections would be equal and it would appear that the cone looked like a cylinder, as if it was made up of equal and not unequal circles, which is extremely far from the truth.'

This passage makes the second supposition on the way Democritus found the volume of the cone more likely: through these rather ingenuous notions appears the idea of a solid constituted by the sum of its sections, according to the concept which was to appear, with a heuristic purpose, in the method of Archimedes, and in the method of indivisibles of Galileo, Cavalieri, and Torricelli.

It would appear that Democritus also studied the angle of contingency (64), on which we shall find an interesting result in Euclid's *Elements*.

Vitruvius testifies more clearly that we owe Agatarcus, Anaxagoras, and Democritus the origins of perspective, particularly in stage design (see §7 of this chapter).

§11. *Socrates* (65).

He was Athenian (470–399 B.C.), a contemporary of Democritus, ten years younger but dying earlier, and Plato's teacher; he shows us a new side of the argument with the sophists. He left no writings, and this makes it particularly difficult to reconstruct his genuine thought. Two writers contemporary with Socrates have handed it down to us: Xenophon and Plato. Xenophon was no philosopher; and in the dialogues of Plato it is hard to distinguish the theories of the historical Socrates from those attributed to him by his great disciple. There are, besides, some important passages of Aristotle which help us to reconstruct Socrates' thought.

His work seems dominated by moral motives which inspired his opposition to the sophists; he sought for truth in concepts to give a solid basis to morality.

The Socratic process is important in the history of science

(64) Enriques–Mazziotti, pp. 202–5.
(65) See Enriques–De Santillana, 1, pp. 174–91.

because it defines concepts. On this subject Aristotle wrote (66):

'As Socrates dealt with moral virtue, at first he tried to define it universally (Democritus had only touched on natural things and thus occasionally defined heat and cold; the Pythagoreans had, before that, defined some things, which their reasoning reduced to numbers: what is the weather, justice, a wedding); he sanely and rationally asked: what thing is?

'Two things, in fact, can be attributed fully to Socrates: inductive argument and universal definition: both pertinent to the principles of science.'

This passage of Aristotle gives us a clue to the history of definition, which in its main lines can be summed up thus:

In the Pythagoreans we find, at an earlier date, definitions by analogy, often inspired by the mysticism of numbers, while later the mathematicians of the school of Pythagoras gave us definitions by division, as in the case of the division of quantity attributed to Archytas of Tarentum: 'Multiform quantity is divided into line, surface, body, space, number and ratio' (67).

But Socrates refuses to accept a definition by division or extension, one which consists in enumerating the kinds or sub-species that constitute a species; he seeks instead for a definition by comprehension in which is expressed what the individuals of the species to be defined have in common. For instance, in Plato's dialogue *Menon*, Socrates ironically answers the man he is talking to, who, in trying to define virtue enumerates several virtues: 'What luck I have had, O Menon! I went looking for *one* virtue, and found you had a swarm of them.'

Socrates' characteristic intention in this matter of definition is to make clear the comprehension of a concept of which the extension is known.

The problem of definition will be taken up again, as we shall see, by Aristotle, and in the modern age considered from another point of view.

§12. *Plato* (428–347 B.C.)

Plato tried to get over the contrast between the Parmenidean

(66) Aristotle, 11, X, 1, 4 (3) and (4), translated by Enriques–De Santillana, 1, p. 184.
(67) See Mullach, p. 573, who mistranslated 'logos' by 'sermo', as Enriques, 14, pp. 50–6, remarked.

conception of being and the Heraclitean conception of becoming. In the following passage (68) Aristotle gives us a rapid summary of his thought: 'After these philosophies came the doctrine of Plato, which in part followed the Pythagoreans, but had theories of its own that were independent of their philosophy. As a young man he was above all influenced by Cratilus, and shared the doctrines of the Heracliteans, believing that all natural things flow all the time and that therefore there can be no science about them; and later, too, he held this opinion. As later, Socrates dealt with moral matters, and took no notice at all of nature in general; and he sought in these moral matters for the universal and was the first to study thoroughly the definitions. Plato welcomed Socrates' thought and understood, in consequence of the doctrine of the Heracliteans, that this defining should be a matter of other things than the sensible, as it was not possible to define as universal any of the sensible things which change continually. And so these other things are called ideas, and the sensibles are outside them.' Plato therefore applied to the sensible world Heraclitus's logic of becoming, and to the world of ideas Parmenides' logic of immutable being. In Platonic thought we thus have an example (perhaps the most ancient) of two logics co-existing in the same philosophical system.

The difference between the universals of Socrates (concepts) and the universals of Plato (ideas) is made clear in another passage of Aristotle (69): 'Socrates did not make these universals separate; neither did he do this to the definitions; others [Plato] separated them, and these universals he called *Ideas*.' That is, the concept of Socrates is not beyond our mind, or even beyond the things it talks about; whereas Plato's idea belongs to a world superior to the sensible universe.

Mathematical thought exercised a profound and wide influence on Plato's theories, and mathematical considerations often appear in his writings. The conception of the world of ideas is inspired mainly by the moral world and by the world of mathematical entities, extending afterwards to all nature (70). According to John Tzetze, an Alexandrine grammarian of the

(68) Aristotle, **11**, A 6; 987 and 29 ff., a passage translated by Manara Valgimigli in his preface to Plato, **10**.
(69) Aristotle, **11**, 4, 1078, b, 27 ff., a passage translated as in (68).
(70) See Windelband, pp. 80–1.

thirteenth century A.D. (71), on the gate of the Academy was written the inscription: 'No one who is not a geometrician can come under my roof'—that is, geometry is an indispensable preparation for understanding Plato's philosophy. And yet a problem exists and has long been discussed (72): are mathematical entities ideas?

We shall examine some of the most significant passages for understanding the position of mathematics in Plato's thought and his conception of the structure of mathematical theories. In the *Republic* we read (73):

'Those who study geometry and arithmetic and similar subjects take as hypotheses the odd and the even, and figures, and three kinds of angles, and other similar things in each different proof. They make them into hypotheses as though they knew them, and will give no further account of them either to themselves or to others on the ground that they are plain to every one. Starting from these, they go on till they arrive at the original object of their proof. . . . Then you know that they use visible squares and figures, and make their arguments about them, though they are not thinking about them, but about those things of which the visible are images. Their arguments concern the real square and a real diagonal, not the diagonal which they draw, and so with everything. The actual things which they model and draw, like shadows and images in the water, these they use as images, seeking to see through them those very realities which cannot be seen except by the idealizing intelligence.'

From this passage it would appear that geometrical entities belong to the world of ideas, where we find 'the square in itself', not perceptible through the senses, of which the sensible square is only an image.

(71) Loria, **2**, p. 110.

(72) See Frajese, **4**, pp. 88–92: the author raises an objection to the interpretation of the Platonic theory, according to which mathematical entities pertain to the world of ideas, whilst, though there is one idea only of a straight line (or a circle), there is not a single straight line (or a single circle). I think that this objection may be overcome: the straight line is only one, when one considers it as an abstract entity; there are many straight lines when one considers them as a part of more complex figures: in such a case these figures are to be considered as Platonic ideas.

(73) Plato, **8**, 510, c, d, e. This passage may be found also in Enriques–De Santillana, **1**, where, at pp. 201–24, a synthesis of Plato's thought is sketched with particular reference to science.

Plato develops a similar idea for numbers, the science of which should be taught not with any practical object in view, but in so far as (74) 'it powerfully draws the soul above, and forces it to reason about the numbers themselves, not allowing any reasoning which presents to the soul numbers that can be seen or touched'.

In another passage of the *Republic* (75) we read: 'None who have even a slight acquaintance with geometry ... will deny that the nature of this science is in exact contradiction to the arguments used in it by its professors. ... They talk in a most ridiculous and beggarly fashion; for they speak like men of business, and as though all their demonstrations had a practical aim, with their talk of squaring and applying and adding, and so on. But surely the whole study is carried on for the sake of knowledge ... of that which always is, not of that which at some particular place and time is becoming and perishing. ... For geometry is a knowledge of that which always is ... and it will draw the souls towards truth.'

In the latter passage the image of geometry as something that draws the mind towards truth is puzzling. But then, in Platonic thought, is geometry not in itself truth? I think that here Plato wishes to say that geometry prepares the mind to rise to other forms of truth, not forgetting the fact that geometry is in itself already truth (as he said, substantially, immediately before in the affirmation that 'geometry is a knowledge of that which always is'), immutable truth, and therefore belonging to the world of ideas.

But again in the *Republic* (76), Plato criticizes the methods of the geometricians of his day: ' ... geometry and suchlike sciences ... dream about being. But they cannot behold it with waking eyes so long as they use postulates and leave them uncriticized, without being able to give an account of them. Really, how may we give the name of science to a discipline which is not aware of its principles and the end and means of which are bound to the unknown?'

This passage is puzzling too; together with the previous one, it suggests that Plato considered that mathematical entities

(74) Plato, 8, 525, c; Enriques, 12, p. 13.
(75) Plato, 8, 527, a, b, reported by Frajese, 4, p. 89.
(76) Plato, 8, 533, c, a passage quoted by Enriques, 2, p. 42. This passage recalls the definition of mathematics given by B. Russell (see ch. XIX, §1).

belonged to a world half-way between the sensible world and the world of ideas.

I think that the difficulty can be overcome as Enriques overcomes it (77), by 'admitting that the inferior position attributed to mathematics in comparison with dialectics refers not so much to pure mathematics, capable of being built up as a science according to Plato's ideal, as to mathematics considered as art'; and I would add that according to Plato mathematics stands below the world of ideas when it is developed in a way that does not conform with the Platonic ideal.

This latter passage of Plato shows us too what his ideal was in the ordering of a deductive science, from which he would like to eliminate the postulates through which the possibility of certain constructions is presumed. According to the Platonic ideal geometry should be based only on definitions and principles as obvious as axioms, which concern innate knowledge. This ideal was to be found again substantially in Leibniz (78).

The conception of innate ideas assumes a mythical form in Plato through the theory of remembrance, according to which the soul will have known the truths of the world of ideas before its earthly life, thus keeping a vague memory in this world which it must make more alive. Plato tries to demonstrate this theory in *Menon* by showing how Socrates, with suitable questions, makes an ignorant slave recognize a particular case of the theorem of Pythagoras. The passage is interesting because it allows us to see Plato's geometrical language and still more because it is a luminous example of the application of the Socratic method of bringing a pupil to knowledge of the truth, which, quite apart from the theory of remembrance, is of very great importance from a didactic point of view. Here, then, is the passage in question (79):

Socrates turns to Menon, who has asked for a demonstration of the theory of remembrance, and says:

'To tell you the truth it's not easy to prove it; but for love of you I'll try. Call someone in your household for me to show it to you.

(77) Enriques, **2**, p. 13.
(78) Enriques–De Santillana, **1**, p. 219.
(79) This passage from the *Menon* is reported by Loria, **2**, pp. 115–20: he gives Plato's text. and moreover, the letters indicating the points that Socrates is supposed to draw on the sand.

Menon: Very well. (*To a slave*) Here you! Come here.

Socrates: Is he Greek? Does he speak Greek?

Menon: Perfectly. He was brought up in my house.

Socrates: Then listen and see which seems right—whether he
    remembers or whether he learns from me.

Menon: I will listen most carefully.

Socrates: Tell me, young man, do you know what a square is?
    Is it a figure [see figure 4] like this one [*ABDC*]?

Slave: Yes, it is.

Socrates: Is it then a quadrangular figure with its four sides
    [*AB, BD, CD, AC*] equal?

Slave: Certainly.

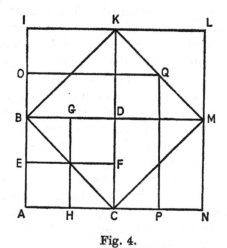

Fig. 4.

Socrates: Are not the lines [*EF* and *GH*] which pass through the
    centre in it equal too?

Slave: Yes, they are.

Socrates: Isn't it true that this figure could be either larger or
    smaller?

Slave: Certainly.

Socrates: Now suppose that the side *BD* was two feet long, how
    many square feet would the whole thing be? Consider it like
    this for a minute: if [*BD*] were two feet long, and [*BA*] only
    one foot, wouldn't the figure contain once two [square] feet?

Slave: Certainly.

Socrates: Now when this *BA* is two feet long as well, won't it contain twice two [square] feet?

Slave: Yes, it will.

Socrates: So we now have a figure that is twice two [square] feet? Work it out and tell me the answer.

Slave: Four.

Socrates: Could we not draw another figure now, double of this *ABDC*, but the same, having all its sides equal?

Slave: Yes, we can.

Socrates: And how many [square] feet would there be in that?

Slave: Eight.

Socrates: Very good. Now try and tell me what length each side of this figure would have. In the first square every side is two feet. Now how long would they be in this square, which is double the other?

Slave: Obviously, Socrates, double.

Socrates (*to Menon*): Do you see, Menon, I'm not teaching anything, I'm only asking. And he firmly believes he knows the length of the side of a square containing eight square feet. Do you agree?

Menon: Yes, I do.

Socrates: Does he really know it?

Menon: No, he doesn't.

Socrates: He thinks it's double. Now listen carefully and you'll see that he remembers another thing as he should remember it. (*To the slave*) Now tell me, did you say that from a line [*IA*], double of *BA*, a double figure is born? But I'm not speaking of a figure *AIKC* with *AI* long and *AC* short, but of one which has all its sides equal just like this one *ABDC* and which contains double of this one, that is eight square feet. Now look, do you think that this comes from the double side [*AI*]?

Slave: Well, yes, I think so.

Socrates: Now doesn't this line become double its length if a second line *CN*, equal in length, is added to it?

Slave: Certainly.

Socrates: So you say that from this one, *AN* results a figure of eight [square] feet, making the sides [*AN, NL, LI, IA*] equal, and that therefore this figure which you say is a square contains exactly eight [square] feet?

Slave: Yes, I say so.

Socrates: What size then will the figure [*ANLI*] be? Isn't it true that it would be the quadruple?

Slave: How could that be otherwise?

Socrates: Then is the quadruple the same as double?

Slave: No, by Jove! It is not.

Socrates: But then what multiple is it?

Slave: Quadruple.

Socrates: Well then, my dear young man, from a double size one obtains a square which is not double but quadruple.

Slave: That's true.

Socrates: Because four by four makes sixteen, isn't that true?

Slave: Yes, it is true.

Socrates: From what size is born a square of eight [square] feet? Certainly not from this one,[*AN*], because from that is born a quadruple square [*ANLI*].

Slave: Yes, I admit it.

Socrates: But from this one [*AC*], which is only half of [*AN*], is born that of four square feet.

Slave: Yes, it is.

Socrates: But the square of eight square feet is double [*ACDB*] and half [*ANLI*]?

Slave: Certainly.

Socrates: Will it therefore be born from a line greater than [*AC*] and less than [*AN*]? Or not?

Slave: I think so.

Socrates: Answer only what you think. And tell me: Is this line *AC* not two feet and this line *AN* not four feet?

Slave: Yes, they are.

Socrates: The side of the figure of eight square feet must therefore be more than this [*AC*], which is two feet long, but less than that [*AN*] which is four feet long?

Slave: Yes, it must.

Socrates: Now try to tell me how long it should be.

Slave: Three feet.

Socrates: Now if it should be three feet, let us add to this [*AC*] its half, [*CP*], and we shall get a line of three feet. Because this [*AC*] is two feet and this [*CP*] is one foot. Similarly for the side [*AB*]; this [*AB*] of two feet and this [*BO*] of one foot. And will this [*APQO*] be the figure you are thinking of?

50

Slave: Yes, it will.

Socrates: But if this whole figure [*APQO*] is three feet long along [*AP* and *AO*], will it not contain three feet by three feet [squared]?

Slave: Obviously.

Socrates: But how much is three times three feet [squared]?

Slave: Nine.

Socrates: And the figure double [of *ACDB*]—how many [square] feet should it contain?

Slave: Eight.

Socrates: Then a line with a three-foot side does not get a square of eight feet either.

Slave: No, that's true, it doesn't.

Socrates: Then what does? Try to get it right and if you cannot express it in numbers show it with a line.

Slave: But, by Jove, I really don't know.

Socrates (*to Menon*): Have you seen how he has gone ahead on the way to remembering? At first he didn't know what the side of a square with the surface of eight [square] feet was, a matter he still doesn't know. But before he thought he knew it and answered casually, like a learned man, without feeling at all perplexed. But now he feels uncertain, he doesn't know it, and doesn't even think he knows it. Now see how, from this uncertainty, with my help he will search and find, while I ask and do not teach. Watch carefully to see if you find me teaching and explaining, or if I don't just ask him how he sees it. (*To the slave*) Is this *ABDC* not our figure of four feet? Do you understand?

Slave: Yes, I do.

Socrates: Can we add to it any equal one, this *BDKI*?

Slave: Yes, we can.

Socrates: And now a third one here, [*KLMD*], equal to these two?

Slave: Yes, we can again.

Socrates: To complete the figure may we not put, in the angle [*CD̂M*], a [*DCNM*]?

Slave: Very well.

Socrates: Will we not thus obtain four equal figures?

Slave: Yes, we will.

Socrates: Well then? How many times is the whole figure [*ANLI*] more than the figure [*ABDC*]?

Slave: Four times.

Socrates: But we wanted it to be only twice as big, don't you remember?

Slave: Certainly.

Socrates: Now these lines [*CB, BK, KM, MC*], going from one angle to another, do they not divide each of these squares in half?

Slave: I see they do.

Socrates: And do these four lines [*CB, BK, KM, MC*] circum-scribe this figure [*CBKM*]?

Slave: Yes, they do.

Socrates: And now look at it for a moment and see how big this figure [*CBKM*] is.

Slave: I don't know.

Socrates: Of these four quadrangular figures [*ACDB, BDKI, DMLK, CDMN*], haven't these four lines [*CB, BK, KM, MC*] cut out half?

Slave: Yes, they have.

Socrates: Then how many halves are there in this figure [*CBKM*]?

Slave: Four.

Socrates: And in this one [*ABDC*]?

Slave: Two.

Socrates: And what is four in relation to two?

Slave: Four is double of two.

Socrates: Then how many square feet does this figure [*CBKM*] contain?

Slave: Eight square feet.

Socrates: And from what line is it born?

Slave: Is it from this one [*CB*]?

Socrates: Do you mean, therefore, from a line that goes from an angle [$\widehat{ACD}$] of the square [*ABDC*] of four [square] feet to another angle [$\widehat{ABD}$]?

Slave: Yes, I mean so.

Socrates: Now the scientists call this line [*BC*] the diagonal; as a result, having said this is the diagonal, from this said diagonal, O slave of Menon, you will obtain the double quadrangle.'

According to a tradition handed down to us by Diogenes Laertius and Proclus, Plato introduced analysis into geometry (80), teaching it to the mathematician Leodamas (remembered by Proclus in his summary). Yet we can assume that analysis was already known to earlier mathematicians. The process of analysis, applied to a problem A, leads it to a problem B, such that having once solved B we can solve A, and so on until we reach a problem which we can solve. This process leads to problems more general than the given one and can find the solution of a problem, but not the demonstration of this solution.

Applied instead to a mathematical proposition A, analysis seeks the principle from which A can be deduced.

But in Platonic thought, science, as definitely constituted, goes from principles to consequences (81).

It is not part of our plan to give a complete list of the mathematical notions contained in the works of this great philosopher; among them is that of regular polyhedra, called Platonic, because they have an important part in the cosmological system expounded in the *Timaeus*. Of what he contributed to the solution of the problem of duplicating the cube we will speak in the next section.

Plato's philosophy exercised a profound and lasting influence on the development of mathematics in the ancient world, on the geometricians who lived in contact with the academy, like Theaetetus and Eudoxus, and on the authors of the mathematical masterpieces of the Alexandrine period and the centuries after it; while even today his influence cannot be said to be completely dead.

The Platonic conception of the objectivity of mathematics finds its lucid expression in the proposition of the *Eutidemus* (82): 'Geometricians . . . are hunters, they do not create the geometrical figures themselves, but go in search of those that exist.'

§13. *The problem of the duplication of the cube*

This has been called the problem of Delos because of its

---

(80) Diogenes Laertius, III, 24; Proclus, p. 211. Brunschvicg deals clearly with this question, pp. 53–5. See also Sabbatini, pp. 3–6.
(81) Plato, 8, 510 B.
(82) Plato, 3, 290, b, c. Concerning Plato's realism and the objectivity of mathematics, see also Frajese, 3.

legendary origin. In a letter attributed to Eratosthenes (third century B.C.) sent to King Ptolemy III, we read an interesting account of the subject (83).

'Eratosthenes to Ptolemy, greetings.

'They say that one of the ancient tragic poets made Minos appear in the act of building a tomb for Glaucus, and that Minos, realizing it was a hundred feet long on each side, said: "This is really a small space that you have allowed for a king's tomb. Double it, keeping the cubic form, double all the sides of the tomb right away." Now it is clear that he was mistaken. In fact by doubling the sides a plain figure is quadrupled, while a solid figure is multiplied eight times. Then the geometricians wondered how you could double any given solid figure, while keeping its shape. And this problem was called *the duplication of the cube*.

'After much hesitation, first of all Hippocrates of Chios found that if between two straight lines, of which the bigger is double the smaller, two mean proportionals are inserted, the cube will be doubled; and thus he exchanged one difficulty for another, not lesser, difficulty.

'It is also said that later the Deleans, urged by the oracle to double a certain altar, fell into the same difficulty. Ambassadors were sent to the geometricians who came together with Plato to the Academy, urging them to work out what was asked. They studied it diligently and it is said that, having tried to insert two mean proportionals between two lines, Archytas succeeded with a semi-cylinder and Eudoxus instead with curved lines. These were followed by others, in making the demonstration more perfect, but they could not effect the construction and use it in practice, except perhaps Menaechmus, and he with great difficulty. . . .'

Some of these views were confirmed and completed in the *Platonicus* of Eratosthenes (84), which we quote:

'When the god announced to the Deleans through the oracle that if they wanted to free themselves from the plague they must build an altar which was double the one that existed, their engineers were greatly perplexed, not knowing how to find a way of constructing a solid double of another similar solid.

(83) Conti, pp. 325–6.
(84) Enriques–De Santillana, 1, p. 249.

So they went to ask Plato's advice; and in his answer he explained the oracle's meaning: it was not that the god wanted an altar twice as big, but that by asking the Greeks to fulfil this task, he wished to make them ashamed of their ignorance of mathematics and their scorn of geometry.'

The problem to which Hippocrates of Chios, according to historians, reduced the problem of Delos, can be expressed in modern language thus: given two segments measuring a and b, determine two segments measuring x and y so that the proportions are:

(I) $$a : x = x : y = y : b$$

From (I) we get:

(II) $$x^2 = ay$$
(III) $$y^2 = bx$$

That is: $$x^4 = a^2 y^2$$
$$x^4 = a^2 b x$$
$$x^3 = a^2 b$$

In particular, if we write:

$$b = 2a$$

we get $$x^3 = 2a^3$$

This equation is in fact the one on which depends the problem of the duplication of a cube with an edge a, while x is the edge of the cube wanted, of double volume.

The insertion of two geometric means between two given segments is in general a problem that cannot be solved with ruler and compasses, as was demonstrated in modern times (85), and geometricians of the ancient world in fact thought out various constructions that fall outside the field in which it is possible to work with these classical instruments, which are the only ones used in Euclid's *Elements* (86).

Archytas of Tarentum (430–365 B.C.), a friend of Plato, solved the problem through an ingenious spatial construction in which

(85) See Conti, pp. 329–31.
(86) We say that a problem can be solved by ruler and compasses, when its solutions may be analysed as a succession of operations, each of which may be one of the following: given two points, to draw the line joining them; to draw a circle, given its radius and centre; to determine the point or points of intersection of two straight lines, of a line and a circle, or of two circles. On this topic, see Castelnuovo, **3**, and Enriques, **7**, 2nd part.

three surfaces of rotation intersect: cylindrical, conical, and toric (87).

We have no precise information about the solutions of Eudoxus, referred to in the letter of Eratosthenes.

A solution of the problem obtained by means of a suitable instrument is attributed to Plato (88). The instrument (see fig. 5) is made of three fixed rods r, s and t, placed like three sides of a rectangle, and of a fourth rod q which runs up and down in such a way as to be always parallel to s. The two

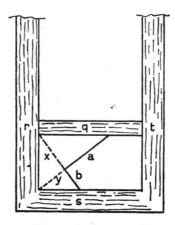

Fig. 5.

segments a and b between which it is wished to insert the two geometric means x and y, are set at right-angles and in such a way that their extensions pass through the points of intersection of the rods respectively qr, rs (making the rod q run up to join the indicated position).

By a well-known theorem (Euclid's 2nd) we immediately get: $a:x = x:y = y:b$; which we wished to obtain.

A contemporary of Alexander the Great, called Menaechmus, to whom is attributed the discovery of the conic sections, also solved the problem of Delos (89).

Bearing in mind that the geometrical properties of cones known to the ancients can be expressed in modern language

(87) See Conti, pp. 331–3, and Enriques–De Santillana, 1, pp. 250–1.
(88) See Conti, pp. 356–8. For another solution of the problem of Delos, by Eratosthenes, see pp. 358–9.
(89) See Conti, pp. 333–5.

through equations (see Chapter VII of this book), we will, for reasons of simplicity, consider the two parabolas (see fig. 6) having the equations (II) and (III). The real point P, different from the vertex of the two parabolas, in which these meet, furnishes with its distances from the axes the two geometric means required.

Other solutions discovered later were based on the use of curves of an order higher than the second; of these we shall speak in §3 of Chapter VII.

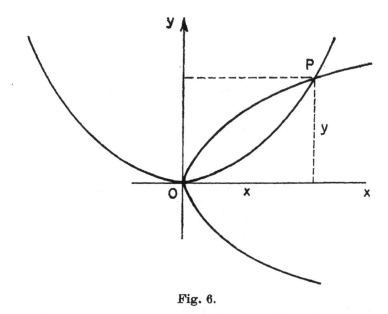

Fig. 6.

§14. *Curve of Hippias, trisection of the angle and squaring of the circle*

The duplication of the cube, the trisection of the angle, and the squaring of the circle, are called the classical problems of geometry. In the modern age it has been shown that it is impossible to solve these problems with only ruler and compasses; the first and the second problem are of the third degree, while the third is transcendental. The second and third of these can easily be solved if one considers a transcendental curve, the discovery of which is attributed to the sophist Hippias in the fourth century B.C.

To describe the curve in question (90) let us consider a square $ABCD$ (see fig. 7) and impart to the side AB a uniform translation towards the position CD, and to the side CB a uniform rotation with a centre C, towards the same final position CD. The two motions are regulated in such a way that the two mobile sections leave at the same time and arrive contemporaneously at the final position. The curve of Hippias is the geometrical locus of the intersection E, of the two mobile segments.

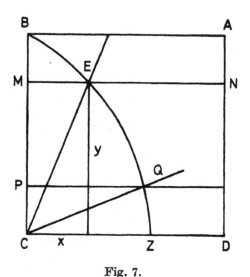

Fig. 7.

Giving unit value to the side of the square, and taking as our x and y axes, respectively, the lines CD and CB, by the given definition of the curve in question we shall have:

$$\frac{y}{1} = \frac{\widehat{ECD}}{\frac{\pi}{2}}, \qquad \frac{y}{x} = \tan \widehat{ECD}$$

from which

$$y = x \tan \left(\frac{\pi}{2} \cdot y\right)$$

(90) See Chisini, 1, p. 71. Loria, 2, pp. 70–1, and 160–2 quotes the passages of Pappus concerning the quadratrix.

The curve is easily drawn by making successive bisections of the segment BC and the angle $\widehat{BCD}$.

If we suppose that the curve of Hippias has already been drawn, it is easy to divide an angle into n equal parts and in particular to trisect an angle (91).

To trisect a given angle $\widehat{ECD}$ (see fig. 7) let us draw for E, the line parallel to CD, to meet BC at the point M, and divide CM into three equal parts. Let P be a point of CM such that:

$$CP = \tfrac{1}{3}CM$$

Let us draw to CD through P the parallel that will meet the curve at a point Q. For the definition of the curve of Hippias we will have:

$$\widehat{QCD} = \tfrac{1}{3}\widehat{ECD}$$

Other solutions of the problem of the trisection of an angle, which turns out to be of the third degree, in general irreducible, were preserved by Pappus. Some of these solutions are based on the use of conic sections.

We owe to Dinostratus the application of the curve of Hippias to the problem of squaring the circle, and for this reason the curve of Hippias was called the squarer (92).

Using modern language for the sake of simplicity, we see that it is easy to calculate as the limit the abscissa of Z, the point of meeting of the curve with the axis of x:

$$\lim_{y \to 0} x = \lim_{y \to 0} \frac{y}{\tan\left(\frac{\pi}{2}y\right)} = \lim_{y \to 0} \frac{2}{\pi} \cdot \frac{\frac{\pi}{2}y}{\tan\left(\frac{\pi}{2}y\right)} = \frac{2}{\pi}$$

Therefore

$$\overline{CZ} = \frac{2}{\pi}$$

Once CZ is known it is easy to rectify the circumference and square the circle, by finding proportions between segments.

Archimedes was to study cyclometry more deeply.

(91) Conti, pp. 378–9.
(92) Chisini, 1, p. 72.

CHAPTER IV

The Logic of Aristotle

§1. *The world of the universals of Aristotle and the fundamental principles of logic*

The dualism established by Plato between the world of the senses and the world of ideas failed to satisfy his great disciple Aristotle (384–322 B.C.) (1), who worked out a criticism of the Platonic ideas, observing, among other things, that if ideas constitute a world completely separate from the sensible one, they cannot be its essence, or furnish an explanation of it. According to Aristotle, universal concepts and ideas form the essence of things, and the said ideas are taken from the mind, having their origin in sensible data. We cannot delay over a minute examination of the Aristotelean theory on the origin of ideas; it will suffice to say that this theory is neither innate, nor simply empirical, and it is happily expressed in the formula of the Renaissance Aristotelian Patrizi: 'Cognitio omnis a mente primam originem, a sensibus exordium habet primum'(2).

These ideas constitute the world of the intelligible on which the logic of Aristotle works, contained especially in the *Analytica priora* and in the *Analytica posteriora*.

Aristotle has been called the father of Logic (even if it existed quite some time before him), in so far as he formulated the statement of the fundamental principles of logic, the theory of definition, of conversion, of syllogism, and the codification of the criteria to be used in the ordering of a deductive science.

The systemization given to logic by Aristotle, whom Trendelenburg called the Euclid of Logic, was inspired by mathematical theory (3).

(1) Trendelenburg has collected Aristotle's passages concerning logic (Greek text, Latin translation and commentary). On Aristotle's thought, see Calogero. On formal logic, according to the Thomists, see Maritain.
(2) See also Aristotle, 5, 19, 99, b, 20 ff. Calogero reports Patrizi's words.
(3) See Aristotle, 5, I, 2 (6).

60

For the beings of the world of the intelligible, of the universals, which according to Aristotle constitute the real essence of reality, the supreme principles of logic can be used: identity, non-contradiction, the excluded middle. Here too we will not go deeply into the precise meaning of these principles according to Aristotle, in whose work the first of them is not expressed clearly and explicitly (4).

The principle of non-contradiction was expressed by Aristotle in this form (5): 'It is impossible that the same property should hold and not hold for the same object in the same sense. . . . This principle is among the most certain of all. In fact it is impossible to suppose that the same thing should be and not be. . . . Therefore all who wish for demonstrations should refer to this certainty.'

'What is true should agree entirely with itself' (6). This latter affirmation is particularly important in making us understand Aristotle's attitude, and that of ancient mathematicians in general in the face of the problem of the non-contradiction or coherence of mathematical theories. From the classical point of view these theories, reflecting objective truths of the world of the intelligibles, which the human mind cannot master as it likes, cannot contain intrinsic contradictions (by contradiction is meant the affirmation and negation of the same judgment). As we shall see later, the mental position of the modern mathematician is very different (see especially Chapters XIX and XX of this book).

The principle of the excluded middle is expressed clearly and explicitly (7): 'Given an affirmation and a negation (of the same judgment) . . . one of these is true and the other is false.' Aristotle therefore builds up the bivalent classical logic, that is, one in which every proposition possesses one and one only of two values: true, false. All the same he leaves open the door to a trivalent logic in which there are these possible values of a proposition: true, false, doubtful. Through a useful definition

(4) Some passages from Aristotle, which seem to express the principle of identity (see *Metaph.* V, 9; VII, 12) have a different meaning.
Some commentators interpret *Metaph.* IV, 3 as a synthesis of the two principles, of identity and of non-contradiction. See Carlini, pp. 75–6 (footnote).
(5) Aristotle, 11, IV, 3; Trendelenburg, §9, pp. 3 and 26.
(6) Aristotle, 4, 1, 32; Trendelenburg, §9, pp. 3, 26–7.
(7) Aristotle, 2, c. 10; Trendelenburg, §10, pp. 4 and 27.

he shows us the field covered by the logic he developed (8):
'Not every speech is a judgment, but only those which may
be true or false.'

In Aristotelean thought the possibility of a proposition that
is neither true nor false is presented in connection with future
contingent propositions, for rigorous application of bivalence
would have led to rigid determinism; and this, since he championed the freedom of the will, Aristotle rejected (9).

This opening which Aristotle left for possible new systems of
logic was to be important in history and in present-day thought,
especially to the Epicureans in their opposition to the Stoics, to
the scholastics, particularly Occam, and finally to the polyvalent logic of Łukasiewicz. Another branch of the mathematical logical thought of our time in which the principle of the
excluded middle is not used is intuitionism, in which all the
demonstrations by reduction to the absurd are forbidden, in so
far as they are based on this principle (see Chapter V, §1;
Chapter X, §4; Chapter XIX, §3; Chapter XX, §2 of this book).

§2. *Concepts, their extension and comprehension*

In concepts we distinguish extension and comprehension.
The extension is made up of the set of elements to which the
concept in question is applied. Comprehension results from the
groups of characters or notes of the concept. Comprehension
diminishes as extension grows and vice versa.

Given two concepts in which the extension of the first contains the extension of the second, it is said that the first is a
genus in relation to the second, which is a species in relation to
the first.

Two operations are possible in regard to the extension and the
comprehension of a concept: division and definition; the first,
which Plato (10) had already dealt with, explains the extension,
the second the comprehension of the concept.

§3. *Definition* (11).

According to Aristotle, the definition expresses what is the

(8) Aristotle, **3**, ch. IV (4).
(9) For this topic, see above all ch. IX of *De interpretatione*.
(10) Plato, **7**, p. 262, cited by Trendelenburg p. 147; Vailati, **1**, pp. 678–9.
(11) Peano, **7**; Enriques, **4**; Piccoli.

essence of a thing, or else it explains the meaning of a name (12). In the first case we have the *real* definition, in the second the *nominal* definition. The distinction is already found in Aristotle, but he finds only real definitions important, since they answer the mind's aspiration to get hold of the essence of a being which we think of as existing outside ourselves in a sensible or intelligible world; while nominal definitions try to make precise a concept which is built up arbitrarily through given terms.

Aristotle recognizes that the definition of a being as itself does not guarantee the logical possibility of the being itself, and referring to Platonic geometry he writes (13): 'A geometer will indicate by means of a definition what the word triangle means, but that a triangle exists or that it is possible to construct one, and that one may therefore obtain results from the fact of having constructed it, is a truth which is neither admitted nor proved by means of the definition and which must be supposed or demonstrated separately.'

In Socrates, as we have seen, the problem of definition can already be posed in modern terms, by saying that it is a matter of determining the comprehension of a concept, the extension of which is known.

In this process we can find implicit the presupposition that to every naturally given class or set corresponds an idea which expresses its unity. From this conception is born Plato's theory of ideas, while for Aristotle the terms whose definition we search for are conceived in a natural order, in a hierarchy of genera and species.

Aristotle's procedure of definition is described by the scholastic formula: 'definitio fit per genus proximum et differentiam specificam' (14). It means that to get a definition, one must state what is the genus of minimum extension which contains the object to be defined, as a species, and afterwards explain the difference between the object and the other species contained in the above-mentioned genus.

Other types of definition were later discovered.

In modern thought an increasing importance is given to the

(12) Aristotle, 5, II, 3 and 7; Trendelenburg §54 and §55, pp. 16–17 and 39.
(13) Aristotle, 5, II, 7. Geymonat (2, p. 31) reports this passage.
(14) Aristotle, 6, 1, 8; Trendelenburg, §59, pp. 18 and 41.

nominal definition, and we stress the conception of mathematics considered as a construction of human thought: the definition seals with a name the result of a constructive process of thought.

For instance Candalla (who left a commentary on Euclid, in 1566) and Pascal recognize that all mathematical definitions are nominal.

In a certain sense Leibniz maintains the distinction between real and nominal definitions. Real definitions, to him, are those which include a judgment of possibility (or logical reality) of the definite, and nominal definitions those which do not include such a judgment. In these cases we have a nominal definition, *as well as* a postulate or a demonstration of existence.

According to the analysis of J. D. Gergonne (1771–1859) (15) the modern mathematical logicians consider the primitive concepts implicitly defined from the postulates that link them, while the other concepts successively introduced are defined explicitly in a nominal way.

All the same, as Enriques observed, the Socratic and Aristotelean problem of definition still corresponds to a need: the need to search for definitions which have some value for the progress of science: this is the historical and dynamic aspect of the problem of definition.

§4. *The categories of Aristotle*

According to Aristotle's conception of a really existing hierarchy of genus and species, in the process of definitions, moving towards ever more general concepts, we must reach concepts which can be considered only as genera but not as species.

Thus we have the '*summa praedicationis genera*' (16), the ten Aristotelean categories.

The first of these refers to 'being', the *substance* (that which is in itself), the other nine refer to the *accidents* (that which can be only in another being) (17). 'Each simple expression indicates either substance or quantity or quality or relationship or place or time or placement or possession or activity or passivity. Substance is, for instance, like man, horse; quantity, like two,

(15) Enriques, **2**, ch. III, n. 21.
(16) Trendelenburg, p. 56.
(17) Aristotle, **2**, c, 4; Trendelenburg, pp. 1 and 24.

three cubits; quality, like white, grammatical; relationship, like double, half, larger; place, like at school, in the forum; time, like yesterday, the previous year; placement, like lies, sits; possession, like is shod, is armed; activity, like saws, burns; passivity, like is sawn, is burnt.'

Later logicians composed the following mnemonic verse to remember the categories of Aristotle:

> 'Arbor sex servos calore refrigerat ustos
> cras ruri stabo sed tunicatus ero.'

§5. *Judgments and rules of conversion*

While concepts are not in themselves true or false, truth and falsehood appear, according to Aristotle, in judgments (18). These are classified into affirmative and negative, or else into universal, particular and indefinite (19). The universal judgments refer to all or to none, the particular to someone; for the indefinite one cannot say precisely if the judgment is universal or particular. Leaving aside the indefinite type we have four types of judgment.

Mediaeval logicians indicated the affirmative universal judgments with the letter A (*A*FFIRMO), the universal negatives with the letter E (N*E*GO), the particular affirmatives with the letter I (AFF*I*RMO), the particular negatives with the letter O (NEG*O*); they summarized this in the verse (20):

> '*A* affirmat, negat *E*, sed universaliter ambae
> *I* firmat, negat *O*, sed particulariter ambae.'

The relationship between these judgments and their concepts were systematically expressed by later logicians in the 'logical square', which sums up the doctrine on the subject.

Two contrary propositions cannot both be true but can both be false; two contradictories cannot both be true, neither can they both be false, but if one of them is true the other is false. If *A* or *E* is true, according to Aristotelean logic its subaltern is also true, respectively *I* and *O*, but the opposite passage is not generally valid. The sub-contraries *I* and *O* can both be true but cannot both be false (in fact if *O*, for example, is false its

---

(18) Aristotle, **3**, c, 4; Trendelenburg, pp. 1 and 24.
(19) Aristotle, **3**, c. 7; 4, I, 1; Trendelenburg, §6, p. 2 and p. 25.
(20) See Petrus Hispanus, tract. I, 1, 21, p. 7.

contradictory *A* is true and therefore its subaltern *I* is true as well, Q.E.D.).

From these considerations are derived the *rules for conversion*, given us by Aristotle (21), through which one proposition draws out another. For the universal negative proposition the *conversio simplex* is used, according to the scheme:

If 'no pleasure is good'... 'no good is a pleasure'. The universal affirmative propositions are converted into a particular (*conversio per accidens*):

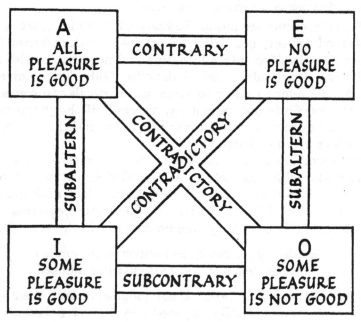

Fig. 1.

If 'every pleasure is good'... 'some pleasure is good'.

Modern logicians have observed that this form of conversion is valid only if the class relating to the subject is not an empty one, that is, contains elements; in our case if no pleasures existed we could not conclude that 'some pleasure is good' (22).

(21) Aristotle, 4, I, 2; Trendelenburg, §14, pp. 5 and 28.
(22) See Peano, 3, ed. 1903, pp. 18 and 317; Padoa, 1, chiefly on p. 79. Maritain, n. 84, pp. 263–73 supports classical logic against contemporary logics.

For a particular affirmative proposition we have the following type of conversion (*conversio simplex*):

If 'some pleasure is good'... 'some good is a pleasure'.

For the negative particulars, as Aristotle observes, you cannot apply this form of conversion:

If 'some animals are not men' it does not follow that 'some men are not animals'.

The examples so far given of conversion are those supplied by Aristotle.

In the mediaeval logicians (e.g. the *Summulae Logicales* of Petrus Hispanus) apart from the types of conversion already indicated we find too the *conversio per accidens* for the universal negative propositions, according to the form 'no man is stone' —'some stone is not man,' and the *conversio per contrapositionem* of the following form: 'every man is an animal'— 'every non-animal is not a man', and 'some men are not stones' —'some non-stones are not non-men' (that is some non-stones are men). The theory of the conversion of propositions is summed up in the following mediaeval verses which form the continuation of the preceding passages above:

'Simpliciter *feci* convertitur *eva* per acci,
Asto per contra, sic fit conversio tota.'

The rules of conversion apply in the theory of syllogism.

§6. *Syllogism*

All knowledge, according to Aristotle, derives from the syllogism or from induction (23). In the demonstrations (which are in fact made up of syllogisms) one proceeds from a general knowledge to a more particular one; inversely we have induction, through which from knowledge of particular cases we reach general laws (24).

Aristotle defined the syllogism as follows (25): 'Syllogism is a speech in which, some premises having been set, by the fact that these have been set, a result is necessarily derived from them', a result different from the premises, without the intervention of other extraneous notions.

(23) Aristotle, 4, II, 23; Trendelenburg, §20, pp. 7 and 30.
(24) Aristotle, 5, I, 18; Trendelenburg, §20, pp. 7 and 30.
(25) Aristotle, 4, I, 1; Trendelenburg, §21, pp. 7 and 30.

Aristotle called terms those concepts which intervene in a proposition: subject and predicate (26).

A fundamental principle of the syllogism is the following: 'What is said of the predicate is also said of the subject' (27). This principle (28) was taken by logicians after Aristotle as a foundation of syllogism, expressed under the form '*nota notae est etiam nota rei*', or also under the form which was called '*dictum de omni et de nullo*': 'Quidquid de omnibus valet, valet etiam de quibusdam et singulis, quidquid de nullo valet, nec de quibusdam et singulis valet.' Aristotle (29) observes that the two principles are equivalent: the first refers to comprehension, the second to extension.

In a syllogism there appear three terms in three propositions, of which the first two are the premises and the third is the conclusion. The middle term, the key to the syllogism, should appear in both premises.

Later logicians formulated in Latin verses eight rules to apply to syllogism; the most important of these, from which the remaining can be found, are the three following (30):

'Terminus esto triplex: major mediusque minorque

. . . .

utraque si praemissa neget, nihil inde sequetur

. . . .

nil sequitur geminis ex particularibus unquam.'

The various positions which can be occupied by the middle in the premises give place to the figures of syllogism which we shall represent through a scheme in which *M* is the middle term which appears in both the premises, *T* is the term which appears only in the first premise, *t* is the term which appears only in the second premise. The premise in which *T* appears will be called the major, the premise in which *t* appears will be called the minor. *Sub* and *prae* indicate respectively if the middle term acts as a subject or as a predicate.

The IV figure of the syllogism, also called the first indirect, is due not to Aristotle but to the physician Galen (A.D. 131–200).

(26) Aristotle, 4, I, 1; Trendelenburg, §22, pp. 7 and 30.
(27) Aristotle, 2, ch. 5; Trendelenburg, §23, pp. 7 and 30.
(28) Trendelenburg, p. 93.
(29) Aristotle, 4, I, 1, quoted by Trendelenburg, p. 93.
(30) Maritain, n. 72, pp. 219–24.

	I figure sub-prae		II figure prae-prae		III figure sub-sub		IV figure prae-sub	
Major Minor	M t	T M	T t	M M	M M	T t	T M	M t
Conclusion	t	T	t	T	t	T	t	T

If for every proposition of the premises of the four figures we substitute propositions of the type $AEIO$, we obtain in all 64 types of syllogism (in fact if we also vary the type of conclusion we get 256); but of these, bearing in mind the rules, only 19 give rise to possible cases in which it is allowable to form conclusions.

The 19 forms of syllogism can be remembered by means of the following mnemonic formula invented by the mediaeval logicians (31), where the vowels $AEIO$ of the first three syllables indicate respectively the type of premises and of conclusion in each form:

> Barbara, celarent, darii, ferion, baralipton.
> Celantes, dabitis, fapesmo, frisesomorum.
> Cesare, camestres, festino, baroco, darapti.
> Felapton, disamis, datisi, bocardo, ferison.

There the first four forms belong to the 1st direct figure, the next five to the 1st indirect, the next four to the 2nd, the last six to the 3rd. For the form of the 4th figure they also used this mnemonic formula: Bamalip, Calemes, Dimatis, Fesapo, Fresison.

To the 1st direct figure are applied directly the principles '*nota notae*' or '*dictum de omni et de nullo*'. The validity of the conclusions for the other figures is shown by reducing their forms to the 1st, applying the rules of conversion. For reasons

(31) Petrus Hispanus, 4–17, p. 41.

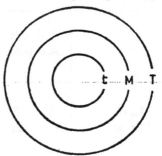

I. Syllogism in *barbara*

    every M is T
    every t is M
    ———————————
    every t is T

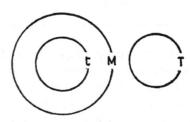

II. Syllogism in *cesare*

    no T is M
    every t is M
    ———————————
    no t is T

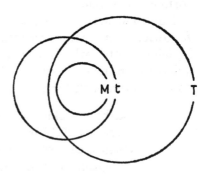

III. Syllogism in *darapti*

    every M is T
    every M is t
    ———————————
    some t is T

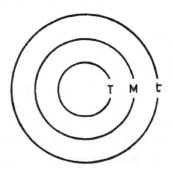

IV. Syllogism in *bamalip*

    every T is M
    every M is t
    ———————————
    some t is T

Fig. 2.

70

of brevity we will omit this demonstration and instead give an example of the scheme of syllogisms for each one of the four figures (see page 70).

For each of these schemes we also give the relative symbolic representation of the extension of the terms through the so-called circles of Euler, who thus showed the subordination of the concepts in the *Lettres à une Princesse d'Allemagne*. But this representation had already been used by Leibniz's teacher Joachim Jungius, and still earlier by Lodovico Vivès (1555) (32).

Modern logicians have observed that of the 19 forms of syllogism of classical logic, four of them need a supplementary existential hypothesis for their validity. These forms are those in darapti, bamalip, felapton, fesapo (33).

To clear up the matter let us take two examples of syllogism in darapti, in the first of which the existential hypothesis is satisfied, whereas in the second it is not:

> Every owl can fly,
> Every owl is a mammal,
> Therefore some mammals can fly.

The conclusion is valid since owls exist.

> Every mermaid has a fish's tail.
> Every mermaid is a good singer
> Therefore some good singers have fish's tails.

This time the conclusion is not valid since mermaids do not exist.

§7 *Aristotelean observations on various forms of reasoning*

Aristotle observes (34) that while it is not possible to deduce false conclusions from true premises, on the other it is possible to get true results from two false premises. It is easy to give examples of this last assertion: let us consider the two false premises: every quadruped is a polygon; every triangle is a quadruped; from these we get the true result: every triangle is a polygon.

(32) Enriques, **2**, p. 132–3.
(33) Padoa, **1**; Hilbert and Ackermann, ed. 1949, p. 48; Maritain (pp. 274–7) explains and supports the positions of classical logic on this point. On the 'conversio per accidens' and the syllogisms in 'darapti', etc. see Bocheński, **1**, pp. 139–45.
(34) Aristotle, **4**, II, 2; Trendelenburg, §32, pp. 10 and 32.

Aristotle fully analyses fallacious arguments; for the sake of brevity we will mention only his observations on defective reasoning by begging the question (35). This defect of reasoning appears when, to demonstrate a certain affirmation which is not self-evident, the affirmation itself is used in the reasoning. Aristotle examines various forms of this, starting with the pure and simple admission of what it is wished to demonstrate (possibly masked by synonyms) and going on to the case of a man who, for example, to demonstrate that the diagonal is incommensurable with the side of the square, admits that the side is incommensurable with the diagonal.

Aristotle examines reasoning *ad absurdum* (36) (which had already appeared in the school of Elea), presenting the reasoning itself according to the following scheme: we wish to demonstrate that A is not B; let us suppose instead that we have: A is B. Besides this, we have that B is C. Then we deduce from that that A is C. If by chance it is known that this result is false, then we get from it that A cannot be B; therefore A is not B, Q.E.D.

It is interesting to observe that Aristotelean logic always refers to aggregates made up of a finite number of elements. Aristotle does not admit the actual infinite and infinitesimal, but only the potential infinite and infinitesimal. He is careful to say precisely that the negation of the actual infinite and infinitesimal does not destroy the mathematicians' arguments (37).

'This reasoning [that denies the actual mathematical infinite] does not destroy the mathematicians' investigations, by denying that the infinite exists in this way, that is that it can increase without ever reaching a limit; since mathematicians have no need of the infinite and do not make use of it, but only ask [that is, postulate] that the finite [straight line] may become as long as they wish. And it is still allowed to divide another quantity of the greatest size in a given ratio, whatever that may be. Thus for them [mathematicians] there will be no difference in the demonstrations.'

(35) Aristotle, 6, VIII. 13; Trendelenburg, §42, pp. 13 and 36.
(36) Aristotle, 5, I, 26; Trendelenburg, §44, pp. 14 and 37.
(37) Aristotle, 7, 207, b, 27–33. This passage is reported in Rufini, 2, pp. 70–1. This work by Rufini deals extensively with: 'Infinite and infinitesimal after Aristotle'.

The question will be taken up again in Chapter VI, §2 of this book.

§8. *A brief glance at modal logic*

The classification of propositions according to the four types $AEIO$ (see §5 of this chapter) is not the only one Aristotle gives us.

Modal logic has its origins in another classification (38).

As Aristotle observes in his *Analytica Priora*, 'Every proposition means: either what it is, or else what it necessarily is, or else what it is in a contingent way'. Among the propositions thus characterized logical relations exist which Aristotle deals with in *De interpretatione* 13, and elsewhere. The scholastic were to take up these ideas again and they were to inspire Łukasiewicz in the construction of trivalent logic (see Chapter XX, §2 of this book).

§9. *Ordering of a deductive science according to Aristotle* (39).

'All rational knowledge,' says Aristotle (40), 'whether acquired or taught, always derives from previous knowledge. Observation shows us that this is true of all the sciences: in fact this is the way mathematics and all the arts proceed, without exception.' This passage clearly shows the influence exercised by mathematical thought on the logical structure of Aristotle's conception of rational theories.

Now from the very concept of knowing 'it necessarily follows,' Aristotle (41) continues, 'that deductive science proceeds from true principles, from immediate principles, better known than the conclusions, of which they are the cause and which they precede.'

Aristotle refutes the objections of two types of opponents of this doctrine, according to which:

(1) Either there are no principles and therefore the demonstration is impossible, giving place to an infinite regression, or

(38) Bocheński, 4, pp. 94 ff., pp. 260 ff.
(39) On this matter, see chiefly Enriques, 2, pp. 14 and ff.
(40) Aristotle, 5, I, 2 (6), translation by Enriques, 2, p. 15.
(41) See n. 40.

(2) the process of the demonstration is relative in so far as it can depart equally well from principles to prove the conclusions or vice versa.

We owe the first view, perhaps, to the empirical anti-mathematical philosophers. The second, which might come from the Megarics or from Democritus, gives us analogies with modern ideas on the arbitrariness of the choice of postulates as the basis of a mathematical theory (hypothetical deductive system: see Chapter XIX, §1, of this book).

But in Aristotle's system a relativism of this kind cannot occur, since the metaphysical theory of universals supposes an absolute order of truth on which the whole science must be based.

The principles of the science cannot therefore be demonstrated and are of various kinds (42):

(1) *terms or definitions*, that is to say assumptions of undefined concepts and definitions truly so called;

(2) suppositions of existence of the objects represented by the terms introduced;

(3) *axioms*: immediate propositions which it is necessary to know to learn anything;

(4) *postulates*, through which we ask the disciple to admit the existence of something.

In the fundamental propositions of his *Elements* Euclid, as we shall see, was inspired by these Aristotelean concepts.

Aristotle first used letters to indicate numbers and geometrical elements, and Euclid and later mathematicians followed him (43).

In Aristotle's enormous synthesis of knowledge mathematics had its place and two mathematical works which have been lost are attributed to him; in the works that remain to us, there are many interesting passages on mathematics, but for reasons of space we will recall only some of the subjects he dealt with.

He criticizes a theory of parallels, which preceded Euclid's systemization, finding that it begged the question. Hypo-

---

(42) Aristotle, 5, I, 2; Trendelenburg, §67, pp. 21 and 67; Aristotle, 5, II, ch. 9, ch. 12; Enriques, 2, pp. 17 and 18.
(43) Aristotle, 7, VII, ch. 6. For notes and bibliography concerning Aristotle's mathematical passages, see Loria, 2, pp. 131–6; Heath, 2.

theses that are probable but not certain have been made concerning this theory (44).

He gives the following definition of the continuum: 'A thing is continuous when two contiguous parts of it have the same boundary' (45) and in substance, in spite of the rather imprecise form, he also formulated the principle that is known as the postulate of Archimedes (46): 'By adding continuously to a finite size I will pass any limited size and by subtracting I will in the same way leave one which is smaller than any other'.

In Aristotle we also find traces of a definition of the equality of ratios, which precedes Euclid's (47). According to this definition, two ratios are called equal when they give rise to the same development in the process of finding the highest common divisor.

For centuries thinkers were persuaded by the perfection of Aristotle's logic that an edifice as formidable as his must be for ever immutable. In the *Critique of Pure Reason*, Kant was still saying that since Aristotle's time logic had not taken a single step forward or backwards (48).

Yet we shall see that, especially in our own day, through the influence of the development of mathematical thought, our view of logic has widened remarkably.

(44) Rufini, 1; Frajese, 4, pp. 108–14.
(45) See Clifford, p. 307.
(46) Aristotle, 7, VIII, 10, 266, b, 2, translated by Rufini, 1, p. 72.
(47) This matter is dealt with by Frajese, 4, p. 116.
(48) See Kant, p. 17, preface to the 2nd edition (1787) of the *Kritik der reinen Vernunft*.

CHAPTER V

Euclid's Elements in Hellenistic Culture

§1. *The Hellenistic world and the post-Aristotelean philosophies* (1).

After the great philosophical and scientific synthesis which Aristotle left us in his vast encyclopaedic work, science became separated from philosophy and the various sciences were divided up: this was specialization.

The conditions in which culture developed in the Alexandrine age must be seen in relation to the political and military events following the conquests of Alexander the Great, whose early death (323 B.C.) ended his efforts to unify the Greek and oriental worlds, which underwent a profound transformation. After the struggles that followed his death, a few large states were formed; the territorial state took over from the urban state. Ambitious princes made their courts centres of the development of art and culture, with all the advantages and disadvantages of patronage.

Ptolemy I (323–283 B.C.), contemporary with Euclid, founded the Museum of Alexandria where learned men could live and teach, using a fine library and scientific material. In these surroundings Euclid and many other scientists worked. Thus at Alexandria and other Hellenistic centres specialized sciences developed which had mostly broken the links with philosophy. This separation at first favoured the development of these specialized sciences, giving scientists a feeling of freedom from wearisome bondage. But with scientific activity divorced from living contact with the fundamental problems of being, of thought and of action, the sources of science began to dry up,

(1) Enriques–De Santillana, 1, pp. 347–56 and pp. 513–40. On the philosophers Geminus and Posidonius, dealt with in this ch., see pp. 385, 444–46, 526–8 of the above work.

intuition to wither, and before long decadence appeared. The first signs of scientific decadence in the Hellenistic period, for instance, can be seen in the indifference with which the school of Alexandria greeted the brilliant heuristic infinitesimal methods of Archimedes.

There were political and economic, as well as cultural, reasons for this scientific decline. Hellenic culture developed within the circle of the small town, in a highly specialized society in which the individual predominated. In the Hellenistic world, however, we find the same cosmopolitan crowd in the centres of Egypt, of Mesopotamia, of Gaul, producing mixed marriages of ill-digested ideas. In the Roman world the situation of science was to be no better.

Just when ancient society saw its ideals of liberty vanish, the post-Aristotelean philosophers, less interested than their predecessors in scientific problems, were mostly turning to morality.

Epicureans, Stoics and Sceptics, although their theories differed so widely, all aspired to find an individual ethic for the citizen of the world, to whom they taught the serenity of the wise man and the independence of the soul in the face of the trials of life. These views influenced the logical and epistemological conceptions of the philosophers of the Alexandrine period (2).

The Epicurean school was founded in 306 B.C. by Epicurus, who held that the object of life was pleasure, but pleasure understood in a high sense. Although the Epicureans partly carried on the rationalist traditions of Democritus, they thought that atoms fell from high to low (no longer moving in all directions, as, according to Democritus's conceptions, they had been), and attributed a slight deviation from the vertical to the atoms themselves (the '*clinamen*' of Lucretius) which would have made possible the meeting of atoms . . . and free will.

This anti-deterministic conception had an influence on the logic of the Epicureans, who asserted the '*fatis avorsa voluntas*' of Lucretius. Like Aristotle (see Chapter IV, §1 of this book) the Epicureans refused to apply the principle of the excluded third to all reality, as is shown in a passage of Cicero's *De Fato*,

(2) On the logic of the Epicureans, of the Stoics, and of the Sceptics, see Enriques, **2**, pp. 30–46, Bocheński, **3**.

which expressed his contrary opinion: 'Contrary to the opinion of Epicurus, it is necessary that of two contradictory propositions . . . one shall be true and the other false: as "Philoctetes shall be wounded" was true before all time, and "he shall not be wounded" false. Unless we wish to follow the Epicureans in saying that these propositions are neither true nor false; or else, ashamed to say so, say what is even worse, that the disjunctions of contradictory propositions are true, but that neither of the two propositions is true.'

The Stoic school was founded about 308 B.C. by Zeno of Citium, who identified the greatest good with virtue. Yet the Stoic cosmological conception is dominated by a rigid determinism which extends even to the field of morality: as the poet Manilius (3) said: 'Fata regunt orbem, certa stant omnia lege'. The unconditioned application of bivalence was upheld in particular by the philosopher Crisippus (head of the Stoic school between 232 and 204 B.C.). On this subject (4) Cicero says: 'We hold then to what was sustained by Crisippus: everything pronounced is true or false'.

For these reasons Łukasiewicz observes that polyvalent logics should not be called non-Aristotelean, but non-Crisippian (5).

Łukasiewicz (6) recognized the Stoic dialectic as the ancient form of the modern calculus of propositions, with which we shall deal particularly in Chapter XVIII, §5.

The Stoics studied the development of logical formalism and the analysis of language. The Stoic Ariston of Chios despised the dry formal schematism, however, and said: 'Those who study dialectics are like crab-eaters who for a mouthful of meat waste time over a whole heap of claws.'

The Sceptics, in the face of the Stoic and Epicurean dogmatism, developed a free criticism of the criteria of truth. The term 'sceptic' derives from a word which means 'I discern'

(3) Manilius, bk. IV, vv. 114.
(4) Cicero, 2 (38); 3, II, 95.
(5) Łukasiewicz, pp. 51 ff. On many-valued logics, see ch. XX, §2 of this book.
(6) Łukasiewicz, p. 77. *Plebe* compares Aristotelian and Stoic logic, in an essay, where the following 'scheme of deduction' by Crisippus is reported:
        'If the first, then the second.
        The first.
        Then the second.'
where 'first' and 'second' are propositions.

and brings with it the idea of methodical doubt. Scepticism, which does not form a school, but rather a movement of thought, began with Pirro of Elide (365–375 B.C.); and Arcesilaus and Carneades introduced it into Plato's Academy.

Carneades (who went to Rome as ambassador in 155 B.C.) held that any demonstration was impossible, because the demonstration would necessarily give place to a *'regressum in infinitum'*, since every premise would have to be reduced from another premise, and there would not exist premises from which to start reasoning, since the premises would derive from sensations and the uncertainty of sensations would react on the intelligence. But Carneades did not limit himself to purely destructive criticism; although he denied the existence of absolutely certain criteria of the true and the false, he attributed a probable value to knowledge; he examines what criteria can make such knowledge more probable: the evidence, the coherence of reasoning.

In the second century A.D. Scepticism found a supporter in Sextus Empiricus, author of a work *'Adversus Mathematicos'* (7). He quotes this sentence of the ancient Sceptics: 'People who speak obscurely are like those who shoot arrows in the dark.'

§2. *Preliminary notes on the Elements of Euclid and their contents*

Euclid's scientific work (8) was carried out in Alexandria at the beginning of the Hellenistic period around 300 B.C., and must be placed in relation to the critical movement revising the principles of geometry which took place in the Platonic Academy, and referred especially to Eudoxus of Cnidus, who is considered the founder of the method of exhaustion.

Euclid benefited from the results of this movement. From Proclus we know that he was younger than Plato's scholars and older than Archimedes.

Here is a brief summary of the subjects treated in Euclid's *Elements.*

(7) Enriques A.
(8) Heiberg has published a critical edition of the *Elements* of Euclid. For an English translation, see T. L. Heath (Bibliog.) and *History of Greek Mathematics* by the same author. Most translations of Euclid are here taken (with modifications) from T. L. Heath's translation. On Euclid's life and work, see Enriques–De Santillana, 1, pp. 357–8; 363–73.

The first book contains the fundamental principles on which the work is based (terms, postulates, common notions), properties of triangles (criteria of equality, etc.), properties of perpendiculars, theory of parallels and the sum of the angles of a triangle, theory of the equivalence of polygons (theorem of Pythagoras and its inverse).

The second book contains the so-called geometric algebra, that is the expression in geometrical form of the algebraic identities and the solution of quadratic equations.

The third book contains the fundamental ideas on circles: intersections, contacts, inscribed angles.

In the fourth book regular polygons of three, four, five, six and fifteen sides are constructed.

The fifth book contains the general theory of ratios between magnitudes.

In the sixth book this theory is applied to geometric figures, together with the theory of similitude.

The seventh, eighth and ninth books are devoted to arithmetic, to the properties of whole numbers (on Greek arithmetic see Chapter VIII, §§6 and 7 of this book).

The tenth book contains an elaborate classification of quadratic irrationals. This inspired the researches of the sixteenth-century Italian algebraists, who succeeded in solving cubic equations.

The eleventh, twelfth and thirteenth books deal with solid geometry. In the twelfth in particular we find applications of the method of exhaustion; equivalence of pyramids of equal base and height, volumes of pyramids, cones and cylinders, proportionality of the sphere to the cube of its radius.

In the thirteenth book, with the construction of the five regular polyhedra, the crowning glory of his *Elements*, Euclid pays homage to Plato's cosmological vision in the *Timaeus*.

It would be interesting to make a critical examination of Euclid's entire work, but for reasons of space we will study only what seems most relevant to the relationship between mathematics and logic.

§3. *The fundamental principles of Euclid's Elements*

These take the name of terms, postulates, common notions or axioms.

The terms are usually interpreted as definitions, but many of them cannot be considered as real and true definitions, to be totally substituted for the object to be defined, according to Aristotle's criterion which modern mathematical logicians use as well; so we shall, with Enriques, translate as 'terms'.

Here are those of most interest to our subject:

1. 'A point is that which has no parts.'

According to Proclus, this definition (if one may call it that) conforms to the criterion of Parmenides, according to which the negative definitions agree with the principles. Enriques maintains that the first of the terms tries to refute the concepts of the early Pythagoreans relating to the monadic point, which still had extension, and of which we have already spoken.

Frajese (9) prefers to consider the term in relation to Plato's definition of unity: 'What is really one must be said to be entirely without parts'; and with a passage of Aristotle: 'The indivisible in quantity is called unity, if it is indivisible in every way and has no place; but if it is indivisible in every way and yet has a place it is called a *point* ... what is quantitatively not divisible in any way at all is called a *point* and a *unity*: unity has no place and a point has.'

Euclid's definition of a point has, of course, aroused strong criticism: it is not applied in any demonstration in the *Elements*.

Waismann (10) observed: 'pain, for instance, has no part; is it therefore a point?'

2. 'A line is a length without breadth.'

This definition harmonizes with the Eleatic conception of fundamental geometric elements, and comes from the Platonic school (11). Proclus refers to another definition which is found in Aristotle (12): 'It is said that a moving line generates the surface, as the line is generated by the movement of a point.'

3. 'The extremities of a line are points.'

Note that endless lines and closed lines are not examined.

4. 'The straight line is that which lies evenly between its points.'

(9) Frajese, **4**, pp. 61–5; Plato, **5**, 245, a; Aristotle, **11**, V, 1016, b.
(10) Waismann, **2**, p. 111.
(11) Aristotle, **6**, VI, 6, 1436, 11.
(12) Aristotle, **9**, I, 4, 409, a, 4.

This definition can be interpreted in various ways (13).

(a) When any two points have been fixed on the line this line is no longer capable of alteration: that is, the line is perfectly characterized by its two points. We may observe then that axiom 9 becomes superfluous: 'Two lines do not enclose any space.' But it seems very unlikely that Euclid meant to say this.

(b) Having taken any segment of a straight line there are in the whole extension of the line segments equal to that segment; but this property is common to the circle and to the cylindrical-helix. This property of the helix is revealed by Apollonius (third century B.C.), the author of a treatise on conic sections.

(c) The straight line is divided into two equal parts by every point on it. This property is also common to the cylindrical helix, as Apollonius observed. For reasons of space we shall omit any other interpretations.

5. 'A surface is that which has only length and breadth.'

This definition recalls Parmenides' passage on surfaces without thickness and tallies with term number 2.

6. 'The extremities of a surface are lines.'

Observations analogous to those set out for term number 3 can be developed in regard to this.

7. 'A plane surface is that which lies equally in respect of its straight lines.'

Here we find the same obscurity and ambiguity met in term number 4, and we could try to interpret it in much the same way as we interpreted the definition of the straight line (14).

For reasons of space we will omit terms 8–21, except for:

14. 'A figure is that which is enclosed by one or more boundaries.'

This shows Euclid's dislike (15) for the actual mathematical infinite, a dislike which we have already met in Parmenides and in Aristotle. Let us go on to:

22. 'Among quadrilateral figures there is the square, which is equilateral, and the oblong; the oblong is rectangular but not equilateral.'

(13) Amaldi, chiefly pp. 42–4.
(14) Amaldi, p. 47.
(15) For further notice on this topic, see Frajese, **4**, pp 69–70.

Admitting the existence of a quadrilateral with four right angles is equivalent to admitting the validity of Euclid's postulate V.

23. 'Parallels are straight lines of a plane which, when produced both ways, do not meet.'

The definition of parallels as 'equidistant straight lines' appears again when vain efforts were made to eliminate the Fifth Postulate of Euclid. This definition comes from Posidonius, a Stoic philosopher, head of the school at Rhodes after 104 B.C., author of the most perfect ancient theory on tides, and to Geminus (first century A.D.), a philosopher of mathematics who studied the principles which are at the base of Euclid's *Elements*.

This definition makes Euclid's postulate V superfluous, but in its turn hides another postulate equivalent to V: the locus of the points situated on one side of a straight line, and having an assigned distance from this straight line, is still a straight line.

'The fallacy of complex definitions' is made clear by Saccheri (1667–1733).

The terms are followed by the postulates and the axioms or common notions. The distinction between postulates and axioms has been made in various ways by various authors (16).

According to Proclus postulates are the basis of problems, and axioms the basis of theorems; postulates have therefore an essentially constructive character: they guarantee the existence of certain elements which are to be constructed. This constructive character is evident in almost all Euclid's postulates, and with a little good will can be found in the end in all the postulates of the first book of the *Elements*.

We must remember, with regard to this, that, as Zeuthen (17) has shown, to the Greeks a construction meant the demonstration of the existence of figures. The moderns, on the other hand, sometimes assure themselves of the existence of a geometrical entity without constructing it, by applying the postulate of continuity (see, for instance, Chapter XVII, §3 of this book).

Usually Euclid reasons only from figures the construction of

(16) Enriques, 2, pp. 21–2. Frajese deals extensively with this topic in 4, pp. 72–92.

(17) Zeuthen, 1; Vailati, 1, p. 547; Enriques, 3, bk. I–IV, pp. 42–4, and 47–8.

which he gives; but there are exceptions—for instance the one relating to the existence of the fourth proportional, which is at the base of the method of exhaustion (18).

According to another point of view, which we hear about from Proclus, while the common notions are principles common to the various sciences, the postulates have to do with a particular science, and especially geometry.

According to a third view of Aristotelean origin—perhaps the best known, although it is contrary to the spirit of contemporary mathematical logic—there is absolute evidence for the axioms, while the postulates are accepted on a basis of intuition only because we have no demonstration for them.

We know from Proclus that some ancient geometers rejected, as do the moderns, the distinction between postulates and axioms.

Let us now examine the postulates:

1. 'It is asserted that from any point a straight line may be drawn to any other point';
2. 'and that any terminated straight line [we should say segment] can be produced to any length in a straight line';
3. 'and that with every centre and every distance a circle can be described';
4. 'and all right-angles are equal to one another.'
5. 'If a straight line meet two straight lines, so as to make the two interior angles on the same side of it taken together less than two right angles, these straight lines, being continually produced, shall at length meet on the side on which are the angles which are less than two right-angles.'

What is clear about this famous postulate, which was meant to establish the conditions of the existence or the constructibility of a point as an intersection of two straight lines, is its constructive and existential character.

In the first 28 propositions Euclid does not make use of postulate V, even at the cost of giving less significant theorems than those which he could have demonstrated if he had used it.

Of Euclid's theory of parallels we shall speak shortly.

Let us now glance briefly at the common notions.

Aristotle calls propositions of this kind axioms, using a term

(18) Frajese, 4, pp. 91 and 206.

of Pythagorean origin. Enriques (19) maintained that the expression: 'common notions' comes from the philosopher Democritus, who was the author of a book on the elements of geometry.

The first 8 common notions are the following;

'Things which are equal to the same thing are equal to one another.'

'And if equal things be added to equal things the wholes are equal.'

'And if equal things be taken away from equal things the remainder is equal.'

['And if equal things be added to unequal things the wholes are unequal.']

'And doubles of the same thing are equal to one another.'

'And halves of the same thing are equal to one another.'

'Magnitudes which coincide with one another are equal to one another.'

'And the whole is greater than its parts.'

To Euclid the equality of geometrical figures is an equality of extension. For instance, he says that two triangles with two sides and an included angle equal, have equal surfaces (in modern language: they are equivalent); then he adds that the two triangles also have their third pair of sides equal and the angles opposite to those sides equal as well.

According to Proclus, Apollonius tried to demonstrate the 1st axiom of the *Elements*. The following fragment of this effort has been preserved (20): 'If a is equal to b and b is equal to c, I say that a is equal to c. In fact a occupies the same place as b, and thus b occupies the same place as c; therefore a also occupies the same place as c.' As Enriques notes, it appears that this reasoning tries to reduce the concept of geometrical equality to that of the superposability of figures through movement, and perhaps Apollonius deluded himself into thinking that he could reduce the given proposition to identical propositions. But by using ideal experiences concerning movement we see that the axiom in question has a synthetic meaning, and cannot therefore be called true by definition, like an analytical proposition.

(19) See Enriques, **2**, pp. 20–1.
(20) Enriques, **2**, pp. 22–3.

With regard to the axioms of inequality, it is worth noting that, for instance, the axiom, 'The whole is greater than its parts' is no longer valid for infinite aggregates or infinite figures. But as we have seen, Euclid refuses to consider these figures.

The last common notion, the 9th, looks quite out of place and would seem originally to have been among the postulates: 'And two straight lines do not enclose a space'.

In the *Elements* Euclid uses implicitly some principles which he does not set out (21). What is noticeable above all is the absence of postulates relating to the order of points on a straight line, and to the division of the plane by means of straight lines (the elementary properties of angular regions are linked to these considerations). In 1832 Gauss drew the attention of Bolyai to these concepts. We owe to Pasch a logical systemization of geometry in which these postulates are borne in mind.

§4. *The first book of the Elements, in particular the theory of parallels*

The propositions of the first book of Euclid's *Elements* aim at two really significant results: the sum of the angles of a triangle is equal to two right-angles (prop. 32); the theorem of Pythagoras (prop. 47).

The first of these results is based on the theory of parallels and is characteristic of Euclidean geometry. Let us see therefore how Euclid builds his theory of parallels and manages to determine the sum of the angles of a triangle.

The first Euclidean proposition refers to the well-known construction of an equilateral triangle of which we are given one side. After two propositions on the transfer of segments (2 and 3) we find (4) the 1st criterion of the equality of triangles (demonstrated through movement).

There follow the theorems on the angles at the base of an isosceles triangle (5 and 6).

Propositions 7 and 8 take us to criterion III of the equality of triangles.

In propositions 9 and 10 we bisect respectively an angle and a segment; from 11 to 14 we find theorems and constructions

(21) Enriques, **3**, bk. I–IV, p. 51.

relating to perpendiculars; and in 15 is demonstrated the equality of opposite angles at the vertex.

Finally propositions 16 and 17 are highly interesting for the theory of parallels.

16. 'If one side of a triangle be produced, the exterior angle is greater than either of the interior opposite angles.'

(By using postulate V, until now never invoked, this more significant proposition can be demonstrated: the exterior angle of a triangle is equal to the sum of the two internal opposite angles. But it is possible to establish proposition 16, without using postulate V; by using the *unlimited prolongability of a*

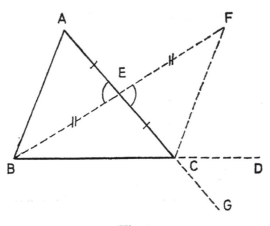

Fig. 1.

segment. In a geometry in which this unlimited prolongability did not exist we should not be able to demonstrate proposition 16.)

Now we come to the demonstration (see fig. 1).

Let us consider a triangle ABC, and produce the line BC to a point D. I say that

$$\widehat{ACD} > \widehat{BAC}$$

as well as

$$\widehat{ACD} > \widehat{CBA}$$

In fact: let E be the mid-point of the segment AC, and let

87

us produce the segment BE beyond E to a point F so that

$$BE = EF$$

Now join F to C and consider the two triangles ABE, CFE equal through the first criterion of equality. In particular we shall have:

$$\widehat{ACF} = \widehat{BAE}$$

But

$$\widehat{ACD} > \widehat{ECF}$$

therefore

$$\widehat{ACD} > \widehat{BAE}, \text{ that is } \widehat{ACD} > \widehat{BAC}$$

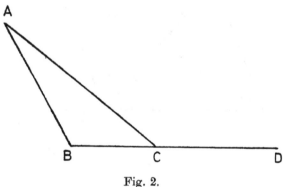

Fig. 2.

Analogously, producing the segment AC to a point G, and bisecting BC, we show that:

$$\widehat{BCG} \text{ (equal to } \widehat{ACD}) \text{ is greater than } \widehat{ABC} \text{ Q.E.D.}$$

From 16, 17 is easily deduced.

'In every triangle any two angles are together less than two right angles.'

Take (see fig. 2) the triangle ABC. Let us produce the side BC to D. We shall have by the preceding proposition:

$$\widehat{ACD} > \widehat{BAC}$$

88

Adding to both the members of this inequality the angle $\widehat{ACB}$ we obtain:

$$\widehat{ACD} + \widehat{ACB} > \widehat{BAC} + \widehat{ACB}$$

that is

$$\widehat{BAC} + \widehat{ACB} < 2 \text{ right angles.}$$

The other inequalities we mentioned could be demonstrated in the same way.

Euclid makes no use of this proposition (22) since it is completely absorbed by proposition 32, but it satisfies the ancient geometer's wish to go as far as possible without making use of postulate V.

Proposition 17 shows us Euclid's dilemma in the face of postulate V, which as far as possible he tries to do without; the proposition is almost a spring-board towards non-Euclidean geometry, as indeed it was to be for Saccheri and Legendre (1752–1833).

In propositions 18 to 25, we find, apart from some constructions, the demonstration of some inequalities relating to angles and sides of triangles: for instance in a triangle the greatest angle is opposite to the greatest side; one side of a triangle is less than the sum of the other two sides, etc.

In proposition 26 Euclid establishes (independently of postulate V) the 2nd criterion of congruence of triangles, both in the case in which the sides given as equal are terminated by angles which by hypothesis are equal, and in the other case, which in modern geometry books is usually established as a result of the theorem that the sum of the angles of a triangle is equal to two right angles (therefore using Euclid's postulate V or another equivalent to it).

Finally we meet the first proposition in which Euclid introduces the concept of parallels, but without yet making use of the famous postulate.

27. 'If a straight line falling on two other straight lines, makes the alternate angles equal to one another, the two straight lines are parallel to one another.'

(22) Frajese, 4, pp. 99–100.

The two straight lines AB and CD are cut by a transversal respectively at the points E and F; the angles $\widehat{AEF}$ and $\widehat{EFD}$ are by hypothesis equal: I say that the two straight lines are parallel.

In fact (see fig. 3), let us suppose that the two lines are not parallel: in such a case they will meet, for instance, when B and D reach G. Thus a triangle EFG will be formed in which the external angle $\widehat{AEF}$ will be equal to the internal opposite $\widehat{EFG}$, which, by proposition 16, is absurd. Similarly it is shown that the two straight lines cannot meet at any other point.

From proposition 27 we can easily deduce (proposition 28)

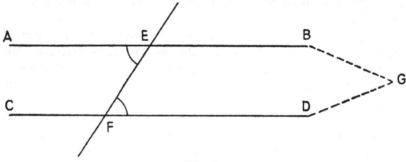

Fig. 3.

that, if a straight line cutting two other straight lines makes the alternate angles equal or the corresponding angles supplementary, then the two straight lines are parallel. But in order to invert proposition 27 postulate V must be used for the first time.

29. 'If a straight line fall on two parallel straight lines, it makes the alternate angles equal to one another, and the corresponding angles supplementary.'

The line EF (see fig. 4) meets the two parallels AB and CD respectively at the points G and H.

I say that the alternate internal angles $\widehat{AGH}$ and $\widehat{GHD}$ are equal. Let us reason indirectly: let us suppose, for example, that $\widehat{AGH} > \widehat{GHD}$. Let us add to both the members of the inequality the angle BGH; we shall obtain:

$$\widehat{AGH} + \widehat{BGH} > \widehat{GHD} + \widehat{BGH}$$

that is

$$\widehat{GHD} + \widehat{BGH} < \text{2 right angles.}$$

Therefore *by postulate V* the two straight lines *AB* and *CD* must meet by the side of *B* and *D*. But since the two lines *AB* and *CD* are by hypothesis parallel, this is absurd.

It is now easy to complete the demonstration of the statement.

30. Transitive property of parallelism: 'Parallels to the same straight line are parallel to each other.'

This is shown by considering the alternate angles and applying propositions 29 and 27.

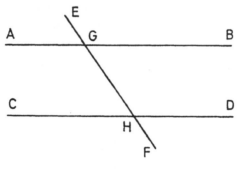

Fig. 4.

Proclus observes that from this proposition we immediately get the uniqueness of the parallel to a given straight line through an assigned point. Vice versa, having admitted this uniqueness it is easy to demonstrate proposition 29, which contains a proposition equivalent to postulate V of Euclid (his contranominal: see §5 of this chapter).

Thus is demonstrated the equivalence of Euclid's postulate V and the uniqueness of the parallel to a given straight line through an assigned point.

31. 'Through a given point to draw a straight line parallel to a given straight line.'

The construction is based on the equality of alternate angles.

32. The sum of the internal angles of a triangle:

'If a side of any triangle be produced, the exterior angle is equal to the sum of the two interior and opposite angles; and the three interior angles of every triangle are together equal to two right angles.'

Given a triangle ABC (see fig. 5), let us produce BC to the point D and draw through C the parallel CE to the line AB. Let us suppose:

$$\widehat{CAB} = \alpha, \widehat{ABC} = \beta, \widehat{BCA} = \gamma, \widehat{ACD} = \delta, \widehat{ACE} = \varepsilon, \widehat{ECD} = \zeta$$

We shall have:

$$\alpha = \varepsilon$$
$$\beta = \zeta$$
$$\alpha+\beta = \varepsilon+\zeta = \delta$$
$$\delta = \alpha+\beta$$
$$\alpha+\beta+\gamma = \delta+\gamma = 2 \text{ right angles}$$
$$\alpha+\beta+\gamma = 2 \text{ right angles} \qquad \text{Q.E.D.}$$

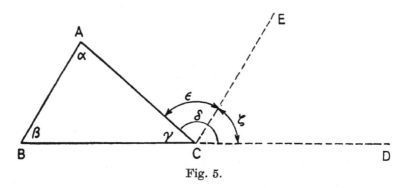

Fig. 5.

Eudemus, quoted by Proclus, attributes the discovery of this theorem to the Pythagoreans.

For many centuries vain efforts were made to demonstrate it without introducing a new postulate. Saccheri and Legendre, as we shall see later (chapter XV, §2 and 3 of this book), made it possible to establish that if, in a single triangle, the sum of the angles is equal to two right angles, this happens in every other triangle, and in that case the validity of postulate V is ascertained.

Substantially, admitting the existence of a single triangle in which the sum of two angles is equal to two right angles is a postulate equivalent to Euclid's V.

But we shall return to this subject in its proper place.

§5. *Application of the laws of the converse in the Elements*

Another reason why Euclid formulated proposition 17, although it is not applied in the *Elements*, depends, according to Frajese (23), on his wish (a wish which we can find in other cases too) to complete the quadrilateral of the propositions *direct, converse, contrary,* and *contranominal*.

To explain the meaning of these terms, let us indicate with H (hypothesis) and T (thesis) two propositions with $\bar{H}$ and $\bar{T}$ their respective negations; while to indicate that H is deduced from T let us write H→T (following the symbolism of Hilbert). We can construct the four following compound propositions:

direct   converse
H → T   T → H

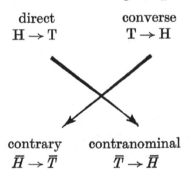

contrary   contranominal
$\bar{H} \to \bar{T}$ $\bar{T} \to \bar{H}$

Let us express proposition 17 in the form: If two straight lines cut by a third form a triangle, the sum of the two internal angles is less than two right angles.

It is now easy to see that if we take 17 as a *direct* proposition we shall have:

as the *converse* proposition the postulate V,
as the *contrary* proposition the 29th (II theorem of parallels),
as the *contranominal* proposition the 28th (I theorem of parallels), since it concerns the conjugate internal angles.

Note that if the direct proposition is true, the converse is not always true, and to be established it needs a suitable demonstration or a new postulate.

But if the direct proposition is true the contranominal is

(23) Frajese, 6. Peano, **3** (ed. 1903, pp. 18 and 317) makes some interesting remarks on the first law of the converse (in reference with the calculus of classes). Peano found such a law in Diogenes Laertius (VII, 80) who attributes it to Zenon the Stoic. This law is clearly expressed by some scholastics, e.g. by Albertus Magnus, t. 1, p. 249.

always true; this logical law is called the *first law of the converses*. It is demonstrated in the following way:

Let the direct $H \to T$ be valid by hypothesis, besides let $\bar{T}$ be valid again by hypothesis; H or $\bar{H}$ should be valid; if H is valid by hypothesis, T should be valid too, but this is absurd because by hypothesis $\bar{T}$ is valid; we therefore conclude that $\bar{H}$, that is $\bar{T} \to \bar{H}$, is valid, Q.E.D.

Observe, besides, that the contrary is the contranominal of the converse, therefore if the first is valid so is the second. Therefore if the direct and the converse are valid all the propositions of the quadrilateral are valid too.

Substantially Euclid follows the same scheme of reasoning as in the general demonstration of the first law of the converse, when he demonstrates, starting from postulate V, proposition 29, while to demonstrate 28 he prefers to go back to 27.

The law we have mentioned recalls the *II law of the converses*: if on a determined subject all the possible hypotheses have been made, $H_1, H_2 \ldots$ from which respectively have been obtained the theses $T_1, T_2 \ldots$ which exclude one another, the theorems converse to those demonstrated are valid too. That is, under the conditions considered, if by hypothesis the theorems are valid:

$$H_1 \to T_1, H_2 \to T_2 \ldots$$

we will also have

$$T_1 \to H_1, T_2 \to H_2 \ldots$$

In fact if by hypothesis T_i, for instance, is valid, either H_1 or H_2 must also hold, because all possible hypotheses have been made. But if we had H_k with $k \neq i$, by hypothesis T_k should hold too, which is absurd because T_i is incompatible with T_k; therefore H_i must be valid, that is $T_i \to H_i$, Q.E.D.

Euclid follows this method of reasoning to obtain proposition 25 from propositions 4 and 24. (Proposition 4 is the first criterion of congruence of triangles, and 24 says: 'If two triangles have two sides of the one equal to two sides of the other, each to each, but the angle contained by the two sides of one of them greater than the angle contained by the corresponding sides of the other, the base of that which has the greater angle

94

is greater than the base of the other.') Proposition 25 says:
'If two triangles have two sides of the one equal to two sides
of the other, each to each, but the base of the one greater than
the base of the other, the angle contained by the sides of that
which has the greater base is greater than the angle contained
by the sides equal to them, of the other.'

In the *Elements* we do not find the statements and demonstrations of the laws of the converse (Euclid did not mean to
write a treatise on logic), but we find the law applied in the
Elements themselves.

On logical procedures applied by Euclid Proclus (24) writes:
'His *Elements of Geometry* is especially admired for the order
which reigns there, and for the choice of theorems and problems
assumed as fundamental, since he has not included all he might
have used, but only those which are capable of functioning as
elements; and also for the variety of his methods of reasoning,
which he used in all possible ways, and sometimes makes their
proofs through reasoning, and sometimes through facts, but
are always irrefutable, precise and learned, and of the most
scientific character. He uses, too, all the processes of dialectics:
the method of *division* to determine the species, that of *definition* for the essential reasoning, the *apodeictical* in proceeding
from principles to the unknown, the *analytical* in proceeding
inversely from the unknown to principles. This treatment
shows us very distinctly the various kinds of reciprocal propositions, sometimes simple and sometimes complicated, the
reciprocity taking place either between the whole and the
whole, or between the whole and a part, or between a part and
the whole, or between a part and another part. And then,
what can we say of his methods of research, of the economy and
order which precedes what is to follow, of the strength with
which every point is consolidated?'

According to A. Guzzo, Euclid realized how arbitrary, at
least within certain limits, were the fundamental principles
set down by geometry as the basis of its rational construction (25).

(24) This passage may be found in the translation given by Loria, **2**,
p. 189.
(25) Guzzo, **2**, pp. 15–60, makes philosophical remarks on the logical
structure of the *Elements* of Euclid.

§6. *The angle of contingency in Euclid and the precursors of non-Archimedean geometry*

The Euclidian definition of an angle, the 8th of Book I, although it does not satisfy the needs of Aristotelean logic or those of modern logic, has the virtue of including curvilinear angles, although it excludes flat angles: 'A plane angle is the inclination of two lines to one another in a plane, which meet together, but are not in the same direction.' The rectilinear angle is defined thus:

'A plane rectilinear angle is the inclination of two straight lines to one another, which meet together, but are not in the same straight line.'

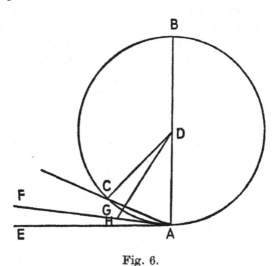

Fig. 6.

Usually Euclid confines himself to considering rectilinear angles, but once, in Book III, proposition 16, he obtains an interesting result regarding the angle formed by a circle and a tangent, which was later, in 1220, called the *angle of contingency* by Giordano Nemorario.

In this proposition it is demonstrated first of all that:

'The straight line drawn at right angles to the diameter of a circle from the extremity of it falls without the circle.'

Let us consider (see fig. 6) a circle *ABC* with a diameter *AB* and a centre *D*, and the perpendicular in *A* on the line *AB*. Reasoning indirectly let us suppose that this perpendicular falls inside the circle, meeting the circumference at a point C. Thus we obtain a triangle DAC, isosceles because $DA = DC$, and therefore $\widehat{DAC} = \widehat{DCA}$; but $\widehat{DAC}$ is by hypothesis a right angle, therefore $\widehat{DCA}$ must also be a right angle; therefore the triangle $\widehat{DAC}$ will have to have two right angles, which is absurd. It is therefore concluded that the perpendicular considered falls outside the circle.

Successively, still in proposition 16 of Book III, Euclid shows that:

'No straight line can be drawn from the extremity between that straight line and the circumference, so as not to cut the circle.'

Reasoning again indirectly, let us suppose that a straight line AF falls between the perpendicular AE and the circumference considered. Let H be the foot of the perpendicular lowered from D on to the line AF. In the triangle DHA the angle in H is a right angle by construction; therefore the angle in A must be acute. Consequently $AD > DH$. But indicating with G the point common to DH and to the circumference, we have $AD = DG$, therefore $DG > DH$, but that is absurd because DG is a part of DH. Therefore no straight line through the point of tangency can intervene between the circumference and the tangent considered.

Finally Euclid demonstrates what is most interesting to us: 'The angle contained by the arc AGC and the straight line [tangent] AE is less than any rectilinear angle.' In fact, Euclid says, 'if there were a rectilinear angle . . . less than the angle contained by the arc CGA and the straight line AE, we could interpose, between the arc CGA and the straight line AE a line that would form . . . an angle less than the angle contained by the arc CGA and the straight line AE. But no such line can be interposed. For this reason no acute [rectilinear] angle can be . . . less than the angle formed by the arc CGA and the straight line AE.'

Thus we see appearing in the mathematical world a kind of magnitude gifted with rather remarkable qualities: if rectilinear, curvilinear, and mixtilinear angles, among them the angles of contingency, are considered as magnitudes of the same kind, the postulate of Archimedes is not valid for this kind of magnitude: given two different magnitudes, there exists a multiple of the lesser which exceeds the greater, and therefore, having admitted the divisibility of the greater one, there exists a sub-multiple of the greater which is smaller than the lesser. In fact in our case no sub-multiple of a rectilinear angle is less than the angle of contingency.

But from his theory of proportion, Euclid wants to banish magnitudes for which he cannot define ratio, as was to happen in the case of a rectilinear angle and an angle of contingency; so he explicitly excludes from his consideration those magnitudes for which Archimedes' postulate (26) is not valid, introducing definition 4 of Book V: 'Magnitudes are said to have a ratio when each of them, when multiplied, is greater than the other' (27).

Archimedes' attitude to magnitudes which we shall call non-Archimedean (that is, magnitudes for which the postulate that takes its name from him is not valid) is similar to Euclid's. Archimedes too does not exclude the logical possibility of these magnitudes, but at a suitable moment he narrows the field of his enquiries, in such a way as to leave them outside his rational construction.

What Archimedes writes in his work on the quadrature of the parabola is interesting with regard to this: 'Let us demonstrate effectively that every segment limited by a straight line and a parabola is greater by a third than the triangle having base and height equal to that of the segment, further admitting, for the demonstration, this assumption: the difference between two different areas can be added to itself until it is greater than every given finite area.

'The geometers who came before us have used this assumption: they used it to demonstrate that the ratio of two circles one to the other is equal to that of the squares of their diameters, and that the ratio of two spheres is equal to that of the cubes of their diameters. Further, they used a similar assumption to demonstrate that every pyramid is a third of the prism having

---

(26) Archimedes formulates this postulate in many passages of his work: e.g. in postulate V of his work 'on the sphere and the cylinder'. Enriques thinks that the postulate may be attributed to Eudoxos of Cnidos.

(27) Frajese 4, pp. 117–21.

base and height respectively equal to those of the pyramid, and that every cone is a third of the cylinder having base and height equal to those of the cone. Yet these theorems are not considered any less real than those considered without this assumption, and I shall be satisfied if those which I now publish are treated with the same degree of certainty.'

As Bortolotti (28) observes, Archimedes admits that his assumption does not express a necessary truth, but rather, as it did for Euclid, a condition which could be applied to the magnitudes for which his theories were valid; it does not, therefore, exclude the possibility of the existence of magnitudes which do not satisfy the assumption. Archimedes thus mentions implicitly the logical possibility of non-Archimedean magnitudes.

On the angle of contingency, which, although it can be increased, diminished, and divided into parts, is always less than a rectilinear angle, subtle arguments (29) took place, with Democritus perhaps taking part in them; and Proclus, in his comment on the 8th definition, hesitates to speak of the angle of contingency as a true angle.

Throughout the ancient world, the Middle Ages, and the modern age, the nature of angles of contingency was discussed at length: can they be considered non-null magnitudes? Is the actual mathematical infinitesimal logically possible?

We shall see later how, through these considerations, we reach non-Archimedean geometry (see Chapter XVII, §3 of this book).

§7. *The theory of proportion*

The discovery of incommensurable magnitudes by the Pythagorean school made it necessary, as we have seen, to construct a new theory of proportion which was valid in their case (30). The rigorous systemization of this theory, which found a place in Book V of Euclid's *Elements*, was attributed by an anonymous commentator of the book to Eudoxus of Cnidus, the mathematician, philosopher and astronomer, who was a contemporary and friend of Plato (but perhaps something of a

(28) Bortolotti, **17**.
(29) Enriques, **3**, bk. I–IV, pp. 216–20.
(30) Vailati, **2**, pp. 143 ff.; on the same argument, especially concerning the structure of Bk. V of the *Elements*, see Frajese, **4**, pp. 115–37; Frajese, **7**.

rival too). Archimedes seems to support this theory by saying that we owe Eudoxus the rigorous demonstration relating to the ratios between pyramids and prisms, cones and cylinders, with equal bases and equal heights (31).

In Euclid's treatment of the theory of proportion, which, logically speaking, is wonderfully delicate, we find, all the same, a first definition of ratio, the 3rd of Book V, which does not satisfy the needs we have mentioned:

'Ratio is a mutual relation of two magnitudes of the same kind to one another in respect of quantity.'

This explanation tells us only that ratio will be defined only for homogeneous magnitudes, and refers to quantity.

In the previous paragraph we have dealt with the 4th definition, which limits the magnitudes whose ratio is to be studied, to Archimedean magnitudes.

Definition 5, on which the Euclidian theory of proportion is effectively based, is the following:

'Two magnitudes A and B are said to *stand in the same ratio* to two others C and D, where for any pair of numbers m and n for which we have:

$$mA \gtreqless nB$$

we also have respectively:

$$mC \gtreqless nD.'$$

Observe that, while Euclid does not give us a definition of ratio according to the Aristotelean rules (*per genus proximum et differentiam specificam*), since we cannot consider the 3rd of Book V a real definition, he gives us, instead of the concept indicated, a *definition* of a new type, called by modern logicians *by abstraction* (32).

In the definition 5 the concept of 'ratio' is defined by saying what is meant by 'equal ratios'. Modern philosophers, such as Grassmann, Helmholtz, Mach, Maxwell, and Vailati, have shown the importance of definitions of this type in physics and economics, as for example when one wishes to define the 'temperature' or the 'value', etc.

But modern logicians hold that, if a given relationship is to

---

(31) Enriques, **8**, p. 24.
(32) Enriques–De Santillana, **1**, pp. 369–70. Enriques, **2**, pp. 144–51, makes interesting remarks on definition by abstraction.

be considered an equality (the abstract concept of which will be defined), this relationship must satisfy the following properties: the *reflexive* $(a=a)$, the *symmetrical* (if $a=b$, $b=a$), and the *transitive* (if $a=b$ and $b=c$, then $a=c$). Now in the case of equality of ratios, the first two properties follow at once, while the third is demonstrated by Euclid, who thus shows he recognizes the requirements of modern logic.

It is instructive to investigate why Euclid laid down the elaborate definitions (at first sight not very natural), which we have examined. To understand his reasons we must compare the definition of equality of ratio between magnitudes with the definition of the equality of ratios between whole numbers (proportion), given by Euclid himself (Book VII, definition 20): 'Four numbers are in proportion when the first and the third of these are obtained respectively from the second and the fourth by multiplying them by the same whole number, or dividing them by the same whole number, or by doing one thing and the other.' That is:

$$a:b = c:d$$

when

$$a = \frac{n}{m}b \text{ while } c = \frac{n}{m}d$$

We have to explain why Euclid does not define proportion between magnitudes in a similar manner (which seems simpler and more natural).

We must remember in the first place that the Greeks generally thought they could reason about a magnitude only when they knew how to construct it. Now it was not always known how to divide a particular magnitude (for instance an angle) into n equal parts with the means offered by Euclid's *Elements* (ruler and compasses); think, for example, of the trisection of an angle. Because of this, instead of dealing with magnitudes which were to be divided into equal parts, Euclid would have to begin by dealing instead with the magnitudes which were to be multiplied by whole numbers (these two relations are substantially equivalent). Thus he would obtain:

$$A:B = C:D$$

when

$$mA = nB \text{ while } mC = nD$$

101

But another more serious difficulty must be overcome: ratios of the type of these two latter equalities are not always valid for the magnitudes, on account of the fact that there exist incommensurables.

If we put aside the other condition which Euclid imposed on himself (the one regarding the divisibility of magnitudes into equal parts) the definition we are examining is equivalent to saying, in modern language, that four magnitudes $ABCD$ are in proportion when for every rational number x:

$$\text{if } A > xB, \; C > xD$$
$$\text{if } A = xB, \; C = xD$$
$$\text{if } A < xB, \; C < xD$$

Putting aside the case in which equality holds (that means the case of commensurable magnitudes) the rational number x, for which $A > xB$, and the rational numbers x', for which $A < x'B$, determine a *section* of the field of rational numbers, which defines (by abstraction) the ratio A/B, an irrational number, according to Dedekind's conception (33).

Euclid's theory of ratios between magnitudes finds application, apart from its place in the theory of similitude, in the method of exhaustion studied by the mathematicians of classical antiquity with the object of rendering the infinitesimal processes rigorous.

(33) On the theory of real numbers, from a historical and critical standpoint, see Enriques, **9**, chiefly pp. 338 ff.

CHAPTER VI

Infinitesimal Methods in Classical Times: Archimedes

§1. *Aspects of the infinite, as considered by classical thinkers*

Since we are now to examine the highest and most perfect expression of the theories concerning the infinitesimal in ancient mathematics, namely that which appears in the writings of Archimedes, we will take the opportunity of glancing rapidly at the many aspects of such theories in the ancient world.

According to a fairly widely held view (1), 'the ancients believed that perfection was limited and finite, and imperfection limitless and infinite', while the Greeks habitually avoided the concept of the infinite, in particular the actual mathematical infinite.

We have already met passages in ancient authors which confirm this point of view (2). Let us recall, for instance, a famous fragment of Parmenides on the limited world, or Aristotle's views on the actual mathematical infinite, or Euclid's definition (14th of Book I) of the figure (conceived essentially as limited).

Yet we should not imagine that Greek thought was confined to a negative attitude to the infinite. According to R. Mondolfo's interesting researches given in *The Infinite in Greek Thought*, we find a wealth of diversity in the attitude of Greek thinkers to problems of the infinite, which are considered under many aspects. A feeling for the infinite is found at the origins of Greek poetry in Homer (3): 'The infinity of space (the ether, the Tartarean abyss, the size of the ocean), the immensity of numbers (myriads of stars), the inherent greatness of natural forces (hurricanes), the immeasurable transcendence of divine

(1) Faggi, cited by Mondolfo, p. 7 footnote.
(2) See ch. III, §4; ch. IV, §7; ch. V, §3 of this book.
(3) Mondolfo, p. 25.

103

power, all of these in Homer give rise to the idea and to the sense of the infinite, of the fascination and the stupor, the fear and the anguish, that sometimes, for some reason or circumstance, the human spirit feels.'

The conception of the infinite in time, the idea of the progenitors of all the gods, appears in Homer, in theogony, and in Orphism (4).

For reasons of space we cannot go into the various conceptions of time held by the Greek cosmologists, who, according to Aristotle, with the exception of Plato always asserted the eternity of time itself. But one, of—it appears—Chaldean origin is worth mentioning, since it was later generally accepted by Greek thinkers (5). From their discovery, in the phenomena of the heavens (eclipses, movements of planets, etc.), of a recurrence that could gradually be more and more precisely established, when longer periods of time could be considered, the Chaldeans, linking their astronomical observations with their astrological theories, according to which human happenings are dominated by the stars, came to the idea of the 'great year' of the 'eternal return', of the universal cycle of formation and dissolution of the cosmos, in which all phenomena are governed according to a cyclic order.

In the ancient world, the representation of temporal infinity (if we may speak of infinity in this case) was generally represented within the limits of cyclicity (6). 'The unrepeatableness of progress, the impossibility (taken from our thermodynamics) of the world going back a second time to something it was before; the infinite progressive way, which all the various modern theories of evolution affirm, towards an ideal end which will never be attained, are [says Duhem] all concepts outside ancient thought, nor could they have been reached without the coming of Christianity.'

We find contrasting opinions among Greek thinkers with regard to the infinity of space. We shall glance only briefly at this too.

We have already mentioned the infinite cosmic material of Anaximandrus (7). The universe of Parmenides, although it

(4) Mondolfo, pp. 29–38.
(5) Mondolfo, pp. 38–45.
(6) Mondolfo, pp. 129–30.
(7) Mondolfo, pp. 223–33.

was finite, expanded in all directions (8). As is shown from the passage of Cicero, already quoted in relation to Democritus' principle of inertia, the atomist philosopher considered the universe spatially infinite.

To Plato's friend Archytas of Tarentum Eudemus attributes the famous argument (9): 'Suppose you reached the end of the celestial spheres, that is, the spheres of fixed stars, could you hold out your hand or your stick beyond it, or not? Not to be able to hold it out would be absurd, but if you could hold it out then there would still be material or space beyond it . . . the material or the space would therefore be infinite.'

Plato and Aristotle (10), however, speak explicitly of the limitations of the universe. But Anaxagoras had conceived of the infinitely small in extended matter.

We shall not reconsider the various ways in which the mathematical infinite appeared to those we have already studied, but will go straight on to the maturer manifestations of Greek thought in this regard.

§2. *The method of exhaustion*

The infinitesimal considerations of Democritus on the sections of the pyramid and the cone, the position taken by Aristotle in regard to the actual mathematical infinite, and their similarity with the development of infinitesimal analysis in the modern age, lead us to suppose that at a certain time Greek mathematicians turned to unrigorous infinitesimal reasoning to solve what could not be solved in finite terms.

But later, with the rise of a more critical spirit, perhaps when it was found that a careless use of the actual infinite and infinitesimal might result in paradox, a rigorous infinitesimal process came into being, known as the method of exhaustion. This term was introduced in 1647 by G. de Saint Vincent (11).

It is generally (12) thought that we owe this method to Eudoxus of Cnidus, since Archimedes attributes to him the rigorous demonstration of the results relating to the volumes of pyramids and cones. But the method has also been attributed

(8) Mondolfo, pp. 278–9.
(9) Mondolfo, pp. 272–3.
(10) Mondolfo, p. 308.
(11) See Tonelli, **2.**
(12) Enriques, **8.**

to Hippocrates of Chios, albeit in particular cases and without the rigour of Eudoxus; this view is based on the fact that Hippocrates uses the proportionality of circles to the squares of their diameters; while this proportionality can be demonstrated rigorously with the method of exhaustion, as is done in Book XII of Euclid's *Elements*. Here the method of exhaustion is systematically applied to demonstrate the theorems on the volumes of pyramids and cones, and the proportions of spheres to the cubes of their respective diameters. Many applications of the method are found in the works of Archimedes, of which we shall speak later.

One of the schemes of the method of exhaustion (13) can be shown in the following way:

To demonstrate the equality of two homogeneous magnitudes Q and Q' which they did not know how to compare directly through decomposition into a finite number of congruent parts, the ancient mathematicians based themselves on the following principle: two homogeneous magnitudes Q and Q' are equivalent when there exist two other (variable) pairs of magnitudes A and A', B and B', such that:

$$A = A', \qquad B = B'$$
$$A < Q < B, \qquad A' < Q' < B'$$

and besides having chosen an arbitrarily small quantity ϵ we can always have

$$B - A = B' - A' < \epsilon.$$

In fact, if $Q \neq Q'$, we should have $Q > Q'$ or $Q < Q'$ (since one postulates that for two homogeneous magnitudes, one only of three relationships is valid: $Q > Q'$, $Q = Q'$, $Q < Q'$).

But it is demonstrated that one cannot have $Q > Q'$, because, if so, supposing that $Q - Q' = \delta$, we should have:

$$B > Q > Q' > A$$

therefore

$$B - A > Q - Q' = \delta$$

that is

$$B - A > \delta$$

Therefore in this case we could not make $B - A$ less than a

(13) Chisini, pp. 63–5.

given small arbitrary quantity, contrary to the hypothesis. Therefore we cannot have $Q > Q'$.

In the same way it may be proved that we cannot have $Q < Q'$. We must therefore conclude that:

$$Q = Q'$$

Note that, in this rigorous process, which reveals a critical mentality, no use is made of the actual infinite or infinitesimal, but only of the potential infinite or infinitesimal.

In the method considered, to establish that $Q = Q'$, it is demonstrated that we cannot have $Q > Q'$ or $Q < Q'$. This demonstration is based on the presupposition, admitted by the ancients as obvious, that every figure has its length, or its area or its volume, and takes the form of the reduction to the absurd so as to avoid using the actual infinite.

§3. *Archimedes and the method of exhaustion*

Archimedes (14) was born at Syracuse about 287 B.C. and died there in 212 B.C., when his city was conquered by the Romans. His work was concerned not only with mathematics, but with physics too, in particular with the statics of solids (the theory of the lever, of mass-centres, etc.) and of liquids (principle of Archimedes). Of his mathematical work we should mention that on cyclometry, on determining the volume of the sphere and the area of its surface; on finding the area of a segment of the parabola or the ellipse, and the volume of a segment of a paraboloid of revolution; on the properties of spirals, on a new method of expressing arbitrarily large numbers.

In his mathematical work known to the modern age before 1906, questions of infinitesimal character were treated with the methods of exhaustion. This is a suitable place to see how the method is applied, through an example.

Let us consider (15) a paraboloid of revolution (see fig. 1) with a vertex A and limited by a plane perpendicular to the axis AD, where D is the centre of the common circle of the plane and the paraboloid, while BC is the diameter of the circle.

---

(14) For an account of Archimedes, see Enriques–De Santillana, **1**, pp. 358–62; 373–7. For the critical edition of Archimedes, see Archimedes, **1**. For an English translation see Archimedes, **2**.

(15) Castelnuovo, **5**, p. 2.

Using the language of modern analytical geometry which translates the geometric properties of the paraboloid, known to Archimedes, let us take a system of Cartesian coordinates in which the z axis coincides with the line AD and the x and y axes are perpendicular to one another and in the plane perpendicular to AD passing through A. The equation of the paraboloid is immediately verified to be:

$$z = x^2 + y^2$$

Let us divide $\overline{AD} = a$ into n equal parts and write:

$$\frac{\overline{AD}}{n} = \frac{a}{n} = h$$

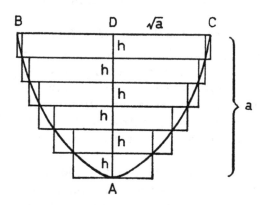

Fig. 1.

On the plane $y = 0$ (in the diagram) the equation of the meridian curve will be the parabola:

$$z = x^2$$

For the points of the division of AD let us draw the chords parallel to BC and form two series of rectangles all of height h and having these chords as bases; the rectangles of the first series contain the arc of the parabola, while the rectangles of the second series are all inside the curve.

Rotating the figure about the axis AD we obtain two volumes

s_n, S_n, which are respectively inscribed and circumscribed to the paraboloid. If V is the volume of the paraboloid,

$$s_n < V < S_n.$$

The bases of the cylinders are: πh, $2\pi h$, $3\pi h$, $\ldots$, $(n-1)\,\pi h$. Therefore:

$$s_n = \pi h^2[1+2+\ldots+(n-1)] = \pi\frac{(n-1)nh^2}{2}$$

(1) $$= \pi\frac{(n-1)n^2h^2}{2n} = \pi\frac{n-1}{2n}a^2$$

$$Sn = \pi h^2[1+2+\ldots+n] = \pi\frac{n(n+1)}{2}h^2$$

(2) $$= \pi\frac{(n+1)n^2h^2}{2n} = \pi\frac{n+1}{2n}a^2.$$

We deduce that:

$$S_n - s_n = \frac{\pi}{n}a^2$$

equal to the greater of the cylinders considered.

Now let us observe that $(\pi/n)a^2$ can be made arbitrarily small. From the equalities (1) and (2) we get immediately:

$$s_n = \frac{\pi}{2}a^2 - \frac{\pi}{2n}a^2$$

$$S_n = \frac{\pi}{2}a^2 + \frac{\pi}{2n}a^2$$

From these we deduce:

$$s_n < \frac{\pi}{2}a^2 < S_n$$

From the two systems of inequalities,

$$s_n < V < S_n$$

and

$$s_n < \frac{\pi}{2}a^2 < S_n$$

by applying the method of exhaustion already mentioned we obtain

$$V = \frac{\pi}{2}a^2 = \frac{\pi a \cdot a}{2}$$

$$= \frac{\begin{array}{c}\text{base of the segment of} \\ \text{the paraboloid}\end{array} \times \begin{array}{c}\text{height of the segment} \\ \text{of the paraboloid}\end{array}}{2}$$

$$= \frac{\text{volume of the cylinder circumscribed by the paraboloids}}{2}$$

$= \frac{3}{2}$ of the cone having the same base πa and the same height a of the segment of the paraboloid.

Using the method of exhaustion Archimedes made important discoveries ranging from elementary geometry to integral calculus: he systematized cyclometry, determined the surface and the volume of the sphere; expressing himself in geometrical language he established that

$$\int_0^c x \, dx = \tfrac{1}{2}c^2, \qquad \int_0^c x^2 \, dx = \frac{c^3}{3}$$

He also found the areas of surfaces limited by arcs of Archimedean spirals having (in our notation) the equation $r = a\theta$. The determination of these areas leads us, as can be seen immediately, to calculating the latter integral (16).

The method of exhaustion applied with such brilliant results by Archimedes seems to have a high degree of logical perfection and to be eminently suitable for the demonstration of theorems. With regard to Archimedes, Enriques and Santillana (17) observe: 'Like the strategist who carefully prepares the blow that will give him victory, we see him clear the ground methodically of every tiny obstacle and set out his forces without letting them be seen: then suddenly comes the decisive theorem. The way to his propositions is hidden but sure, and in the end it leaves even the experienced reader astonished.'

All the same it can be understood that such a process is not the most suitable for heuristic purposes. In fact it poses a

(16) Geymonat, **2**, pp. 48–50.
(17) Enriques–De Santillana, **1**, pp. 360–1.

problem: by what means did Archimedes obtain the results which he demonstrated with the method of exhaustion?

The reply is given by Archimedes himself in a letter to Eratosthenes.

§4. *The mechanical method of Archimedes*

Archimedes' letter to Eratosthenes giving information about his heuristic work disappeared, and for centuries mathematicians knew nothing of it; not even Galileo, Cavalieri, or Torricelli, who, independently of Archimedes, took up his methods again. The letter in question was discovered by Heiberg in 1906 at Constantinople, in a palimpsest belonging to the monastery of the Holy Sepulchre of Jerusalem.

Here is the introduction (18) to the '*Method of mechanical theorems* [letter from] Archimedes to Eratosthenes'.

'Archimedes to Eratosthenes, greetings.

'I wrote to you before about some theorems I had found, and I sent them to you written out, inviting you to find the demonstrations, which I could not then indicate. . . .

. . . .

The demonstrations to these theorems I have now written in this book and I am sending them to you.

'But since I know you, as I have already said, to be a learned and excellent master of philosophy, and, if need be, you can appreciate mathematical researches, I have thought it well to explain to you, in this same book, the particulars of a method through which you will find it possible to gain a certain facility in treating mathematical matters by mechanical means. Besides, I am persuaded that this method will be no less useful for the demonstration of the theorems themselves as well. In fact I myself saw some things for the first time through mechanical means, and then I demonstrated them geometrically; because the research done in this way is not a real demonstration. But it is certainly easier, having in that way gained a certain knowledge of the question, to find the demonstration, instead of seeking it without any preliminary knowledge. This is why with the theorems on the cone and the pyramid, which Eudoxus first demonstrated—that is, that the cone is the third

(18) Rufini, **2**, pp. 105–8.

part of the cylinder, and the pyramid the third part of the prism, having the same base and the same height—a fair share of the credit should go to Democritus, who first declared, without demonstrating it, that the figures had these properties. In my case too, the theorem which I am now publishing was discovered in a way similar to that of the theorem I have mentioned. And on this occasion I have decided to set down the method in writing, both because I had said I would do so and did not want it said that I had made a promise in vain, and because I am persuaded that it will be of some use to mathematics; I think, in fact, that now and in the future, other theorems which I have not yet thought of may be discovered through this method. In the first place I am going to put down one which first came to me by mechanical means, that is that *every segment of a section of a rectangular cone is equal to the four thirds of the triangle having the same base and the same height* (19), and after this some of the other results obtained with this method. At the end of the book I explain the geometrical demonstration of the theorems which I have already told you about.'

As it is useful to see how the same problem is treated by the method of exhaustion and the mechanical method, we will give the *Archimedean mechanical process for determining the volume of a segment of the paraboloid of revolution* (20), a theorem which we have already demonstrated by the method of exhaustion.

'Every segment of a right conoid (paraboloid of revolution), cut from a plane perpendicular to the axis, is one-and-a-half times greater than the cone which has the same base and the same height as the segment. This can be seen in the following way: given a right conoid, we cut it by a plane passing through the axis; let the intersection of this plane with the surface of the conoid be the section of the right cone *ABC* (see fig. 2). Cut it with a second plane perpendicular to the axis, and let *BC* be the common intersection of the two planes. Let the axis of the segment [of the conoid] be *DA*; and let *DA* be produced to *H* so that:

$$AH = DA,$$

(19) A readable description of Archimedes' mechanical device to determine the area of the segment of a parabola may be found, e.g. in Chisini, 1, p. 80 ff.
(20) Rufini, 2, pp. 130-3.

and let us consider *DH* as the beam of a balance, in which *A* is the mid-point. Let the base of the segment be the circle of the diameter *BC* perpendicular to *AD* and consider the cone which has as a base the circle with diameter *BC* and as a vertex the point *A*; besides, let there be a cylinder having as base the circle with diameter *BC* and as axis *AD*. In the parallelogram *EC* draw a line *MN* parallel to *BC*, and from *MN* draw a plane perpendicular to *AD*; this plane will cut the cylinder in a circle of diameter *MN*, and the segment of the right conoid in a circle of diameter *OP*.

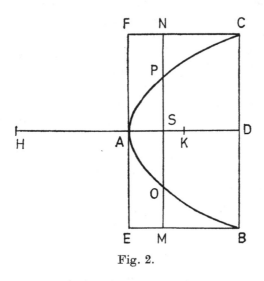

Fig. 2.

'Now since *BAC* is a section of the right cone, *AD* is its diameter, and *OS* and *BD* are two ordinates, we have:

$$DA : AS = \text{quad. } (BD) : \text{quad. } (OS)$$

But

$$DA = AH \text{ [and } BD = MS];$$

and as a result

$$HA : AS = \text{quad. } (MS) : \text{quad. } (SO)$$

We also have:

$$HA : AS = \text{(circ. diam. } MN) : \text{(circ. diam. } OP)$$

'Therefore the circle of diameter *MN*, which is in the cylinder,

8   113

will balance the circle of diameter OP, transported and placed on the beam at H, in such a way that the centre of gravity is H; in fact, the centre of gravity of the circle of diameter OP, transported, is H, and we have in inverse proportions:

$$HA:AS = \text{(circ. diam. } MN):\text{(circ. diam. } OP)$$

'In the same way it can be demonstrated that if in the parallelogram EC another straight line parallel to BC is drawn, and on this straight line is raised a perpendicular plane to AH, the circle which is obtained in the cylinder, remaining in its place, will balance the circle which is obtained in the segment of the right conoid transported to the beam at H, in such a way that its centre of gravity is H.

'Thus, therefore, the cylinder and the segment of the right conoid being filled, the cylinder will balance the segment of the right conoid, transported and placed on the beam at H, in such a way that its centre of gravity is H.

'Then, since the above magnitudes balance each other with respect to the point A, and the centre of gravity of the cylinder is K (since the line AD is divided in half at point K [lemma 8]) and the centre of the transported segment is H, we have the inverse proportion:

$$HA:AK = \text{cylinder}:\text{segment}$$

but

$$HA = 2AK$$

therefore also

$$\text{cylinder} = 2(\text{segment})$$

'On the other hand, the same cylinder is thrice the cone which has as a base the circle of diameter BC and as its vertex the point A; therefore it is clear that:

$$\text{segment} = \tfrac{3}{2}(\text{cone } ABC).\text{'}$$

As far as we know Archimedes' excellent heuristic methods inspired no successor in the ancient world. Enriques and Santillana have, as we have already seen, detected a sign of scientific decadence in the indifference with which these methods, like his other discoveries, were received by the school of Alexandria.

Archimedes himself has something significant to say on the

subject in his introduction to the *Treatise on spirals* (21). He had sent a list of problems and theorems to his friend Conon, a teacher at Alexandria, but Conon died before he could answer, and the other mathematicians at Alexandria took no interest in them. Archimedes wrote: 'Many years have passed since the death of Conon, but I have not heard that anyone has studied a single one of these problems.' So much so that none of the scientists at Alexandria noticed that two of the theorems sent by Archimedes to Conon were wrong, until Archimedes himself corrected his mistake.

§5. *The arithmetical infinite of the 'sand-reckoner'*

In an ode of Horace, dedicated to Archytas of Tarentum, we read:

'Te maris et terrae numeroque carentis arenae
mensorem cohibent Archyta,
pulveris exigui prope litus parva Matinum
munera, nec quicquam tibi prodest
aerias temptasse domos animoque rotundum
percurrisse polum morituro.' (*Carmina* I, 28)

In these verses we find the two motives: the calculation of numbers from grains of sand, and the measurement of the distances of the stars (22).

These subjects were developed scientifically in the work called *The Sand-reckoner* of Archimedes (23), who in the dedication of his treatise to Gelon, son of the King of Syracuse, expresses himself thus:

'Some think the number of grains of sand is infinite, and I am speaking not just of those who are around you in Syracuse and the rest of Sicily, but in any other place, cultivated or uncultivated. Others think that their number is not infinite, but that they could name a greater number.'

Archimedes shows that it is possible to express, with a finite number, how many grains of sand would fill the universe,

(21) Rufini, **2**, pp. 80–3.
(22) On what may be believed concerning Archytas's contribution to these matters, see Mondolfo, pp. 159–62.
(23) Mondolfo, pp. 162–6; Geymonat, **2**, pp. 41–2. On the heliocentric system of Aristarchus of Samos, see Archimedes, **2** (comm. by P. Ver Eecke), pp. 354–5; Enriques–De Santillana, **1**, pp. 299–303.

including the sphere of the fixed stars, attributing to the universe itself the dimensions obtained from the heliocentric theory of Aristarchus of Samos, who was called the Copernicus of the ancient world. He quotes Aristarchus as saying that the length of the world's orbit round the sun is to the distance of the fixed stars as the centre of any sphere is to the whole surface of the sphere. This conception, which would entail putting the fixed stars at an infinite distance, was criticized by Archimedes, who thought the ratio between a single point and a surface made up of infinite points could not be meaningful. He then interpreted what Aristarchus said in the following way: the ratio of the earth, considered as the centre of the world, to the world understood as a sphere having as its radius the distance between the earth and the sun, is equal to the ratio of the said world to the sphere of the fixed stars.

On this basis, and with some hypotheses concerning the distances of the stars, Archimedes calculates the number of grains of sand which could be contained in the sphere of the fixed stars, showing that the number can be expressed by means of a system of numeration planned by himself, a system that allows numbers, however great, to be expressed substantially by the use of powers of 10.

Mondolfo concluded, and we may conclude with him, that Archimedes' criticism 'is useful in stressing the character of infinity given by Aristarchus to the radius of the universal sphere. And, besides, what determined Archimedes' critical attitude was his intention to show that mathematical calculation, which in its way is without limits, manages to deal with any reality. And so he must exclude Aristarchus's hypothesis of an infinite reality, and keep infinity to thought and to mathematical calculation.'

CHAPTER VII

Introduction to Higher Geometry: Apollonius and His Followers

§1. *Conics as sections of a circular cone* (1)

We have seen that Menaechmus, a disciple of Eudoxus, used conic sections (in particular parabolas) to solve the problem of Delos. The introduction of these curves into the Greek mathematical world is attributed to Menaechmus.

Pappus tells us that Euclid, Aristaeus, and Archimedes all dealt with conics.

These authors, who preceded Apollonius, called a cone respectively acute, right-angled, or obtuse, according to whether the angle common to the cone and a plane passing through its axis was acute, right, or obtuse; and in every case considered sections by planes perpendicular to a generator of the cone; thus they obtained as intersections the three types of conic, called respectively: section of the acute cone (afterwards called ellipse), section of the right-angled cone (afterwards called parabola), section of the obtuse cone (afterwards called hyperbola).

From this spatial generation of the conic they obtained, as we shall now see, results in plane geometry equivalent to those expressed by our Cartesian equations.

Let BAC (see fig. 1) be the angle of a circular cone, and let $AB = AC$. Thus we can take a plane perpendicular to the axis which meets the conical surface in to a circle passing through the points B and C. Through a point E of the segment AB let us take a plane perpendicular to the straight line AB. This plane will meet the surface of the cone in a line passing through two points M and N of the circumference considered, and will

---

(1) On this topic, see Bortolotti, **5**, pp. XIII–XVI, Geminus tells us of a theory of conics, preceding the work of Apollonius, in a passage reported by Eutocius in his commentary to Apollonius (see Loria, **2**; p. 154).

cut the axis of the cone at a point F. Let L be the point at which EF meets BC and let H be the point at which the parallel to BC through E meets AC.

A segment (such as LM) normal to the axis of the conic enclosed between the axis and the curve was called an *ordinate* of the conic corresponding to the *abscissa*.

From this figure we get at once: $\overline{LM}^2 = \overline{BL}.\overline{LC}$.

In the case in which the angle $\widehat{BAC}$ is a right angle, we shall have:

$$\overline{BL} = \sqrt{2}.\overline{EL}$$

$$\overline{LC} = \overline{EH} = \sqrt{2}.\overline{EF}$$

therefore

$$\overline{LM}^2 = 2\overline{EF}.\overline{EL}$$

Now supposing

$$\overline{LM} = y, \qquad \overline{EF} = p, \qquad \overline{EL} = x$$

we obtain

$$y^2 = 2px \qquad\qquad (I)$$

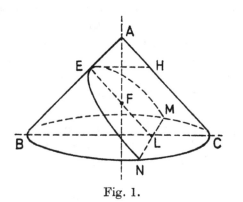

Fig. 1.

This translates Proposition XX of Book I of Apollonius: 'In the parabola the squares of the ordinates are proportional to the corresponding abscissae.' For a conic with acute or obtuse sections the formation of the analogous equations is based on the following lemma:

'If from any point L of the plane of the angle $\widehat{BAC}$ two straight lines *BLC* and *ELD* are drawn, parallel to two fixed straight lines, the ratio

$$\frac{\overline{BL}.\overline{LC}}{\overline{EL}.\overline{LD}}$$

of the segments intercepted between the point *L* and the sides of the angle is constant.' The demonstration is immediate if we consider figure 2 for $\widehat{BAC}$ acute and figure 3 for $\widehat{BAC}$ obtuse,

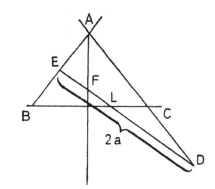

Fig. 2.

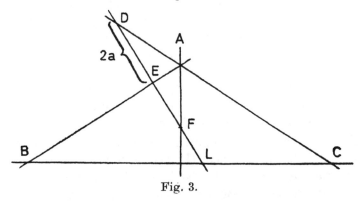

Fig. 3.

and observe that the ratios $\overline{BL}/\overline{EL}$ and $\overline{LC}/\overline{LD}$ remain constant.

Supposing that $\overline{ED} = 2a$, and indicating with *p* a constant segment (identified with *EF*), we shall have:

$$\frac{\overline{BL}.\overline{LC}}{\overline{EL}.\overline{LD}} = \frac{p}{a}$$

119

That is, remembering that $\overline{BL}.\overline{LC} = \overline{LM}^2$ we obtain:

$$\frac{\overline{LM}^2}{\overline{EL}.\overline{LD}} = \frac{p}{a}$$

Supposing that $\overline{LM} = y$, $\overline{EL} = x$, $\overline{ED} = 2a$,
in the case of the acute angle,

$$\overline{LD} = 2a - x$$

in the case of the obtuse angle,

$$\overline{LD} = 2a + x$$

Therefore the equation of the section of the acute cone will be:

$$\frac{y^2}{x(2a - x)} = \frac{p}{a}$$

that is

$$y^2 = \frac{p}{a}x(2a - x) \tag{II}$$

while the equation of the section of the obtuse cone will be:

$$y^2 = \frac{p}{a}x(2a + x) \tag{III}$$

Equations (II) and (III) are the analytical translations of Proposition XXI of Book I of Apollonius's treatise on conics.

Supposing that $p/a = m$, the equations (I), (II) and (III) of this paragraph can be summarized by the single equation:

$$y^2 = 2px + mx^2$$

Where

> $m = 0$ for the section of the right-angled cone,
> $m < 0$ for the section of the acute cone,
> $m > 0$ for the section of the obtuse cone (2).

If we interpret y^2 as a known area and x as the measure of a segment to be determined, (I), (II) and (III) turn out to be quadratic equations which have been solved in geometrical form in Euclid's *Elements* (Books II and VI) in the problems called *application of area*.

§2. *The conics of Apollonius* (3)

From the previous considerations arise the three names

(2) See Frajese, **4**, p. 156 ff.
(3) A critical edition of Apollonius' work has been published by Heiberg: see Apollonius **1**; for the English translation by Heath, see Apollonius **3**. On Apollonius' work, see Bortolotti, **5**, pp. XVIII–XX; Enriques–De Santillana, **1**, pp. 362–3, 377–81.

parabola, ellipse, and hyperbola, introduced by Apollonius instead of the old names stated above.

In fact in the case of the equation $y^2 = 2px$ we have a simple application: from a Greek term we obtain *parabola*.

In the case of the equation $y^2 = 2px - mx^2$, the square of the area y^2 is less than the rectangle $2px$: from which we get *ellipse*.

In the case of the equation $y^2 = 2px + mx^2$, y^2 is greater than the rectangle $2px$: from which we get *hyperbola* (4).

We owe to Apollonius the consideration of conics obtained by means of a section by any plane of the cone obtained by projecting a circle from any point.

Apollonius's treatise on conics, one of the masterpieces of the Alexandrine age, consisted of eight books, of which seven have survived, four in the Greek text, and three in an Arabic translation. It was rediscovered by the Italian humanists of the Renaissance, and later translated into Latin, and had an important influence on Descartes, who includes Apollonius among the ancient mathematicians for whom he felt the highest admiration (5). This is easily explained, since the work of Apollonius, by giving us geometrical writings which can immediately be translated into analytical terms, foreshadowed Descartes' analytical geometry.

Apart from the equations of conics, already dealt with, which are expressed substantially in Propositions XX and XXI of Book I, Apollonius gives us in geometrical form the equation of the hyperbola referred to its asymptotes, in Proposition XII of Book II. In Book III we find the properties of poles, of polars, and of foci. Apollonius expresses the properties of the tangents in a form suitable for characterizing the curves as the envelopes of tangents.

Book IV deals especially with the reciprocal positions and the intersections of conics, Book V with the normals to a conic from a point, reaching, substantially, the equation of the evolute, where a new solution is found to the problem of Delos and problems of maxima and minima are solved. Book VI deals with the congruence and similitude of conics. In Book

(4) See Pappus, **2**, t. II, pp. 503–4.
(5) Baillet, t. II, pp. 481–2. This passage is quoted in Descartes, t. X, p. 481.

VII we find dealt with, among other things, conjugate diameters of conics.

In Apollonius's treatise on conics we find a great number of the metrical properties of these curves (of course the projective properties, at least in an explicit form, are lacking), and yet there are some unexpected gaps: he does not speak of the focus of the parabola or of the directrix of a conic.

Apollonius's work, interest in which was aroused when Kepler discovered that the planets describe elliptical orbits, forms, as Enriques and Santillana observe, an introduction to modern higher geometry (6).

§3. *The last geometric studies in the Alexandrine age* (7)

Classical mathematics reached its most perfect level of maturity with the works of Euclid, Archimedes and Apollonius. At this point in the development of mathematics, any other important advance became difficult without a radical renewal of ideas, such as took place in the modern age.

'The very perfection of the models,' observe Enriques and Santillana (8), 'the abstract character of the rigorous treatment which conceals the line of thought, caused first a halt, and then a decline.'

There was no lack of interesting discoveries in this period, though; among them new curves, by means of which it was possible to solve the classical problems of the insertion of two geometric means between two given segments, and of the trisection of an angle: the conchoid of Nicomedes which is applied to the two problems, and the cissoid of Diocles which is applied to the first of these (9). It is said that these two geometers lived between 250 and 100 B.C.

The conchoid, which can be constructed by using a simple

(6) Enriques–De Santillana, 1, p. 380.
(7) Enriques–De Santillana, 1, pp. 382–4.
(8) Enriques–De Santillana, 1, p. 382.
(9) In his *Collection*, Bk. III, Pappus deals of the conchoid of Nicomedes and examines the problem of the duplication of the cube, intuitively seeing that it is not solvable by ruler and compasses. P. Ver Eecke maintains that Wantzel, pp. 366–72, cited in Pappus, 2, p. XVII, was the first who proved that the problems of the duplication of the cube and the trisection of an angle are not solvable with ruler and compasses. On the conchoid, the cissoid, and their application to classical problems, see Conti, pp. 343–9, 352–6, 388–90.

mechanical apparatus attributed to Nicomedes, according to analytical geometry, is a curve of the 4th order having the equation:

$$(x^2 + y^2)(y - a)^2 = b^2 y^2$$

while the cissoid is of the 3rd order, having the equation

$$y^2 = \frac{x^3}{2r - x}$$

The relative chronology of various mathematicians of the ancient world is very uncertain.

It is doubtful when Perseus lived; he studied the sections of a toroidal surface by a plane parallel to the axis of rotation of the generating circle; but presumably he belongs to the Alexandrine period (10). It is said that Zenodorus (11) belongs to this period too; to him we owe the first systemization known to us of the isoperimetrical problems, of which we shall speak in relation to Pappus (third century A.D.), who left an exposition of the same theory. Later we shall speak of Hero too (12). The Greek Alexandrine period, with which we have dealt in Chapters V, VI and VII, ends suitably with the foundation of the empire of Augustus, and gives place to the period which Tannery calls Graeco-Roman (13).

(10) Other notes on Perseus may be found in Loria, **2**, pp. 415–18.
(11) Loria, **2**, pp. 418–20.
(12) Cantor and Hultsch maintain that Hero lived in the first century B.C., Diels that he lived in the second A.D., Tannery, Heyberg and Heath in the third A.D.: see Enriques–De Santillana, **1**, p. 386, footnote.
(13) Bortolotti, **20**, p. 621.

CHAPTER VIII

Mathematics in the Roman World

§1. *Philosophy, science and law*

Mucius Scaevola, pontifex maximus, master and friend of Cicero, once said: 'There are three kinds of religion: that of the poet, that of the philosopher, and that of the statesman. The first two are futile, or superfluous, or even dangerous, and we must refute them. Only the last is acceptable' (1).

The mentality of the ancient Roman statesmen, shown in this quotation, extended to the sciences as well; in the Roman world only the practical applications of science were as a rule appreciated, especially those which referred to the government of the people. As Virgil realizes (*Aeneid* VI, 847–53):

> 'Excudent alii spirantia mollius aera
> Credo equidem; vivos ducent de marmore voltus,
> Orabunt causas melius coelique meatus
> Describent radio et surgentia sidera dicent;
> Tu regere imperio populos, Romane, memento:
> Hae tibi erunt artes, pacisque imponere morem,
> Parcere subiectis et debellare superbos.'

There is a great deal to say on the various aspects of the development of techniques in the Roman world, but we cannot linger on the subject, which is rather outside our field. We will confine ourselves to recalling that in Vitruvius's *Architecture* (written towards the end of the first century A.D.) we find an explicit mention of the vanishing point of perspective in stage scenery, and so this author is one of the forerunners of descriptive geometry (2).

For reasons of space we shall not linger on the encyclopaedias

(1) Enriques–De Santillana, 1, p. 549; *Science in Roman Civilization*; pp. 541–607.
(2) On technology in ancient Rome, and particularly on Vitruvius, see Bortolotti, 20, pp. 623–6.

of Varro (116–27 B.C.), Celsus (who lived at the time of Augustus and Tiberius), Seneca (3 B.C.–A.D. 65), and Pliny the elder (A.D. 23–79), who died nobly during the eruption of Vesuvius, with the double object of bringing help to the stricken people and of observing what was happening.

The philosophy of the Roman world has above everything else a moral and human meaning, as for example in Cicero (106–43 B.C.), whose eclecticism we note in passing. His influence in the diffusion of culture was very important, both in relation to philosophy properly so-called, and in relation to science. We have already referred to passages of Cicero on the principle of Democritean inertia, and on the principle of the excluded middle in the Epicurean and Stoic doctrines. Copernicus says that he was led to construct his theory by a reference which Cicero made to the ideas of the Pythagorean Icetas on the movement of the earth (3). For reasons of space we shall not examine Cicero's interesting discussion in *De finibus bonorum et malorum*, on his mechanical and teleological ideas, and on the astronomical information found in his works, for instance in the *Somnium Scipionis*. He made useful contributions to the formation of the Latin philosophical language; but we find no evidence, in classical antiquity, of a Latin mathematical language, with the exception of that of the Roman geometers, whose work related to technology more than to science, and the late translation of mathematical works by Apuleius and Boetius.

The philosophy of Epicurus is expressed in poetic form and with speculative vigour by Lucretius (97–55 B.C.) in *De rerum natura*. Lucretius admired Epicurus as a liberator of the mind from the terrors of superstition and death, but he does not follow him slavishly; he meditated too on the writings of Empedocles, on the historical views of Thucydides, and on the wisdom of the Stoic Posidonius. He tried to explain all natural phenomena through the atomistic theory of Epicurus (which, as we have noted, means a regression from the concepts of Democritus). But in Lucretius there is only the echo of Greek science: the interest in the scientific explanation of phenomena has now vanished, as it already had for Epicurus. What strikes the scientist far more than the mistakes Lucretius

(3) Maddalena, p. 237.

makes is the obvious indifference with which he gives, as equally probable, rational explanations and fantastic explanations of natural phenomena (for instance, eclipses). But his aim was not so much to explain these phenomena as to explain his own serene and severe ideal of life. The poem *De rerum natura* kept alive through the centuries the memory of the atomistic theory, which was to arise again, renewed, in the modern age.

With Lucretius's poem is associated, through its similarities and contrasts, the astronomical and astrological poem of Manilius (4) which was inspired by the Stoic philosophy. Manilius expresses his great faith in human reason which '*se quaerit in astris*'. The poem shows the influence of oriental astrological ideas on the relationship between the microcosm (man) and the macrocosm (the universe). And a profound gnoseological need, which reveals itself under various aspects in the history of thought, is expressed in the verses of the poet-astrologer who wonders:

'Quid mirum noscere mundum
si possint homines, quibus est et mundus in ipsis
exemplumque Dei quisque est in imagine parva?'

The Latin genius appeared outstandingly in the construction of Roman law, which appears in the form of rational theories of outstanding logical perfection (5). Stoic thought influenced Roman law both morally and logically. Among the Stoic philosophers of the Roman world were Seneca, Epictetus (A.D. 50–93), and the Emperor Marcus Aurelius (A.D. 121–180).

Other currents of thought originated in the Graeco-Roman period and influenced the development of science, among them the neo-Pythagoreans and the neo-Platonists.

The neo-Pythagoreans revived the Pythagorean school's ancient speculations on numbers, and mystical ideas relating to them. Among the neo-Pythagoreans was the senator Nigidius Figulus (6). Of neo-Pythagorean mathematicians we shall speak later.

(4) Carruccio, **3**. The passages of Manilius we report belong to the IV book of *Astronomica*, v. 910 and vv. 893–5.
(5) Even in our times there have been studies on the formal theory of law with reference to logic: see Bobbio.
(6) On the neo-Pythagoreans, see Enriques–De Santillana, **1**, pp. 551, 622–3.

The neo-Pythagorean school flows into Neo-Platonism (7), founded by Ammonius Sacca (A.D. 175–240) and developed by his disciple Plotinus (A.D. 204–270), one of the richest of mystical philosophers.

Plotinus takes as his springboard Plato's theory of ideas, but is not satisfied with contemplating this world of ideas outside his own mind; he wants to raise himself towards the absolute One in which thought and what is thought of coincide, which can be reached by the human mind only in mystical ecstasy. He considers the One as infinite, too. But this infinite must not be confused with the mathematical infinite. In these concepts we find religious influences, Jewish and Christian.

'The spirit of the late Hellenistic age,' observes Enriques (8), 'drawing away from classical attitudes, is shown in the new expressions of art with the romantic sense of the indefinite. Eyes which had longed for what was clear and distinct—for the concept of order and measure—now turned fascinated to mystery, to the unknowable.'

Yet the tradition of Hellenistic science, in particular of mathematics, was preserved in the unified world of Rome (9) during the Graeco-Roman period which lasted from the foundation of the empire of Augustus to the political and religious revolution of Constantine. There are still interesting discoveries in the period, which foreshadow the developments of modern mathematics, but no works comparable to those of Euclid, Archimedes, and Apollonius.

Through a decline in pure aesthetic idealism, or through a more mature realization of the needs of practical life, mathematics turns to its practical applications.

§2. *The origins of trigonometry* (10)

In the Graeco-Roman period astronomy reaches its maturity in the works of Ptolemy, whose observations were made between A.D. 127 and 151.

But as early as the time of Archimedes, Aristarchus of Samos, the Copernicus of the ancient world, had formulated the helio-

(7) On the neo-Platonists, see Enriques–De Santillana, 1, pp. 623–7.
(8) Enriques, 12, p. 130.
(9) Vacca, 13.
(10) Enriques–De Santillana, 1, pp. 388–9. On the origins of trigonometry, see Theodosius of Tripolis, 1 and 2; Menelaus; Ptolemy.

centric hypothesis for the planetary system, and Hipparchus (second century A.D.), the discoverer of the phenomenon of the movement of the equinoxes, as far as we know was the first to undertake the calculation of the chords of circular arcs. Let us note that the chord $2x$ (which subtends an arc $2x$) is a function of this arc, while $\sin x = \frac{1}{2}$ chord $2x$. The substitution of the sine for the chord of an arc was to be made by the Indians and the Arabs.

The spherical figures of Theodosius of Tripoli (first century A.D.) and of Menelaus, who flourished, it appears, about the end of the first century A.D., were likewise studied for astronomical reasons.

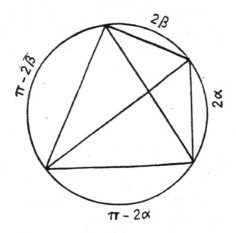

Fig. 1.

Similarities and differences between plane geometry and spherical geometry are revealed by Menelaus, who showed that the sum of the interior angles of a triangle having as its sides the arcs of great circles of a sphere is greater than two right angles. Thus in classical antiquity we find an interpretation of the non-Euclidean geometry of Riemann (11).

Ptolemy ordered his geocentric system in the mathematical composition, afterwards called by later astronomers 'the greatest', from which term in Greek was derived the Arab term *Almagest* (12).

(11) Guzzo, 1.
(12) Enriques–De Santillana, 1, p. 408.

The calculation of chords was undertaken by Ptolemy, who with his theorem on the quadrangle inscribed in a circle gave, in the language of chords, the addition theorem for the circular functions (13).

In fact Ptolemy's theorem says: 'The sum of the products of the pairs of opposite sides of a cyclic quadrilateral is equal to the product of the diagonals.'

Let us consider (see fig. 1) a circle of radius r on which we take in order the consecutive arcs 2α, 2β, $\pi - 2\beta$, $\pi - 2\alpha$. Having constructed the quadrilateral whose sides are the chords which subtend these arcs, and their diagonals, we shall have, applying the theorem of Ptolemy: $2r.\text{chord } (2\alpha + 2\beta) = \text{chord } 2\alpha . \text{chord } (\pi - 2\beta) + \text{chord } (\pi - 2\alpha) . \text{chord } 2\beta$.

If $r = 1$, dividing its two members by 4, remembering that $\frac{1}{2}$ chord $2x = \sin x$, we obtain:

$$\sin (\alpha + \beta) = \sin \alpha \cos \beta + \cos \alpha \sin \beta.$$

§3. *Hero and a property of light rays*

Hero of Alexandria (14), whose name is linked with the famous formula for the area of the triangle:

$$\triangle = \sqrt{p(p-a)(p-b)(p-c)}$$

where p is the semiperimeter and a, b, c are the sides, was also the inventor of ingenious machines, such as the steam turbine. Furthermore, he deduced the laws of the reflection of light from the hypothesis that the path taken by light rays from the eye to the mirror and from the mirror to the object is the shortest possible. (Note that according to Hero light was sent out from the eyes (15).)

Here is a translation from the Greek of Hero's original passage on the subject:

'Since in fact it is commonly admitted by all, that nature does nothing in vain, nor troubles in vain, if we did not admit that reflection comes with equal angles, nature would be troubled in vain with unequal angles, and the visual ray,

(13) Zappelloni.
(14) Hero. The passage we quote is in the second volume, first part, p. 368 ff. On Hero's work see Enriques–De Santillana, 1, pp. 386–8; on the reflection of light according to Hero, see Mach, p. 398; Chisini, 2, p. 217.
(15) Ronchi, pp. 27–9.

instead of arriving at the object by the shortest path, would obviously come to the object itself by the longest way. In fact the sum of the segments which from the eye led to the mirror and from there to the object form unequal angles, is found to be greater than the sum of those which form equal angles. And that this is true is shown thus:

'Let there be [see fig. 2]) a segment AB which supports a mirror, let G be the visual centre, D the object seen, and E the point of the mirror at which the visual ray is reflected towards the object seen; we join G to E and E to D. I say that $\widehat{AEG} = \widehat{DEB}$. If in fact the equality is not verified, let Z be another

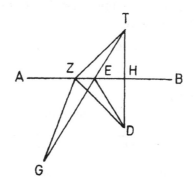

Fig. 2.

point of the mirror in which the visual ray forms reflecting unequal angles; and let G be joined to Z and Z to D. It is obvious that $\widehat{GZA} > \widehat{DZE}$ [we should say: suppose, to fix the ideas, that $\widehat{GZA} > \widehat{DZE}$]. If GZ and ZD form unequal angles with AB, while GE and ED form equal angles with AB, I say that $GZ + ZD > GE + ED$. Take the perpendicular from D to AB, let H be the foot of the said perpendicular and let the segment be prolonged as far as T. It is obvious that the angles at H are equal, because they are right angles. Let $DH = HT$ and let T be joined to Z and T to E. This is the construction. Then since $DH = HT$, but since also $\widehat{DHE} = \widehat{THE}$, and the side HE is common to the two triangles,

$TE = ED$ too, and the triangles HTE and HDE are equal, and the remaining sides are respectively equal to those which are opposite to equal angles. Therefore $TE = ED$. Besides, since $HT = HD$ and $\widehat{DHZ} = \widehat{THZ}$ and HZ is common to the two triangles DHZ and THZ, these are equal, then TZ equals ZD, and the triangles ZHD and THZ are equal too. TZ is therefore equal to ZD. And since $TE = ED$ let us add the common segment EG; therefore $GE + ED = GE + ET$.

'Therefore the whole segment $GT = GE + ED$ [since $\widehat{AEG} + \widehat{GED} + \widehat{DEH}$ is equal to two right angles; but $\widehat{AEG} = \widehat{DEH} = \widehat{HET}$, therefore $\widehat{GED} + \widehat{DEH} + \widehat{HET}$ is equal to two right angles, and since the points GET are in line $GE + ET = GT$]. And since in every triangle the sum of two sides, whichever they are, is greater than the remaining side, in the triangle TZG, $TZ + ZG > GT$. But $GT = GE + ED$. Therefore $TZ + ZG > GE + ED$. But $TZ = ZD$. Therefore $ZG + ZD > GE + ED$. And GZ and ZD enclose unequal angles.

'Therefore the sum of two segments which form unequal angles is greater than the sum of the segment which form equal angles.'

From this we derive that the angle of incidence and the angle of reflection (complementary to the angles $\widehat{AEG}$ and $\widehat{BED}$ considered by Hero) are equal too.

§4. *Isoperimetrical theory in ancient times*

The problems of maxima and minima are very important in the history of scientific thought, both for the mathematical studies which originate in these problems (infinitesimal analysis, questions of existence) and for the suggestiveness of the language of maxima and minima in relation to the laws of physics (principle of Fermat in optics, principle of least action . . .) (16), and lastly for their application to engineering.

In ancient times Euclid, Archimedes, and Apollonius treated various problems of maxima and minima by synthetic processes, but for reasons of space we shall not go into other problems of this kind and will consider only those concerning isoperimetrical theory.

(16) Enriques–De Santillana, **2**, pp. 381–2.

The object of this theory is to find the maximum area for figures of a given perimeter satisfying given conditions, or the minimum perimeter for figures of a given area satisfying given conditions. Between the two problems, obviously, a strict relationship exists which can be expressed in the following way:

Law of reciprocity (17): the figure F of maximum area S for those figures which have the same perimeter p and satisfy certain conditions C invariant under similitude, coincide with the figure of minimum perimeter for those which have the same area and satisfy the conditions C.

In fact, reasoning *per absurdum*, let us suppose that there exists a figure F' of area S and perimeter $p' < p$, and that F' satisfies the conditions C. Let us transform by similitude the figure F' so that its perimeter becomes p. Then the area of F' thus increased will become $S' > S$, while the perimeter will be p. Therefore F will not have the maximum area among all figures of perimeter p and satisfying conditions C, contrary to the hypothesis. F therefore must have a minimum perimeter among those figures which have an equal area and satisfy the conditions C.

The above theory, developed by Zenodorus, who has already been mentioned, was again expounded and perfected in the *Collection* of Pappus (who lived about 300 A.D.), who gathered together varied and interesting types of mathematical and mechanical reasoning (18). Space will not allow us to examine Pappus's work very carefully, so we will confine ourselves to parts of the *Collection* which are especially significant in the development of modern mathematics. Let us read the introduction to Book V, which deals with this theory.

'God, my worthy Megetius, gave men a very high and perfect understanding of science and mathematics, but he gave a part of it too to the unreasoning animals. He therefore provided that reasoning men should work out everything through arguments and demonstrations, while the animals, who could not reason, were allowed, through a natural disposition, to get what is necessary and useful in life. This has been foreseen in very many species of animals, but not least in the bees; whose

(17) Padoa, **2**, pp. 153–60; Chisini, **4**.
(18) Pappus, **1** and **2**. In P. Ver Eecke's introduction we find much information on Pappus's work.

discipline and obedience to the queen is wonderful, and whose diligence and cleanliness in collecting and caring for the honey, and providence and economy, are even more wonderful. In fact persuaded by the gods, so it seems to me, to prepare for men, expert in the liberal arts, a specimen of ambrosia, they do not wish to spill it on the earth, or in wood or some other ugly and untidy material; but among the sweetest flowers of the earth they pick the loveliest and from them prepare the cells destined to hold the honey, called honeycombs, all equal, similar and held together in the figure of a hexagon.

'But let us note that they do this according to a certain geometrical foreknowledge. In fact bees have to compose and join their figures in such a way that heterogeneous bodies, meeting at their joints, do not waste their labour. And three figures can satisfy these conditions, regular, as I say, equilateral and equiangular: irregularity does not suit the genius of the bees.

'The equilateral triangle therefore, the square and the hexagon, without any supplement, can be connected by common sides.

'In fact . . .

'Therefore, of the three figures themselves which might fill a space around a point, that is the triangle, the square, and the hexagon, the bees wisely chose the polygon having the greatest number of angles since it could contain the greatest amount of honey.

'And the bees knew only this which was useful to them, that is that the hexagon is greater [for a given perimeter] than the triangle and the square and can contain more honey, in equality of material used by the depositor, but we who think we have greater wisdom than the bees, seek for something much more subtle. In fact, of plane figures with equal perimeters, equilateral and equiangular, the one with more angles is always greater, and the maximum of all is the circle, when it has the same perimeter' (19).

Let us now see how Pappus develops the theory of iso-perimeters. As far as the logical rigour of the reasoning is concerned we should note in his treatment the so-called 'traditional lacuna'. That is, some of Pappus's arguments

(19) For a more detailed account of the studies on bee-cells, see Carruccio, 4.

require, in order to be complete, that the existence of the figure of maximum area in the class of figures considered should be demonstrated or postulated (20).   Now, as is well known, a bounded aggregate of numbers can lack a greatest member : consider, for instance, the real numbers less than unity.

More precisely, let us examine the following scheme of argument : let m be a magnitude belonging to an aggregate G ; let us proceed in such a way that every element of G that differs from m is increased while m remains unaltered, and so we conclude that m is the greatest member of G.   The conclusion is correct if it is established that G admits a maximum. Otherwise the reasoning is fallacious.   As O. Perron noted (*Jahresbericht der deutscher Math. Vereinigung*, vol. 22, 1913) with reasoning of this type we could demonstrate that the number 1 is the greatest of all positive integers . . . in fact the operation of squaring increases all the numbers considered, with the exception of 1.

For Pappus the existence of the maximum is intuitive, and besides, as is easily seen, it can be demonstrated on the basis of the theorem : 'Every continuous function $f(x_1 \ldots x_n)$ in a finite closed and continuous field C admits a maximum and a minimum' (21).

In Pappus (22) the theory of isoperimeters is based on the following two assumptions :

First lemma.   Among all triangles with a given base and a given perimeter the isosceles has the greatest area.

This can be demonstrated by considering the minimum property of the path of light, in the case in which, in Fig. 2 of this chapter, GD is parallel to AB, and applying the law of reciprocity.

Second lemma.   Given (see fig. 3) the similar isosceles triangles ABC and BDE, and the isosceles triangles ABC' and BDE' which are not similar to each other ; and suppose that :

$$BC + BE = BC' + BE'$$

We must show that

$$\text{area}(ABC) + \text{area}(BDE) > \text{area}(ABC') + \text{area}(BDE')$$

(20) Padoa, **2**, p. 130.
(21) Enriques, **11**, p. 335.
(22) Chisini, **2**, pp. 204–9.

In fact we note that C and C' are on the axis of AB and E and E' are on the axis of BD.

Since

$$BC + BE = BC' + BE'$$

if C' is external to the triangle ABC, E' will be internal to the triangle BDE and vice versa.

To fix the ideas, let us suppose that C' is external to ABC, and let us distinguish the two cases:

1. $AB \leqq BD$

2. $AB > BD$

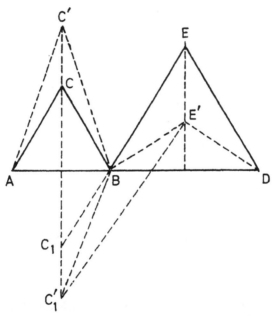

Fig. 3.

For reasons of space we will confine ourselves to considering the first case, which is the one used to develop the theory.

Let C_1 and C_1' be the images of C and C' with respect to AB. By hypothesis:

$$\widehat{ABC} = \widehat{EBD}$$

therefore C_1B and E are points in line and

$$C_1B + BE = C_1E$$

Besides, by hypothesis

$$A\widehat{B}C' \neq E'\widehat{B}D$$

therefore $C_1'B$ and E' are not in line.   We shall have:

$$EC_1 = EB + BC = E'B + BC' = E'B + BC_1' > E'C_1'$$

therefore

$$EC_1 > E'C_1'$$

Again, the straight lines CC' and EE' being parallel we have

$$EE' > C_1C_1'$$

and also

$$EE' > CC'$$

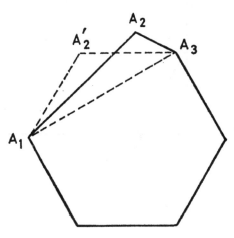

Fig. 4.

Therefore (comparing the two triangles $CC'A$ and $EE'B$, considering bases and heights):

$$\text{area}(ABC') - \text{area}(ABC) < \text{area}(BDE) - \text{area}(BDE')$$

therefore

$$\text{area}(ABC') + \text{area}(BDE') < \text{area}(ABC) + \text{area}(BDE)$$

Q.E.D.

The theorem of Zenodorus was demonstrated in ancient times by admitting implicitly the existence of the maximum among all the isoperimetrical polygons with n sides.

Theorem. Among all isoperimetrical polygons with n sides the maximum is the regular polygon.

Let (see fig. 4) $P = A_1 A_2 \ldots A_n$, be the polygon of maximum area; we shall show that it must be regular.

If the given polygon is not equilateral, at least two consecutive sides must be unequal; let these sides be $A_1 A_2$ and $A_2 A_3$.

It is easy (Assumption 1) to construct a triangle which will replace $A_1 A_2 A_3$, giving a triangle of greater area, and with the same perimeter; but then the polygon considered would not be the maximum, which would be contrary to the hypothesis.

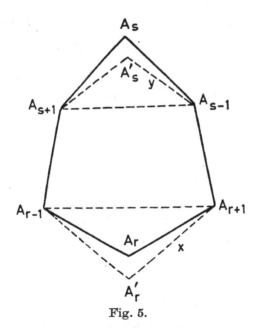

Fig. 5.

It must still be demonstrated that the polygon which attains the maximum is also equiangular.

When $n = 4$, we get an immediate result since it is easy to see that a rhombus with unequal angles has an area smaller than that of a square.

Let us suppose therefore that $n > 4$. We show that if the polygon is not equiangular there must always be two angles which are different but not consecutive. Let us suppose that this is not the case. Then all angles of even position and also those of odd position must be equal among themselves; but besides this, the non-consecutive angles A_1 and A_4 will be equal; since one of these is of even place and the other of odd place, all the angles will be equal, which is contrary to the hypothesis.

Let $\hat{A}_r$ and $\hat{A}_s$ be non-consecutive unequal angles (see fig. 5). To fix the ideas, let

$$\hat{A}_r > \hat{A}_s$$

Let us consider the triangles:

$$A_{r-1}A_rA_{r+1} \quad \text{and} \quad A_{s-1}A_sA_{s+1}$$

Since the sides of the given polygon are all equal we shall have:

$$A_{r-1}A_{r+1} > A_{s-1}A_{s+1}$$

Let us now divide the segments $2A_rA_{r+1}$ into two parts x and y proportional respectively to A_rA_{r+1} and $A_{s-1}A_{s+1}$

$$x+y = 2A_rA_{r+1}$$
$$\frac{x}{y} = \frac{A_{r-1}A_{r+1}}{A_{s-1}A_{s+1}}$$

Now let us construct the isosceles triangle $A_{r+1}A_r'A_{r-1}$ in which

$$A_{r+1}A_r' = A_r'A_{r-1} = x$$

and the other isosceles triangle

$$A_{s+1}A_s'A_{s-1}$$

in which

$$A_{s+1}A_s' = A_s'A_{s-1} = y$$

Now in the given polygon let us substitute for the triangles $A_{r-1}A_rA_{r+1}, A_{s-1}A_sA_{s+1}$ respectively the triangles $A_{r-1}A_r'A_{r+1}$, $A_{s-1}A_s'A_{s+1}$.

In this way we get a polygon with the same number of sides, with the same perimeter, and with a greater area, since by Assumption 2 the sum of the areas of the new triangles is greater than the sum of the areas of the triangles of the given polygon. But this is contrary to the hypothesis; therefore the polygon is equiangular. We have already demonstrated that it is equilateral; so we can conclude that the polygon is regular.

Corollary. The regular polygon with n sides has a smaller area than the regular polygon with $n+1$ sides which has the same perimeter.

Pappus demonstrated the corollary by showing that the apothegm, or in-radius, of the n-gon is smaller than that of the $(n+1)$-gon.

But the corollary can be demonstrated more simply: take a point A_{n-1} on the side A_1A_n of the n-gon. Then the regular n-gon can be considered as an irregular $(n+1)$-gon which will have a smaller area than the regular $(n+1)$-gon of equal perimeter.

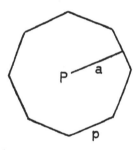 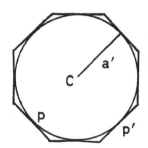

Fig. 6.

Isoperimetric property of the circle

Theorem: The circle has a greater area than any polygon P with the same perimeter p.

By the theorem of Zenodorus it suffices to demonstrate the theorem in the case in which the polygon is regular.

P has a perimeter p and an apothegm a; let C be the circle with a perimeter p and a radius a'. Let P' be similar to P, circumscribed by C, with a perimeter p' and apothegm a' (see fig. 6).

We shall have $\dfrac{a}{a'} = \dfrac{p}{p'}$. But $p < p'$, therefore $a < a'$.

Now: $P = \tfrac{1}{2}ap$, $C = \tfrac{1}{2}a'p$

So we conclude that $C > P$.

Zenodorus and Pappus also guessed that among all the surfaces of a given area the spherical surface encloses the maximum volume.

The problems of maxima and minima were destined to be

solved in the modern age, as we shall see, by using the processes of infinitesimal analysis. In particular the isoperimetric problems were to be treated systematically in the calculus of variations.

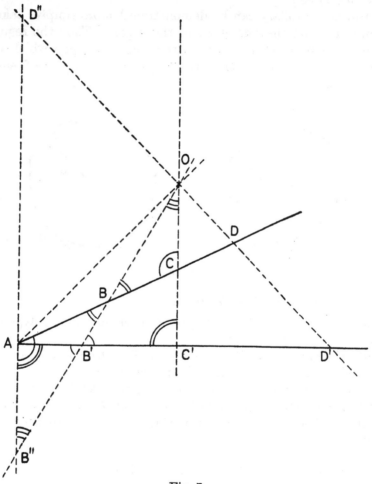

Fig. 7.

In the nineteenth century the Swiss mathematician Steiner (23) was to revive the synthetic method.

§5. *Beginnings of projective geometry in the work of Pappus*
Important developments in modern mathematics were

(23) Steiner.

foreshadowed in the ancient world, and among these we shall consider those shown in Book VII of Pappus's *Collection*, which can be translated into modern language as theorems of projective geometry (24).

In proposition 129 of Book VII Pappus demonstrates the invariance of the cross-ratio of four points under projection and section in the case in which the two sets of four points have an element in common.

Let $ABCD$ (see fig. 7) be four points in line and $AB'C'D'$ their projections from a point O on a straight line passing through A.

Pappus shows that

$$\frac{AB.CD}{AD.BC} = \frac{AB'.C'D'}{AD'.B'C'}$$

In modern language this is expressed by:

$$(BDAC) = (B'D'AC')$$

Let us draw through A to the straight line OC the parallel which will be met by the straight lines OD and OB respectively at the points D'' and B''.

Pappus's proof consists in establishing that the two expressions in question are both equal to the ratio AB''/AD''.

Considering the two pairs of similar triangles ABB'' and CBO, and CDO and ADD'', we shall have:

$$AB:BC = AB'':OC$$
$$CD:AD = OC:AD''$$

From this we get:

$$\frac{AB.CD}{AD.BC} = \frac{AB''}{AD''}$$

In the same way, considering the two pairs of similar triangles $AB''B'$ and $C'OB'$, and $C'D'O$ and $AD'D''$, we obtain:

$$AB':B'C' = AB'':OC$$
$$C'D':AD' = OC':AD''$$

From this we get:

$$\frac{AB''.C'D'}{AD'.B'C'} = \frac{AB''}{AD''}$$

(24) Enriques–De Santillana, 1, p. 385; Frajese, 2.

We conclude that:

$$\frac{AB.CD}{AD.BC} = \frac{AB'.C'D'}{AD'.B'C'}$$

Q.E.D.

Note that, having once demonstrated the invariance of the cross-ratio under the operations of projection and section for two tetrads of points having an element in common, it is easy to extend the result to the general case. In fact (see fig. 8) let

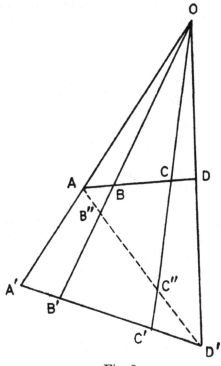

Fig. 8.

$A'B'C'D'$ be the projections on any straight line of points $ABCD$ in line from a point O. Draw the straight line AD' which meets the straight lines BB' and CC' respectively at the points B″ and C″. By Pappus's result, we have:

$$(ABCD) = (AB''C''D') = (A'B'C'D')$$

that is

$$(ABCD) = (A'B'C'D')$$

Note that in this demonstration, considered from a modern

142

point of view, a cross-ratio is reduced to a simple ratio when one of the elements goes to infinity; this case is treated separately in proposition 137.

Pappus also considers the harmonic set of points. He shows that harmonic conjugates of two given points are those which divide in the same ratio (internally and externally) the segment joining the two given points.

That is, when $ABCD$ are in line, with C between A and B, and D on the extension of AB to the right, we have the absolute value $AB/BC = AD/BD$ (considering the sense of direction, according to modern use one of these ratios should be preceded by a minus sign).

In proposition 145 Pappus shows that harmonic sets are invariant under projection and section.

Proposition 130 gives us, though in a language different from that of today, the theorem of Desargues on the involution determined by a complete quadrangle: 'the three pairs of opposite sides of a complete quadrangle are cut by a transversal not passing through any vertex, in three pairs of points belonging to an involution.'

Proposition 131 gives the particular case in which the transversal passes through the intersection of two pairs of opposite sides of the quadrangle: thus in Pappus we get the construction of the harmonic set by means of the complete quadrangle. Proposition 132 gives us the particular case in which a diagonal of the quadrangle is parallel to the transversal, and so we get the particular harmonic set in which C is the middle point of the segment AB and the point D goes to infinity.

Other propositions of Book VII relate to discoveries which, translated into modern language, give us important properties of involutions (propositions 22, 30, 32 and 34); while 37 and 38 give us the double points of an involution. Propositions 39 and 40, translated into modern terms, say substantially that: 'If in a projectivity the elements of a pair A and D correspond doubly, then the elements of another (generic) pair correspond doubly.'

In Pappus we also find the particular case of Pascal's theorem on conics, when the conic degenerates into two straight lines.

143

Propositions 138, 139, 141 and 143 of Book VII deal with this subject. In the first of these the degenerate conic is made up of two parallel straight lines, and in the next we have the case of two incident straight lines.

In Book VII too we find a theorem on the volume of the solids of rotation which anticipates by thirteen centuries the theorem of Guldin (1577–1643) (25).

§6. *The arithmetic of the neo-Pythagoreans and neo-Platonists* (26)

The Pythagoreans hoped, with the theory of monads, to base geometry on arithmetic. But the discovery of incommensurables made arithmetic take second place to geometry. Yet in his *Elements* (Books VII, VIII and IX) Euclid dealt with divisibility, and with prime numbers, the infinite number of which he demonstrated, and gave a rule for the construction of perfect even numbers (27); while in his *Sand-reckoner* Archimedes dealt with the way in which numbers however large could be expressed. Besides, arithmetic was studied for its applications (geodesy, astronomy), and Hero left us processes for the extraction of square and cube roots.

But it was the neo-Pythagoreans and the neo-Platonists who rediscovered a theoretical interest in arithmetic (and the mystique of numbers). At the end of the first or the beginning of the second century A.D. Nicomachus of Gerasa wrote his *Introduction to Arithmetic*. There we find developments which do not appear in Euclid on figurate numbers, on progressions, and a theorem from which it is easy to obtain the sum of the cubes of natural numbers—the latter found in the Arcerian Codex, which contains works by Roman geometers (28). But

(25) Guldin, Bk. 2, Ch. VIII, prop. 3. For a comparison between the formulation by Pappus, and the one by Guldin, see Pappus, **2**, introduction by P. Ver Eecke, p. XCV.

(26) Enriques–De Santillana, **1**, pp. 389–90; Cipolla, pp. 95–101: these works also give an account of the Greek numeration, of logistics, and of arithmetic before Diophantus.

(27) Euclid's rule for obtaining perfect numbers N may be expressed by the formula: $N = (2^p - 1)\, 2^{p-1}$, $2^p - 1$ being a prime. Euler has proved that all even perfect numbers are given by Euclid's formula. Odd perfect numbers are not known, but it has not been proved they do not exist; see Cipolla, p. 58.

(28) Bortolotti, **19**, p. 265; Cipolla, p. 100. This formula, which may be proved by induction (see ch. XVIII, §3), is:

$$1^3 + 2^3 + \ldots + n^3 = \left[\frac{n(n+1)}{2}\right]^2.$$

Nicomachus's demonstrations are less rigorous than Euclid's.

For the study of ancient arithmetic, the *Exposition of things useful for reading Plato* by Theon of Smyrna (second century A.D.) is interesting—not to be confused with Theon of Alexandria, a neo-Platonist of the fifth century who wrote a commentary on the Almagest, and was the father of Hypatia, the woman commentator of Apollonius, Ptolemy, and Diophantus, who died tragically in A.D. 415.

Lastly there is the *Collection of Pythagorean doctrines* of Giamblicus of Calcides, who died about A.D. 330. There we find the first pair of amicable numbers, 220 and 284, which presumably had already been noted by the Pythagoreans; the second pair, 17296 and 18416, was discovered in 1636 by Fermat (29).

§7. *Diophantus Alexandrinus* (30)

Greek arithmetic reached its highest perfection with the work of Diophantus, who flourished about the middle of the third century A.D., and can be considered the most notable forerunner in classical antiquity of two important branches of modern mathematics: the theory of numbers and algebra.

We hear of his life in an epigram of the *Palatine Anthology*, a collection of arithmetical questions attributed to Metrodorus in the third century A.D. From a simple equation we can discover the length of Diophantus's life. How accurate this is we cannot tell, but it is an example of a problem of the first degree on numbers, set down by the Greeks.

'In this tomb, wonderful to contemplate, lies Diophantus. By means of the art of arithmetic it shows the measure of his life. God gave him his childhood for a sixth of his life; after another twelfth his beard covered his cheeks; after a seventh he lit the nuptial flame, and after five years he had a son. Alas! The unlucky child, though so much loved, having reached barely half the years of his father's life, died. Four years later, consoling his own sorrow with the science of numbers, Diophantus reached the end of his life.'

(29) Enriques–De Santillana, **1**, p. 628. The Arab mathematician Thabit Ibn-qurrah (second half of the ninth century) gave a rule for finding pairs of amicable numbers.

(30) Critical edition: Diophantus, **1**. See also: Enriques–De Santillana, **1**, pp. 390, 392; Bortolotti, **19**, pp. 628–9; Cipolla, pp. 101–5.

From this we deduce the equation:

$$x = \tfrac{1}{6}x + \tfrac{1}{12}x + \tfrac{1}{7}x + 5 + \tfrac{1}{2}x + 4$$

from which we get that the life of Diophantus lasted $x = 84$ years.

But putting curiosities of the kind aside and considering the work of Diophantus, the newness of his methods and the scale of the arithmetical problems he solved, without recourse to authors who preceded him, and whom we know, make us consider the problem of his historical antecedents. There have been various views on the subject. Some people think Diophantus throws light on a little-known aspect of classical mathematics, the relevant documents of which have been lost: the numerical treatment of equations, a stage of progress which the early Pythagoreans had reached. It has also been thought that he was influenced by Indian or Babylonian culture. However, his arithmetical genius could also have developed in a fairly autonomous way in his own cultural surroundings. These points of view do not, of course, exclude one another.

Diophantus originally wrote 13 books on arithmetic, of which six and a fragment of a seventh remain. The work contains the solution of equations and of determinate and indeterminate systems of equations, where he systematically looks for solutions given by absolute rational numbers. Although he did not yet use negative numbers, Diophantus realized that in the development of $(a-b)$ and $(c-d)$ one must let $(-b)(-d) = +bd$ (31).

His work is important too in the history of arithmetical and algebraic symbolism. For fractions, Diophantus used a notation similar to ours but with inverted terms, for example:

$$\frac{\delta'}{\varepsilon'} = \frac{5}{4} \quad \text{(where } \delta' = 4 \text{ and } \varepsilon' = 5\text{)}$$

He does not use a sign to indicate the sum (he writes additions one after the other without any sign in between), while for subtractions he uses a sign which is like a Ψ upside down ($\pitchfork$).

In the case of a 1st degree equation with a single unknown Diophantus's process for solving it very much resembles that used today: to one member are added the terms containing the unknown and to the other the known terms; in the end the

(31) Enriques, **12**, pp. 23–4.

value of the unknown is found, by a division or by the determination of a fourth proportional.

For example, the first problem of the First Book: 'Divide a number into two others which have a given difference', is solved by this process.

A characteristic example of Diophantus's way of working is problem 8 of the Second Book, to which Pythagoras and his followers Plato and Euclid had already given some solutions.

'Divide the square of an integer into two squares' (32).

'We want to divide 16 into two squares. If the first is x^2, the other will be $16 - x^2$ and must be a square. I make the square of any multiple of x, diminished by as many units as are contained in the root of 16 [that is 4], let it be $2x - 4$, the square of which is $4x^2 + 16 - 16x$. This expression is equalized to $16 - x^2$; the denied (subtracted) quantities are added to the two members, and similars are taken from similars. Then we shall have $5x^2 = 16x$, where $x = \frac{16}{5}$. So one of the squares will be $\frac{256}{25}$, the other $\frac{144}{25}$, the sum of which is $\frac{400}{25}$, or 16, and each is a square.'

It is easy to give the general form of Diophantus's processes for solving the Pythagorean equation in positive whole numbers.

$$z^2 = x^2 + y^2$$

Let us put $y = mx - z$, therefore:

$$y^2 = m^2x^2 - 2mxz + z^2$$

that is

$$z^2 - x^2 = m^2x^2 - 2mxz + z^2$$

from which

$$x = \frac{2m}{m^2 + 1} z$$

therefore

$$y = \frac{m^2 - 1}{m^2 + 1} z$$

Putting $m = a/b$ (with a and b whole numbers which are relatively prime) we get:

$$x = \frac{2ab}{a^2 + b^2} z, \qquad y = \frac{a^2 - b^2}{a^2 + b^2} z$$

(32) A passage reported by Cipolla, p. 103, with its generalization. On p. 104 is found the above information concerning Fermat's great theorem.

Assuming that $z = a^2 + b^2$ we obtain the complete solutions:

$$x = 2ab, \quad y = a^2 - b^2, \quad z = a^2 + b^2$$

This latter result was known implicitly to Diophantus as a result of question 22 of the third book, but it was given explicitly by the Indian mathematician Bramagupta (about A.D. 598).

It is interesting to compare Diophantus's solution of the Pythagorean equation with the method suggested by F. Klein (33).

Dividing by z^2 the two members of the equation

(I) $$x^2 + y^2 = z^2 \quad ,$$

the equation assumes the form:

(II) $$\xi^2 + \eta^2 = 1$$

To every solution in rational numbers of (II) corresponds a solution in whole numbers of (I) obtained by multiplying the two members of (II) by the highest common multiple of the denominators which compose it.

(II) represents a circle with a centre at the origin and radius 1. It suffices to find the rational points on the circumference. These points (and these only) are the residual intersections of the lines passing, for instance, through the point (0, 1) with the circle. The equation of any such line is:

$$\eta - 1 = \lambda \xi \quad (\lambda \text{ rational})$$

Substituting the value of η in (II) we get:

$$\xi^2 + (\lambda \xi + 1)^2 = 1$$

Rejecting the root $\xi = 0$, we get:

$$\xi = \frac{-2\lambda}{1 + \lambda^2} \quad \text{while} \quad \eta = \frac{1 - \lambda^2}{1 + \lambda^2}$$

Putting $-\lambda = b/a$, where m and n are whole numbers, we get:

$$\xi = \frac{2ab}{a^2 + b^2} \qquad \eta = \frac{a^2 - b^2}{a^2 + b^2}$$

therefore $x = 2ab, \ y = a^2 - b^2, \ z = a^2 + b^2$.

(33) Scarpis, pp. 50–2.

It is worth noting that Diophantus's process is substantially equivalent to intersecting a circle of radius z with a straight line determined by a rational parameter and passing through the point $(0, -z)$.

The problem considered relating to the Pythagorean equation suggests the generalization: decompose an nth power into the sum of 2 nth powers. With regard to this, Fermat (1601–65) writes in his notes in the margin of the edition of Diophantus edited by C. G. Bachet de Meziriac:

'It is absolutely impossible to divide a cube into two cubes, a bi-square into two bi-squares, and in general any power into two powers, except the square into two squares. With regard to this I have found a marvellous demonstration, but the margin is too narrow for me to reproduce it.'

Fermat's assertion has been demonstrated for $n = 3$, $n = 4$, $n = 5$, and for an infinity of other values of n, but the general demonstration has not yet been found, in spite of the efforts of the greatest mathematicians, efforts which have now lasted for about three centuries.

Did Fermat really know the demonstration, or was he deluding himself? Doubt has even been cast on the demonstrability of the proposition (34). At least for the moment we cannot exclude the possibility that Fermat's theorem belongs to the class of undecidable propositions.

(34) Waismann, **2**, p. 178.

CHAPTER IX

The Decline of Ancient Science

§1. *St. Augustine*

After the third century A.D. (1) the profound ideological, political, social and economic crises of ancient society grew more acute in the Roman empire. The founts of classical thought were drying up, and men's minds were moving from the exact sciences to the occult. And yet, in the midst of this declining culture, new philosophical and mathematical elements were appearing, yielding seeds from which later were to spring new and flourishing branches of modern mathematics.

The mathematical Middle Ages can be said to begin with the appearance of these elements in the deep meditations of St. Augustine.

The idea of starting a new period in the history of scientific thought with St. Augustine is justified not only by the position which mathematics occupied in his philosophy (2) but more precisely by the overcoming of scepticism, and by the recognition of the extra-sensory character of mathematics and its ascetic value, in demonstrations of the immortality of the soul and the existence of God. But leaving aside all these subjects, we will concentrate for a moment on aspects of St. Augustine's thought that foreshadow certain developments in modern mathematics.

It is worth remembering that, after the first Pythagorean efforts to arithmeticize mathematics, following the discovery of incommensurables, the mathematicians of the classical world

---

(1) Enriques–De Santillana, 1, pp. 611–19, 352–5, 624–51.
(2) The following passage from St. Augustine may be misinterpreted: 'A good Christian may beware of the mathematicians, and of all men that are devoted to impious forebodings, chiefly if these forebodings prove to be true, lest such folk, in agreement with the devils, deceive his soul, and bind his soul in a devilish covenant.' (Augustine, 4, bk. II, XVII–37, cited by Enriques, 12, p. 19.) Here 'mathematicians' means 'astrologers'. St. Augustine's attitude towards the mathematicians is a very different one.

turned towards a prevalently geometrical view of mathematics. While the writings of the neo-Pythagoreans, of the neo-Platonists, and of Diophantus, as we have seen, reaffirm an interest in the properties of numbers, the philosophical conception of numbers considered as fundamental in all mathematics was clearly expressed by St. Augustine (3): 'Hinc [ratio] est profecta in oculorum opes et terram coelumque collustrans, sensit nihit aliud quam pulchritudinem sibi placere, et in pulchritudine figuras, in figuris dimensiones, in dimensionibus numeros.'

Thus St. Augustine turned decisively towards the reduction of geometry to analysis, foreshadowing Descartes' conceptual systemization with analytical geometry.

Another point of interest is the actual mathematical infinite, which, as we have seen, Aristotle refused to face. Christian thought affirmed the philosophical infinite in the conception of the perfect Being, and the actual mathematical infinite found a supporter too in St. Augustine, who proved that the totality of infinite whole numbers is thinkable and logically possible, since it is known to the divine mind, which certainly does not stop at any finite numbers (4). St. Augustine's reasoning has been pointed out by G. Cantor, the founder of the modern set theory (5).

St. Augustine also foreshadows modern developments in the philosophy of mathematics with regard to the problem of expressing rational thought through a symbolism, a language. As we have seen, in *Menon* Plato tries to show that mathematics is not taught, but that suitable questions from his master arouse in the pupil the memory of what he has already learnt in a former life in contact with the world of ideas (the theory of remembrance). St. Augustine illumines the Platonic theory of ideas, which he considers existing in the mind of God (6), and thus affirms the objectivity of mathematics.

(3) Augustine, **3**, bk. II, ch. XV, 42.

(4) 'Ita vero suis quisque numerus proprietatibus terminatur, ut nullus eorum par esse cuicumque alteri possit. Ergo at dispares inter se atque diversi sunt, et singuli quique finiti sunt, et omnes infiniti sunt. Itane numeros propter infinitatem nescit omnes Deus; et usque ad quandam summam numerorum scientia Dei pervenit, caeteros ignorat? Quis hoc etiam dementissimus dixerit?' (Augustinus, **5**, bk. XII, ch. 18.)

(5) 'Energischer, als dies hier von St. Augustin geschieht, kann das *transfinitum* nicht verlangt, vollkommener nicht begründet und verteidigt werden.' (Cantor, pp. 401–2.)

(6) See Augustine, **7**, quaestio XLVI: De Ideis.

As it was for Plato, so truth, according to St. Augustine, and in particular mathematical truth, was not something to be taught. He reached this conclusion in a work called *De Magistro*, through a subtle and complex discussion on the signs through which we seek to express thought. 'In fact,' observed St. Augustine (7), 'if I am given a sign, and I am ignorant of its meaning, it cannot teach me anything, but if I already know it what does the sign teach me?' He does not, however, interpret knowledge as the memory of a hypothetical former life (8), but concludes that if we want to know the truth we must consult the eternal wisdom which lives in the depths of our soul: 'De universis autem quae intelligimus non loquentem qui personat foris, sed intus ipsi menti praesidentem consulimus Veritatem' (9). Guzzo expresses St. Augustine's thought on the subject by saying: 'No expression, no sign with which others communicate or signify their concept, has the power to make us understand anything without an act of our own, a completely personal act of understanding' (10).

These thoughts of St. Augustine can at once be applied to the teaching of any mathematical theory, which is learnt by the pupil only when he reconstructs it in his own mind with a personal thinking act of his own.

St. Augustine's theory of signs is linked, too, to questions relating to the analysis of language, which the modern mathematical logician, who tends to express a hypothetical deductive system in symbols, meets too. Towards the end of this book we shall study modern theories, which will bring us back, substantially, by other ways, to the concepts of *De Magistro*.

'In the Middle Ages,' Enriques and Santillana (11) observe, 'scientific thought had to start its development again, moving, not as in Greece, from the contemplation of nature, but from the inner anguish of the religious and moral conscience.'

§2. *Doxographists and encyclopaedists* (12)

Chronologically later than St. Augustine, but nearer to

---

(7) Augustine, **1**, ch. X, 23.
(8) To know how Augustine outgrows the Platonic theory of recollection, see Augustine, **6**, bk. 12, ch. XV–24.
(9) Augustine, **1**, ch. XI, 38.
(10) Augustine, **2**, introduction by Guzzo, p. 10.
(11) Enriques–De Santillana, **2**, p. 226.
(12) For further information, see Bortolotti, **20**, pp. 630–1.

classical civilization, was Proclus Diadocus (420–485) (13), a characteristic figure of the last phase of classical civilization, the phase of the doxographists, that is of the collectors and commentators of the works left from antiquity. The work of these doxographists, even if it contains only a pale reflection of what Greek science was in its highest moments, is valuable for the information it gives us, through which the historian can try to reconstruct ancient mathematical and philosophical thought; particularly with regard to pre-Euclidean mathematics in its relation with philosophical speculation. To Proclus, as to others of his time, the exact sciences were confused with mysticism and the occult.

We have already noted (see the notes on Chapter III, §2), Proclus's passage which shows us the rise and fall of scientific civilization as seen through an astrological conception of cosmic cycles.

Other ideas of the doxographists, through the transcription of ancient texts and the re-evocation of myths, show us highly dramatic moments in the history of thought, relating to the early rational construction of mathematics. There is the scholium of Euclid's Book X, attributed in fact to Proclus, which we quoted in Chapter III, §3, where we are shown, not just the tragic fate of the man who disclosed the secret of incommensurable magnitudes, but the whole agony of Greek thought when faced with the inexpressible.

We shall mention Proclus's contribution to the criticism of Euclid's postulate V, a criticism which in some ways foreshadows the thought of Lobachevsky, again in Chapter XV, §1.

In the fifth century, with Europe overrun by barbaric hordes, and learning in full decline, whatever was preserved of classical science was recorded in the encyclopaedias of Marcianus Capella (about A.D. 450) and of Boetius (470–524). With his translation of, and commentary on, part of the *Organ* of Aristotle and the *Isagoge* of Porfirius, Boetius made Greek bivalent logic known to the Latin world, in the syllogistic and propositional calculus (14), although he remembered Aristotle's

(13) Enriques–De Santillana, 1, chiefly at pp. 393, 394, 630–1.
(14) Dürr.

153

observation that leaves open the door to trivalent logic (15).

In 529, by order of the Emperor Justinian, the School of Athens founded by Plato was closed: thus ended a long and luminous period in the history of thought. The philosophers of the Academy, with Simplicius, the commentator of Aristotle, chased out of Athens, took refuge at the court of the King of Persia, Cosroe: and the Greek tradition was successfully grafted on to the oriental. This union was to have important developments in the history of mathematical thought.

Meantime, in the West, what was left of culture was preserved in the cloisters and in the shadows of the Church: there were the collections of Cassiodorus (490–585), who lived his last years in a monastery he himself founded, and of Isidore, Bishop of Seville (570–636). The preservation of Western culture in the high Middle Ages is due mainly to the Benedictine monks.

§3. *The Byzantines* (16)

The School of Architecture in Byzantium must not be forgotten, for its work in preserving ancient science, from which, at the Renaissance, modern science had its origins.

This school began because of the difficulties met with in building the dome of Santa Sophia in Constantinople in the sixth century. As the engineers of the time had lost the mathematical tradition, in particular Archimedes' work on statics, the cupola twice fell down. Finally a school of geometry and mechanics was opened by the architects Artemius of Tralle, Isidore the elder of Miletus, Isidore the younger, and Metrodorus, learned commentators of the classics of Greek mathematics, who managed to finish the cupola successfully.

Eutocius of Ascalona, who left interesting comments on the work of Archimedes and Apollonius, was a disciple of Isidore of Miletus.

The preservation of ancient learning was very valuable in the development of mathematics, when cultural relations were established between scholars in the East and the humanists of Western Europe, especially in two historic circumstances: The Council of Florence, which ended with the (short-lived) return of the Church of Constantinople to unity with the

(15) Łukasiewicz, p. 76.
(16) See Bortolotti, **20**, pp. 629–30; Enriques, **12**, p. 110.

Catholic Church (1439); and the fall of the Eastern Roman Empire to the Turks in 1453, which sent many learned men from Byzantium into Italy with their wonderful Greek codices. And thus the highest aspects of Greek civilization were revived in the most auspicious surroundings.

CHAPTER X

Mathematics and Logic in the Middle Ages

§1. *Indian and Arabic mathematics* (1)

To reconstruct the development of algebra during the Renaissance, which was of fundamental importance to modern mathematics, we must go back not just to Greek mathematics, but to Arabic and Indian mathematics as well.

India boasts one of the most ancient civilizations in the world; its literature includes among its sacred books its most ancient writings. The *Vedas* (sacred books), and the poems *Mahabarata* and *Rahamajana* are at least twenty-five centuries old. But it has not been established with any certainty if Indian mathematics had an autonomous development independent of the Greek.

The most ancient work of Indian mathematics known to us is the *Sulva-sutras*, which some people maintain goes back to the fifth century B.C., and others to the eighth; there we find geometrical rules not rationally justified, for the construction of altars, and also the statement of the theorem of Pythagoras, but without any general demonstration (2).

A second period of Indian mathematics lasted from the fifth to the twelfth century A.D.

Indian geometry is, as a rule, only a collection of practical metric rules, which are not always exact. The demonstration or the explanation of geometrical propositions is found in the figure. For instance, the demonstration of Bhaskara Carya (born 1114 A.D.) of Pythagoras's theorem is found in Fig. 1: from which figure we get that:

$$a^2 = (b-c)^2 + 4\frac{bc}{2} = b^2 + c^2$$

(1) Enriques, **12**, pp. 22–7; Bortolotti, **20**, pp. 634–5; Cipolla, pp. 113–23.
(2) Brunschvicg, p. 46.

The Indians, as this shows, were more interested in calculation than in geometry, and they perfected their methods by introducing the decimal position in numeration, which they may have invented. The large numbers which appear in the Indian cosmogony, in which cycles of millions of years recur, are among the reasons that led Indian mathematicians to represent numbers simply and expressively (3). Their positional representation, in which nine significant digits and zero are used, appears in an Indian treatise of the fourth or fifth century A.D.

The Indians have too a sense of formal development, which gives rise to the use of negative numbers: Bramagupta (born

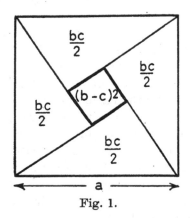

Fig. 1.

A.D. 598) gives practical rules for the addition of credits and debts, introduces an algebraic symbolism, and solves general equations of the 2nd degree, of which particular cases had been solved by Aryabhata (born A.D. 476). Bhaskara Karya writes a treatise on arithmetic which deals with a calculation of roots, quite proficiently used. He observes that if in the fraction a/b the denominator is evanescent, the fraction itself assumes a value superior to any given number.

For reasons of space we will not linger over the interesting subject of the Indian mathematicians, who in trigonometry substitute the sine of the arc for the chord.

In the Islamic empire, in which people of varying origins were united, a culture developed that can be compared on the

(3) Enriques–De Santillana, **2**, pp. 245–6.

one hand with Indian civilization, and on the other with the Hellenistic tradition. In Baghdad where, after 529, philosophers fled when Justinian closed the School of Athens, the Arabs' interest in Hellenistic culture is shown in translations from the Greek of the works of Euclid, Ptolemy, Diophantus, Hero, Archimedes, and Apollonius. But the Arabs' most important contribution to modern mathematics consists in the development of algebra, with the theory of 1st and 2nd degree equations, residuals of the geometrical doctrines of the Greeks.

'Algebra' (Al giabr) is a word of Arabic origin which indicates the transference of a term from one member of the equation to another with a change of sign, while 'algorithm' derives from the name of the Arabic mathematician Al-Khuwarizmi. Nisir Eddin made a memorable contribution to the criticism of Euclid's theory of parallels (to which we shall return), the Jew Abraham Savasarda worked on infinitesimal processes and Omar Khayyam on cubic equations solved by using conics, with which Arabic mathematicians also solved 3rd degree problems in the construction of the regular heptagon (4).

In the history of the philosophy and science of the Islamic world we should remember Avicenna (980–1037, who lived in Persia), and Averroes (1126–98, a Mussulman of Spain), whom Dante recalls for his commentary on Aristotle (*Inferno*, canto 4, verse 144). Both these men had an important influence on scholastic thought (5).

§2. *Mathematics in the late Middle Ages in the West* (6)

The work of Cassiodorus, mentioned in the previous chapter, reached the Venerable Bede (675–735), an English Benedictine monk and historian, who worked too on digital calculation and problems relating to movable feasts, through the writings of Isidore, Bishop of Seville (570–636). From his school came Alcuin (735–804), who was charged by Charlemagne to found the Palatine Academy, and to whom is attributed the work *Propositiones ad acuendos iuvenes*, which contains elementary but interesting and curious problems. There we find the famous problem of the man who has to cross a river carrying a

(4) Schoy, pp. 21–40; Carruccio, **6**.
(5) Nallino, **1** and **2**.
(6) Enriques, **12**, pp. 21–2; Bortolotti, **20**, pp. 631–4.

wolf, a goat, and a bundle of cabbages, one at a time, and without leaving the wolf with the goat or the goat with the cabbages. The fact that a problem of this kind is found with others which are typically mathematical makes it seem that the traditional conception of mathematics as a science of quantity had been widened; that is, that with Alcuin purely logical problems, although only very elementary ones, are brought again into the field of mathematics. The intellectual revival which originated in the Academy of Charlemagne lasted a short time and ended towards the end of the ninth century with Scotus Erigena, a neo-Platonic philosopher.

The cultural tradition was kept alive in the abbeys of Monte Cassino, Bobbio, Fulda, St. Gallen, Auxerres, Cluny, and Aurillac.

New movements of thought originated about the year 1000. In the writings of the mathematician Gerbert, who became Pope under the name of Sylvester II, we find a still rather rudimentary use of Indo-Arabic digits, and the treatment, sometimes rather naïve, of some geometrical questions. We owe Gerbert the definition of an angle considered as a plane region, a definition which substantially agrees with the modern concept.

§3. *Leonardo Pisano* (7)

A wonderful synthesis of classical and Indo-Arabic mathematics was made in the thirteenth century by Leonardo Pisano, who had travelled in Algeria and the countries of the Middle East. In the preface to his *Liber Abbaci*, written in 1202, we read:

'All that was studied in Egypt, in Syria, in Greece, in Sicily, and in Provence . . . with various methods belonging to these places, where I wandered as a merchant, I investigated very carefully and . . . having very accurately studied the ways of the Hindi [algebra], instructed by my own enquiries, and adding what I was able to take from Euclid, I wanted to write a work of fifteen chapters, with nothing capital left without a demonstration; and this I did, so that the science might be easily understood, and the Latin people should no longer be deprived of it.'

(7) Leonardo Pisano, Enriques, **12**, pp. 27–8. Bortolotti, **20**, pp. 646–7; Frajese, **4**, appendix.

So the national exigency again gained ground, after long centuries of scientific decadence in the West.

The *Liber Abbaci* collects Arabic arithmetic and the system of decimal numbers based on the position of the digits and on zero, and deals with questions of Diophantean analysis.

Chapter XIV of the book was important for the later progress of algebra, for there Leonardo dealt with square and cube roots, showing that cube roots cannot be reduced to the forms classified by Euclid in Book X of the *Elements*.

These studies were, according to Bortolotti and G. Vacca (8), the point of departure for the algebraic solution of cubic equations by the Italian algebraists of the Renaissance.

One of the questions examined by Leonardo Pisano in a short work called *Flos* is of interest. It was a matter of establishing if the solution of a cubic equation

$$x^3 + 2x^2 + 10x = 20$$

can be presented in a rational form or in the form of one of the irrationals studied by Euclid in Book X of the *Elements*. The conclusion is negative. Then Leonardo thinks of solving the equation approximately and gives a solution that is remarkably close to the true one. We cannot know for certain how this solution was calculated.

In the *Practica geometriae* (1220), where problems of areas and volumes are solved, the classical inspiration of Euclid and of the comments on Archimedes of Eutocius are more obvious than in the *Liber Abbaci* (9). It is worth noting that Leonardo Pisano introduces unit segments into geometrical constructions. This introduction, which appears in the writings of Rafael Bombelli too, was to be one of the essential elements of Cartesian geometry.

§4. *Traditional logic in the scholastic philosophy* (10)

In Scholasticism, which was an ordered and coherent systemization of Christian philosophical thought, logic in which formal technique was pre-eminent developed consider-

(8) Bortolotti, **8**, or **11** ; Vacca, **6**.
(9) Carruccio, **7**.
(10) Enriques, **2**, pp. 46–50. The passage from Occam is reported from G. Vacca, **3**.

ably. St. Thomas Aquinas (1227–74) gives the following definition of logic:

'Ars directiva ipsius actus rationis, per quam scilicet homo in ipso actu rationis ordinate et faciliter et sine errore procedat.' (*In Analyt. post.* Book I, lect. 1).

One of the best treatments of the Aristoteleanly-directed scholastic logic is found in the *Summulae logicales* of Petrus Hispanus, who became Pope under the name of John XXI (died 1277) (11). The scholastics' efforts to improve Aristotelean logic have been examined in Chapter IV of this book. The calculus of propositions had already appeared in the Stoic School (see Chapter III, §1) and we shall now mention some of its elements as they appear in scholastic logic. In Petrus Hispanus's *Summulae Logicales* (1–22 and 1–23), the propositions resulting from the fundamental logical operations—implications, logical conjunction, logical disjunction—are respectively indicated by the terms: *conditionalis, copulativa, disiunctiva,* and characterized in the following way:

'1–22 . . . *Conditionalis* est illa, in qua coniunguntur duae categoricae per hanc coniuctionem "si", ut "si homo currit, homo movetur"; et illa categorica, cui immediate coniungitur haec coniunctio "si" dicitur "*antecedens*" alia vero "*consequens*". *Copulativa* est, in qua coniunguntur duae categoricae per hanc coniunctionem "et" ut "Socrates currit et Plato disputat". *Disiunctiva* est illa in qua coniunguntur duae categoricae per hanc coniunctionem "vel" ut "Socrates currit vel Plato disputat".

'1–23. Ad veritatem conditionalis exigitur, quod antecedens non possit esse verum sine consequente, ut "si homo est animal est"; unde omnis conditionalis vera est necessaria et omnis conditionalis falsa est impossibilis. Ad falsitatem eius sufficit quod antecedens possit esse vera sine consequente ut "si homo est, album est". Ad veritatem copulativae exigitur utramque partem esse veram, ut "Deus est et homo animal est." Ad falsitatem eius sufficit alteram partem esse falsam, ut "homo est animal et equus est lapis." Ad veritatem disiunctivae exigitur alteram partem esse veram, ut "homo est animal vel corbus est lapis" et permittitur quod utraque pars ipsius sit vera, sed non ita proprie, ut "homo est animal vel equus est

(11) Petrus Hispanus.

11   161

hinnibilis". Ad falsitatem eius exigitur utramque partem esse falsam, ut "homo non est animal vel equus lapis".'

Note the character of necessity in '*conditionalis*', a character which is not found in the material implications (see Chapter XVIII, §5), but which is found in the strict implication of Lewis; see Vaccarino, *2*, page 251: $p \rightarrow_3 q = \infty \diamondsuit (p \infty q)$, where the signs $\rightarrow_3 \diamondsuit \infty$ mean respectively: strictly implies, is possible, not. Meantime the formula shown reads: p implies strictly q means that it is impossible that p and the contradictory of q be true at the same time.

An important property concerning negation, conjunction or disjunction, as Vacca has noted, was expressed by William of Occam (died about 1349) in the following terms: 'Sciendum est etiam quod opposita contradictoria disiunctivae est una copulativa composita contradictoriis ex partium ipsius disiunctivae'. (*Tractatus logicae fratris* Guillermi Ockan, Parisiis, 1488, fol. 53r., col. 1). This property can be put in the form

not (*A* or *B*) = (not *A*) and (not *B*)

While Buridan, as Vacca observes and Enriques mentions, gives us the property

not (*A* and *B*) = (not *A*) or (not *B*).

We shall meet these logical operations and properties when we consider the systemization of Hilbert's calculus of propositions (see Chapter XVIII, §5).

The scholastics knew the possibility of deducing any proposition from two contradictory propositions: the theorem of Pseudo-Scotus: 'ex falso sequitur quodlibet'; example of the scholastics: 'If Socrates exists and Socrates does not exist, then Plato is an ass'. We shall easily demonstrate this theorem by means of Hilbert's calculus of propositions (12).

Rámon Lull (1234–1315) is considered a precursor of mathematical logic because of his efforts to find a mechanical process which would allow the systematic attainment of all deductions, starting from given principles. According to Lull this is attained through concentric rings whirling and divided into parts that represent determined concepts; and, as I. M.

(12) See Bocheński, *1*, p. 81–2. For a modern proof, see Hilbert and Ackermann, 1928, p. 21, or ch. XX, §5 of this book.

Bocheński (13) notes, this doctrine is not just a curiosity in the history of logic but had some influence on Leibniz.

§5. *The question of universals* (14)

This was debated among the mediaeval scholastics and showed various aspects of human thought, which are still reflected in our own day, in the varying attitudes of mathematicians towards the meaning and foundations of mathematics, with special reference to the theory of infinite sets (see Chapter XVII, §1).

It is a matter of establishing if a reality outside the human mind corresponds to universal ideas. The question is asked clearly in a passage of Porphirius (*Isagoge* 1, 3): 'And above all, as far as genus and species are concerned, I will avoid searching to see if they exist in themselves, or if they exist only as pure notions of the spirit; and—admitting that they exist in themselves—if they belong among the corporeal or the incorporeal; and finally if they have a separate existence or only in the sensible things. It is too profound a question, one that needs a different study from this, and very much longer.'

While the nominalists like Roscellino denied the autonomous reality of universals, which they conceived as pure names: '*flatus vocis*', the realists, like St. Anselm of Aosta (1033–1109), who were inspired by Plato's theory of ideas, affirmed the existence of universals themselves.

William of Occam and Buridan held a doctrine somewhere between nominalism and realism: terminism (Abelard's *conceptualism* comes somewhere near it). According to terminism concepts are considered as signs or terms, representing single things and classes of existing things, while logic relates to the relations between the signs (written, spoken, or conceptual).

St. Thomas Aquinas admits the universal '*ante rem*' as a divine idea, '*in re*' as the essence of things, and '*post rem*' as the human concept.

While Aquinas was inspired chiefly by Aristotle, St. Bonaventure, through St. Augustine, was mainly under Plato's influence.

(13) Bocheński, **4**, p. 319.
(14) Enriques, **2**, pp. 48–9.

§6. *Mediaeval forerunners of trivalent logic*

While bivalent logic developed and spread either under the form of syllogism, or in the calculus of propositions, Aristotle's observations on propositions which are neither true nor false, which had been taken from the Epicureans by Boetius and other authors (see Chapter IV, §1; Chapter V, §1; Chapter IX, §2) were considerably developed by the scholastics. In his comment *In libros peri Hermeneias* St. Thomas Aquinas (15) seems particularly interested in these views, as other mediaeval philosophers are elsewhere (16). It is especially interesting, with the new logic in mind, to read Occam on the Aristotelean passage we have mentioned. Occam thinks that Aristotle does not deny the principle of the excluded middle, much less the principle of non-contradiction, but admits that considered in themselves certain propositions are neither true nor false (17). Thus we find again the Epicurean point of view which Cicero considered 'most scandalous' (see Chapter V, §1).

Occam looks carefully at the calculus of propositions (see §4 of this chapter), with special reference to the implication, examining cases in which either of the two given propositions (*antecedens, consequens*), on which one is working, is neither true nor false. He deals outstandingly clearly with the case in which '*antecedens*' being false, and '*consequens*' neither true nor false, the 'consequentia' is valid (18): 'Per praedicta patet quod Philosophus concederet istam consequentiam: Deus scit A fore, igitur A erit, sed diceret antecedens esse simplicitur falsum et consequens nec esse verum nec falsum. Nec est inconveniens, quod ex falso sequatur illud, quod nec est verum nec falsum, sicut ex falso sequitur verum.'

Can we conclude from this that Occam supported trivalent logic? With Boehner, we must answer 'no', since, although Occam examines the possibility of propositions that are neither true nor false, he definitely rejects it, as the following passage (19) shows: ' ... The philosopher's meaning is that in such a contingent future no part of the contradiction is true or

(15) See Thomas Aquinas, 2, pp. 62–7.
(16) Occam, pp. 118–38. On Boehner's work on scholastic logic, see Buytaert.
(17) Occam, pp. 61–110.
(18) Occam, Super I librum Perihermeneias, pp. 112–13.
(19) Occam, p. 111.

false ... and this would mean that God does not know one part of the contradiction more than another, in fact that he would know neither ... all the same according to the truth and the theologians one must say otherwise, because God knows determinedly one part [the alternative that will happen]. How this is, theology must explain.' So Occam does not believe in propositions that are neither true nor false, since with any given event A, God knows which he will verify of the two alternatives: A will be, A will not be.

Occam's position with regard to trivalent logic is rather similar to Saccheri's with regard to non-Euclidean geometry (see Chapter XV, §2). Just as Saccheri is a forerunner of non-Euclidean geometry, for having deduced important theorems from the negation of Euclid's fifth postulate, he ends by affirming the unconditioned validity of the postulate and of Euclidean geometry, so Occam, although he deduced some consequences of the hypothesis of trivalence, which he definitely refuted, deserves to be numbered among the forerunners, but not among the supporters, of the new logics (see Chapter XX, §2).

§7. *The scholastics' views on the mathematical infinite*

The scholastics reaffirmed the distinction, already met among the neo-Platonists, between the philosophical infinite (the Absolute, God) and the mathematical infinite. A great deal could be said about their subtle reasoning on the mathematical infinite, but for reasons of space we will confine ourselves to some of their ideas which are especially relevant to the modern theory of sets.

Roger Bacon (20) (not to be mistaken for Francis Bacon) in his *Opus maius* (London, 1233) raises again the possibility of establishing a one-to-one correspondence (through projection) between the points of a side of a square and the points of its diagonal; and observes that a half-line is equal to another half-line contained in the first. These observations make him think he can deduce that the mathematical infinite is not logically possible. But in spite of these paradoxes Galileo was to reach another conclusion, as we shall see later.

Aquinas several times tackled the problems of the mathe-

(20) Bacon, pp. 66–70, and chiefly pp. 68–70.

matical infinite, and his best-known views on it agree with those of Aristotle, who rejected the actual mathematical infinite (21). But when he refers to the mind of God, he agrees with St. Augustine in recognizing the existence of the infinite objects of divine knowledge: 'Necesse est dicere quod Deus etiam scientia visionis scit infinita' (22).

Aquinas mentions, as Vacca (23) observed, the possibility of one infinity which is greater than another: there are infinite sets, which are themselves parts of other sets. The examples he refers to, though, are of aggregates which we consider of equal power (see Chapter XVII, §1).

Other writings on the mathematical infinite and on the continuum are found in the works of Thomas Bradwardine (1290–1349), Archbishop of Canterbury (24).

§8. *Mathematics from the thirteenth to the fifteenth centuries* (25)

Jordan Nemorarìus, perhaps to be identified with the successor of St. Dominic as head of the Order, in his *Arithmetica* takes up Nicomachus, and in his *Algorithmus demonstratus* expounds the system of decimal numeration and the calculus of fractions, and elsewhere explains through examples the solution of 1st and 2nd degree equations; he studied elementary geometry and statics as well.

Campanus of Novara, who in 1280 translated Euclid's *Elements* from Arabic into Latin, dealt with important matters relating to the angle of contingency and to continuity, and also studied the axioms of arithmetic.

Nicholas Oresme, Bishop of Lisieux (1323–82), precursor of Cartesian geometry, studied natural phenomena by mathematical means, through diagrams that foreshadow the Cartesian diagrams. He gave theorems on geometric series, and proved the divergence of the harmonic series (26). He also introduced

(21) See also Thomas Aquinas, **3**, pars I, q. VII, a. 4.
(22) Thomas Aquinas, **3**, pars I, q. XIV, a. 12.
(23) Thomas Aquinas, **1**. Quodlibet nonum, art. I: 'Nihil enim prohibet illud quod est infinitum per unum modum, superari ab eo quod est infinitum pluribus modis, sicut si esset aliquod corpus infinitum secundum longitudinem, finitum vero latitudine, esset minus corpore longitudine et latitudine infinito . . . infinito enim non est aliquid maius in illo ordine quo est infinitum: sed secundum alium ordinem nihil prohibet aliquid esse aliud maius infinito; sicut numeri pares et impares simul accepti sunt plures numeris paribus.'
(24) Enriques, **12**, p. 29.
(25) Enriques, **12**, pp. 28–9.
(26) Oresme, questio 1–2, pp. 1–6.

the use of fractional exponents, while negative exponents were used by Chuquet (1484).

Fra Luca Paciolo (1445–1514) (27) left, in his *Summa de Arithmetica, Geometria . . .* (Venice 1494), a kind of encyclopaedia of the pure and applied mathematics taught in his day. Leonardo da Vinci, who was a friend of his, drew the plates for another of his works, *Divina proporzione*, in which he dealt with the golden section, its application to various geometrical questions, and odd notions which we might call the mystique of geometry.

Luca Paciolo can be said to have reached the dividing line between mediaeval and modern mathematics. He managed to solve 1st and 2nd degree equations, and generally those with the form $x^{2m} + px^m + q = 0$, and then shows an unsatisfied longing to solve cubic equations: 'In the Chapters on equations of the form

$$x^3 + px + q = 0,$$

or of the form

$$x^3 + px^2 + q = 0,$$

it has not yet been possible to formulate general rules. . . . I should say that the art of mathematics has not yet given us the way of doing so, as it has not yet given us a way of squaring the circle.'

(27) Bortolotti, **19**, pp. 27–33.

CHAPTER XI

The Mathematical Renaissance and the Algebraists

§1. *Art and mathematics in the Renaissance* (1)

Just as the works of the artistic and literary Renaissance are inspired by classical art and literature, but do not simply recreate the ancient world, so the rational constructions of the scientific Renaissance, though based largely on ancient science, conceived in a new spirit, have elements in them which cannot be found in the classical world.

The humanists brought back the masterpieces of the Alexandrine age, influences from the east contributed to the progress of algebra, number took precedence over geometry, the search for infinitesimal analysis was undertaken from the point of view of the mathematical infinite, to which Renaissance mathematicians were powerfully drawn; science and art were thus united in a wonderful synthesis.

The Renaissance feeling for measure and harmony fitted well with the mathematical mentality. A similar spirit appeared in art and in science. The artists of the Renaissance were often scientists as well; among them Brunelleschi, Paolo Uccello, Alberti, Piero della Francesca, Leonardo da Vinci (2), who in mathematics 'found a sense of beatitude as in music and in things which are certain'; and Albrecht Dürer....

The scientific Renaissance, it is worth noting, reached its highest point when the artistic and literary Renaissance was already exhausted: about the time Michelangelo died, Galileo was born.

§2. *The Italian algebraists of the Renaissance* (3)

During the sixteenth century fruits of ideas that had long

(1) Enriques, **12**, pp. 30–1.
(2) Severi, **5**.
(3) This matter has been dealt with by Bortolotti, **12**, **20**. See also Vacca, 8; Enriques, **12**, pp. 33–6.

been maturing were gathered, above all in the field of algebra, with the solution of cubic and quartic equations; the solution of cubic equations can, in fact, be considered the Pillars of Hercules of ancient and mediaeval mathematics, after which we come to modern mathematics. At that time mathematicians shrouded their discoveries in secret, so as to use them in public competitions, on the success of which could depend the acquisition or the loss of a university chair.

At one of these competitions N. Tartaglia of Brescia (died 1559) solved the cubic equation of the form

$$x^3 + px = q \quad (I)$$

(We should write $x^3 + px + q = 0$, but then, to avoid negative numbers, they did not bring all the terms to the same side.)

G. Cardano, mathematician, physician and philosopher (1501–76), wished to know the rule for solving cubic equations, and with difficulty managed to get Tartaglia to tell him, in the form of an obscure triplet, but had to promise not to publish it. For nearly six years Cardano kept his promise, but then he came to know the solution in another way: cubic equations had been solved, before Tartaglia, by the Bolognese mathematician Scipione del Ferro (1465–1526), who told his son-in-law, Annibale della Nave, who in turn had told Cardano; who then felt free of the solemn promise he had made to Tartaglia, and in his book *Artis magnae* published the solution of cubic equations. There too Cardano showed how to reduce an equation of the type:

$$y^3 + ay^2 + by + c = 0$$

to the form:

$$x^3 + px + q = 0$$

(All one needs is to write:

$$y = x - \frac{a}{3}).$$

Cardano himself attributes this transformation to his disciple Ferrari.

To solve the equation $x^3 + px = q$, all that one needs is to interpret Tartaglia's rhyme correctly. Here are the verses,

169

with their meaning expressed in modern algebraic symbols:

Quando che 'l cubo con le cose appresso
Se agguaglia a qualche numero discreto $\qquad x^3 + px = q$
Trovan dui altri differenti in esso $\qquad u - v = q$
Da poi terrai questo per consueto,
Che 'l lor produtto sempre sia eguale,
Al terzo cubo, delle cose neto, $\qquad u.v = \left(\dfrac{p}{3}\right)^3$
El residuo poi suo generale
Delli lor lati cubi ben sottratti,
Varrà la tua cosa principale $\qquad x = \sqrt[3]{u} - \sqrt[3]{v}$

. . . .

Tartaglia's rule for solving the equation $x^3 + px = q$ (I) therefore consists in finding the numbers u and v such that

$$u - v = q \quad \text{(II)}$$

$$uv = \left(\frac{p}{3}\right)^3 \quad \text{(III), putting } x = \sqrt[3]{u} - \sqrt[3]{v} \quad \text{(IV)}$$

It is easily verified that the positions (II), (III) and (IV) identically satisfy equation (I). In fact by substituting expression (IV) for (I), we get

$$(\sqrt[3]{u} - \sqrt[3]{v})^3 + p(\sqrt[3]{u} - \sqrt[3]{v}) = q$$

that is

$$u - v - 3\sqrt[3]{u}\sqrt[3]{v}(\sqrt[3]{u} - \sqrt[3]{v}) + p(\sqrt[3]{u} - \sqrt[3]{v}) = q$$

$$u - v + (\sqrt[3]{u} - \sqrt[3]{v})(p - 3\sqrt[3]{u}\sqrt[3]{v}) = q$$

This will be satisfied identically by writing

$$u - v = q \quad \text{(II)} \qquad p - 3\sqrt[3]{u}.\sqrt[3]{v} = 0$$

that is

$$u.v = \left(\frac{p}{3}\right)^3 \quad \text{(III)} \qquad\qquad \text{Q.E.D.}$$

How did the Italian algebraists of the Renaissance manage to solve cubic equations? Bortolotti and Vacca (4) maintain

(4) Bortolotti, **8**, **11**; Vacca, **6**.

that Leonardo Pisano's work on the irreducibility of cubic irrationals in the form studied in Euclid's Book X started them off in the right direction.

Leonardo observes that according to Euclid

$$\sqrt{a+\sqrt{b}} + \sqrt{a-\sqrt{b}} = \sqrt{2a+2\sqrt{a^2-b}}$$

Now he does not give a similar reduction for the sum of the two cube roots:

$$u+v = \sqrt[3]{a+\sqrt{b}}+\sqrt[3]{a-\sqrt{b}}$$

It therefore seems natural to look for an equation that satisfies the expression $u+v$, which we indicate with x.

Raising to the cube both the members of $x = u+v$, that is

$$x = \sqrt[3]{a+\sqrt{b}}+\sqrt[3]{a-\sqrt{b}}$$

we get

$$x^3 = a+\sqrt{b}+3\sqrt[3]{a^2-b}.\left(\sqrt[3]{a+\sqrt{b}}+\sqrt[3]{a-\sqrt{b}}\right)+a-\sqrt{b}$$

that is

$$x^3 = 3\sqrt[3]{a^2-b}.x+2a$$

This last equation is identified with the general equation of the type:

$$x^3 = px+q \quad ;$$

hence

$$2a = q, \text{ that is } a = \frac{q}{2},$$

and

$$3\sqrt[3]{a^2-b} = p, \text{ that is } b = \frac{q^2}{4}-\frac{p^3}{27}$$

Therefore

$$x = \sqrt[3]{\frac{q}{2}+\sqrt{\frac{q^2}{4}-\frac{p^3}{27}}}+\sqrt[3]{\frac{q}{2}-\sqrt{\frac{q^2}{4}-\frac{p^3}{27}}}$$

If the equation, as in present day use, is written in the form

$$x^3+px+q = 0$$

171

all we need do is substitute, in the preceding expression for x, $-p$ for p and $-q$ for q. Finally we get

$$x = \sqrt[3]{-\frac{q}{2}+\sqrt{\frac{q^2}{4}+\frac{p^3}{27}}} + \sqrt[3]{-\frac{q}{2}-\sqrt{\frac{q^2}{4}+\frac{p^3}{27}}}$$

Quartic equations were solved by Cardano's disciple Ludovico Ferrari, presumably in 1540, and published in the book *Artis magnae*, together with the rule for solving cubic equations.

We must remember that Cardano studied the solution of equations, not just by using algebraic methods, but by considering the problem from the point of view of the numerical evaluation of roots by successive approximations as well (5). The process applied by Cardano to cubic and quartic equations can be used for equations of any degree. Cardano sees that 'per iteratus operationes semper propinquius licet accedere' to the roots required.

§3. *On the development of algebraic symbolism: the contribution of Descartes*

It would be interesting to deal in detail with the evolution of algebraic symbolism from its origins to the present day, but for reasons of space we can only glance at it rapidly.

The Italian algebraists of the Renaissance made an important contribution to the progress of algebraic symbolism, particularly Rafael Bombelli, a mathematician and engineer in Bologna in the sixteenth century, in his book *Algebra* (Bologna 1572), part of which remained unpublished and was discovered by Bortolotti and published in 1929, with an interesting introduction on Bombelli's symbolism, and tables that reproduce several pages of the original manuscript (6).

A decisive step in the perfection of algebraic symbolism was taken by Descartes. In his *Discours de la Méthode* he criticizes the algebraic symbolism used before his day: 'on s'est tellement assuieti ... a certaines reigles et a certains chiffres, qu'on en a fait un art confus et obscur, qui embarasse l'esprit au lieu d'une science qui le cultive' (7).

(5) Cardano, **2**, ch. 38, **3** in **1**, vol. IV, p. 273. Cossali, pp. 316 ff. and Genocchi, p. 166, pointed out the importance of Cardano's discovery.
(6) Bombelli, chiefly pp. 32–6.
(7) Descartes, t. VI, p. 18.

For example the unknown (root), its square and its cube were represented respectively thus:

R, Q, C;

or else, with cossive characters, respectively:

𝓔, ℨ, 𝓒,

which Descartes was still using in 1619.

These characters did not express clearly what was being done with the unknown to obtain the square and the cube, nor were they easily used in algebraic calculation.

Descartes used letters instead of numbers, to perform operations ($a, b, c, \ldots$ for known quantities, A, B, C for unknown, and later he changed these to x, y, z). The successive powers of a given quantity x were represented, as is still done, by

$$x, x^2, x^3, \ldots$$

Therefore the polynomial in cossive characters

P ı 𝓔. P 4 ℨ. M 7 𝓒

became, according to Descartes' symbolism, which is our own as well,

$$x + 4x^2 - 7x^3 \quad (8).$$

But in the field of algebra Descartes did not confine himself to giving us an adequate, lucid, manageable symbolism; he made as well important discoveries concerning the roots of algebraic equations, collected especially in Book III of his *Géométrie*: in particular the methods of lowering the degree of an equation a root of which is known, and of removing the term of degree $n - 1$ in an equation of degree n; and above all the celebrated theorem about the sign of the roots of an algebraic equation of degree n with real coefficients and roots. Descartes says: 'On connoit aussy, de cecy, combien il peut y avoir de vroyes racines, et combien de fausses, en chasque Equation, A sçavoir: il y en peut avoir autant le vroyes que les

(8) Adam, p. 53.

signes + et − s'y trouvent de fois estre changés; et autant de fausses qu'il s'y trouve de fois deux signes +, ou deux signes −, qui s'entre suivent" (9).

With Descartes's contribution to the construction of analytical geometry and to the development of infinitesimal analysis, we shall deal in the next two chapters.

§4. *The irreducible case and imaginary numbers* (10)

The solution of cubic equations was the first decisive step in modern mathematics beyond the limits of ancient science. This opened up the way for new developments.

Cardano realized that when the three roots of a cubic equation are real, the discriminant of the equation is negative, that is, in the expression of these real roots imaginary values appear.

In such a case the roots are real because they are the sum of conjugate complex numbers. But the imaginary cannot be made to disappear algebraically; it can be eliminated, though, by the complex number being expressed in trigonometrical form, as was shown by F. Viète (1540–1603), who perfected algebraic symbolism and did important work on equations.

While imaginary numbers were found only in quadratic equations, senseless symbols appeared in correspondence with impossible problems; and so these numbers did not seem to matter very much. But when imaginary numbers appeared in the solution of cubic equations, they seemed extremely important, just because they corresponded to real roots. As Cardano observed: 'R.m.9 non est 3p nec 3m sed quaedam tertia natura abscondita'. Which means '$\sqrt{-9}$ is not + 3 nor − 3, but a being of hidden nature'.

The rules for the calculation of complex numbers were given by Rafael Bombelli in his *Algebra*, which we have already mentioned.

We owe further developments of the theory to J. Wallis (1616–1703), A. de Moivre (1667–1754), and R. Cotes (1628–1716). But in Leibniz's day the imaginary still had something mysterious about it: in 1702 Leibniz called the imaginary

(9) Descartes, t. VI, p. 446.
(10) Gigli, **1**, p. 133 ff.

number: 'Analyseos miraculum, idealis mundi monstrum, pene inter Ens et non Ens amphibium.'

The logical possibility of imaginary numbers—that is the non-contradiction and coherence of the new theory—was justified by the geometrical representation of the complex plane of Wessel (11) (1745–1818), Argand (1768–1822), and Gauss (1777–1855), and from the theory of number-pairs, which we owe in a sense to Gauss and in a clearer form to W. R. Hamilton (1805–65).

In the first of these theories the complex number is interpreted as a point on a plane, of which the abscissa is the real part and the ordinate the coefficient of the imaginary, but in the second the complex numbers are considered as pairs of real numbers, which have determinate properties.

From the didactic and heuristic points of view it is preferable to have complex numbers introduced through a geometrical interpretation. But from the logical point of view the theory of pairs is much more satisfying, since it shows the consistency of the theory of complex numbers starting from the consistency of the real numbers.

Thus we have the advantage of not needing to use geometry to establish consistency.

In modern mathematics the consistency of the various geometries is established on the basis of the consistency of the analysis; so if the latter were based in turn on the consistency of geometry, we should have a vicious circle.

In the complex field algebraic theorems have a simplicity and an elegance of enunciation not found in the real field. For instance, in the complex field we find the fundamental theorem of algebra of which d'Alembert (1717–83) gave a non-rigorous demonstration, and which Gauss first demonstrated strictly: every algebraic equation of degree n has n roots.

§5. *A glance at the theorem of Ruffini–Abel and at the work of Galois* (12)

After the wonderful development of algebra during the Renaissance, a development that culminated in the solution of cubic and quartic equations in the sixteenth century, for more

(11) Wessel.
(12) Notari; Enriques, **12**, pp. 198–207; Carruccio, **11**.

than two centuries mathematicians tried in vain to solve algebraically equations of a higher degree than the fourth: see Lagrange's reflections (1770) on cubic and quartic equations, the solution of which he places in relation to the groups of substitutions and permutations of three or four letters.

An important mathematical historian, Montucla, comparing the problem to an enemy defending himself in a besieged fortress, wrote: (13) 'Les dehors de la place sont enlevés de toutes parts; mais renfermé dans son dernier reduit, le problème s'y défend en désespéré. Quel sera le génie heureux qui l'emportera d'assaut, ou le forcera de capituler?'

Paolo Ruffini, mathematician and physician from Modena, succeeded in a very different way from that attempted by mathematicians during centuries of research. At the very time when he was dismissed from his university chair and barred from all public office for having refused to swear against his conscience (April 1798–October 1799), he completed his *General theory of equations* (Bologna, 1799), in which the algebraic solution (that is, through rational operations and extraction of roots) of general equations of a degree higher than the 4th is shown to be impossible. This discovery, which was no less outstandingly interesting because it was negative in character, went far beyond the mental capacity of Ruffini's contemporaries; and the most important mathematicians of the day greeted it coolly.

A gap in Ruffini's demonstration was filled by Abel (1802–29) who, as far as we know, discovered the result independently of Ruffini.

In Ruffini and Abel's demonstration the concepts of combinatory analysis are applied, with reference to the substitutions on the roots $x_1 \ldots x_n$ of the given equation. E. Galois (1811–32) went further into the subject, and established fundamental relationships in higher algebra between the theory of groups of substitutions and the theory of algebraic equations, reaching the necessary and sufficient condition that an equation of any degree n can be solved by radicals.

Galois gave his findings without proofs, in a work that was written on the night before the duel in which he was killed. At first his results, expressed in a concise and very original

(13) Montucla, t. III, p. 18, a passage reported by Bortolotti, **1**, p. 23.

form, were not understood by mathematicians; they were finally demonstrated by E. Betti (1823–92).

This work in the theory of substitutions allows us to establish a vital link between algebra and something that is in itself independent of quantity, and of a purely logical character.

CHAPTER XII

Origins of Analytical Geometry and Cartesian Rationalism; Vico's Gnoseology

§1. *The forerunners of analytical geometry*

While early Pythagorean mathematics was based on the concept of numbers, after the discovery of incommensurables classical mathematics developed, as we have seen, mainly in a geometrical way. All the same many geometrical discoveries of the ancients can be immediately translated into analytical terms; that happened, for instance, to Apollonius's work on conics, which Descartes particularly admired, although he does not mention Strabo (1) (about 63 B.C. to A.D. 25), whose geographical co-ordinates foreshadowed analytical geometry, or N. Oresme, with his diagrams; while his mention of Marino Ghetaldi, who is remembered as a forerunner of analytical geometry, does not seem to show the influence of his thought.

The interest in number, revived when mathematics declined in the ancient world, led to the wonderful discoveries of Renaissance algebra which prepared the way for Fermat and Descartes' synthesis of algebra and geometry.

Algebra was now ready to be applied to geometry, not just to the dimensions of figures, but to the position of points, and in such a way that it could solve appropriate problems with its highly developed simplicity and power.

§2. *Fermat's contribution to the construction of analytical geometry*

Fermat's *Isagoge ad locos planos et solidos*, published posthumously in 1674 after Descartes' *Géométrie* in 1637, but shown earlier to a circle of friends, contains the fundamental principles

(1) Bortolotti, **5**, pp. XXX–XXXI.

of analytical geometry: the correspondence between points of the plane and pairs of numbers by means of a system of co-ordinate axes, and the equations of the straight line and of the conics (2).

Descartes' *Géométrie*, however, which we will examine in more detail, is philosophically more interesting, since Fermat uses algebra only as a help to geometry (3), whereas Descartes asserts the logical primacy of number over the geometrical world (4), thus overturning the whole classical conception.

§3. *Philosophical exigencies answered by Cartesian geometry*

We propose to examine briefly the philosophical motives which made Descartes establish the primacy of numerical analysis over geometry (5). Among them was Descartes' wish for clear and distinct ideas above all: he says that the concept of numbers and their use is simpler than that of geometry and so logically they come first (6).

About this he writes to Desargues: '. . . il y a bien plus de gens qui sçavent ce que c'est la Multiplication, qu'il n'y en a qui sçavent ce que c'est que composition de raisons etc.' (7).

The exigencies we have mentioned depend on the first of the four precepts which, in his *Discours de la Méthode*, Descartes substitutes for 'ce gran nombre de precepts dont la Logique est composée' (8).

(2) Fermat, p. 92.
(3) Boutroux, pp. 105–9.
(4) Enriques, **12**, pp. 43–4.
(5) Here it may be useful to remember Descartes' idea of a 'Mathesis Universalis', a science where geometry, astronomy, physics meet, ruled by number: Quod attentius consideranti tandem innotuit, illa omnia tantum, in quibus ordo vel mensurae examinatur, ad Mathesim referri, nec interesse utrum in numeris, vel figuris, vel astris, vel sonis aliove quovis objecto, talis mensura quaerenda sit, ac proinde generalem quandam esse debere scientiam, quae id omne explicet, quod circa ordinem et mensuram nulli speciali materiae addictam quaeri potest, eamdemque, non ascititio vocabulo, sed jam invete-rato atque usu recepto, Mathesim universalem nominari quoniam in hac continetur illud omne, propter quod aliae scientiae Mathematicae partes appellantur (Descartes, t. X, pp. 377–8). See Augustine, **3**, bk. II, ch. XV (this passage is referred to in this book, ch. IX, §1).
(6) Enriques, **12**, p. 44.
(7) Letter of June 16th, 1639; Descartes, t. I, p. 555.
(8) Descartes, t. VI, p. 18. The four logical precepts of the *Discours de la Méthode* (part 2) may be summed up as follows:
    1. One must accept as true only what is evident.
    2. Difficulties must be divided into their elements, to be solved more easily.
    3. Statements must be ordered, beginning with the simpler ones.
    4. Enumeration must be always complete.

It is a question of using evidence, intuition, clear and distinct ideas; and this not only opposes the principle of authority for the admission of fundamental axioms, but says that even deduction is not subject to the rules of formal logic, but acts through a succession of acts of intuition (9). This explains why Descartes, while he rejects traditional logic, 'syllogismorum formae nihil juvant ad rerum veritatem percipiendam' (10), does not substitute precise rules for deduction, like those of contemporary mathematical logic, for instance (rules of substitution and scheme of conclusion) (11), but confines himself to giving general laws like those contained in his *Discours de la Méthode* and in his *Regulae ad directionem ingenii*.

Descartes's point of view is extremely interesting, even though —in fact precisely because—it seems far from the contemporary logical mathematical way of thinking. Mathematicians in our day, however, do sometimes restate—and do so very effectively —the value of intuition in the face of a system of logical rules (12).

There is another, more subtle, reason why Descartes replaces geometry as far as possible by algebra: the irreducible heterogeneity of the *res extensa* and of *res cogitans*. The question is now: how can extension be thought of?

The question can be answered, up to a point. Brunschvicg (13) observes that Descartes, interpreting geometrical relations through relations between numbers, frees mathematics from the observation of sensible objects and from the imagination of figures; for him the notion of quantity is no longer abstracted through the use of the senses, but has a purely intellectual character. An incomplete work, *Calcul de M. Descartes*, tries to construct an abstract algebra without any use of geometrical representations.

But in fact Descartes does not entirely abolish the imagina-

(9) Milhaud, p. 2: 'At vero haec intuitus evidentia et certitudo, non ad solas enuntiationes sed etiam ad quodlibet discursus requiritur.' (Descartes, t. X, p. 369.)

(10) See: 'Regulae ad directionem ingenii', reg. XIV, Descartes, t. X, pp. 439–40.

(11) See Hilbert and Ackermann, ed. 1949, p. 23.

(12) See Severi, **3**.

(13) Brunschvicg, p. 123. Laporte, in his *Le rationalisme de Descartes*, does not agree. (**1**, pp. 126–8.)

tion in his *Géométrie,* bringing to light the correspondence
between the quantities x, x^2, x^3, and determinate segments
(no longer segments, surfaces or solids, as they were for the
ancients); but the decisive step which makes Cartesian geometry no longer a simple application of algebra to geometry,
but real reduction of geometry to algebra, was made, according
to Brunschvicg, by Malebranche (14).

Leibniz criticized this idea; equations in themselves, he says,
have no geometric significance: they need translating to be
applied to spatial relations; for instance when we say that:
$x^2 + y^2 = a^2$ is the equation of a circle, the meaning of x and y
in the figure must be explained, etc. (15).

Leibniz's conclusion was that the: 'synthèse des Géomètres
n'a pu être changée encore en analyse. ... On s'entonnera
peut être de ce que je dis ici, mais il faut savoir que [l'algèbre],
l'analyse de Viète et Descartes est plutôt l'analyse des nombres
que des lignes, quoiqu'on y réduise la géométrie indirectement,
en tant que toutes les grandeurs peuvent être exprimées par
nombres' (16).

The difficulty inherent in the problem of conceiving extension
is linked, in Cartesian philosophy, to the mysterious relationship between mind and body (17). Descartes tried to get over
the difficulty by saying that the mind in contact with the body
gets the notion of extension directly from it (18). This
substantial union of mind and body is an irrational element in
Descartes' philosophy which he himself admitted (19).

And yet, through his thought we reach these conclusions:
the logical relations of geometry are translated into analytical
relations: for instance, a point's belonging to a straight line is
translated into the condition that two numbers (co-ordinates
of the point) satisfy a 1st degree equation in two variables; and
so on.

When, later on, modern mathematical thought came to conceive mathematical theory as a hypothetical deductive system
based on arbitrary postulates from a strictly logical point of

(14) See Brunschvicg, p. 132. Here also Laporte, 1, p. 127, does not agree.
(15) Leibniz, 1, 1te Abth., Bd. II, p. 30; 3, I Bd., p. 580.
(16) Leibniz, 4, p. 181.
(17) See Brunschvicg, pp. 128–9.
(18) Laporte, 2, pp. 257–89.
(19) Descartes, t. III, p. 693; Laporte, 2.

view, and this raised the question of the non-contradictoriness of the postulates of geometry, for instance Euclidean geometry, Descartes' analytical geometry made it possible to reduce the non-contradictoriness of his geometry to that of analysis (20).

And this is one of his most important contributions to mathematical thought.

§4. *Aspects of Descartes' analytical geometry*

The first step in constructing the 'Mathesis universalis', in which, according to Descartes, geometry, astronomy, and physics were dominated by number, consisted in the reform of algebraic symbolism, to make it clear and expressive: we dealt with this subject in the previous chapter.

Another essential step was the representation of any proportion by a proportion between segments expressed through symbols, 'chiffres'. This concept, which was to be systematically applied in his *Géométrie*, was already expressed in the *Discours de la Méthode*, where on the subject of proportion Descartes says: 'Puis ayant pris garde que, pour les connoistre, i'aurais quelques fois, besoin de les considerer chaschune en particulier, et quelque fois seulement, de les retenir, ou de les comprendre plusieurs ensemble, je pensay que, pour les considerer mieux en particulier, ie les devois supposer en des lignes, a cause que ie ne trouvais rien de plus simple, ny que ie pusse plus distinctement representer a mon imagination et a mes sens; mai que, pour les retenir, ou les comprendre plusieurs ensemble, il fallait que ie les expliquasse par quelques chiffres, les plus courts qu'il serait possible; et que, par ce moyen, i'empreunterois tout le meilleur de l'Analyse Géométrique et de l'Algebre, et corrigerois tous les defaus de l'une par l'autre' (21).

In his *Géométrie*, introducing the unit segment of measure as Leonardo Pisano (22) and Rafael Bombelli (23) had already done, Descartes interpreted the algebraic expressions as relations between segments. That is, for instance, when a is the measure of a segment, a^2 and a^3 were represented not as

(20) See Carruccio, **14**; chiefly n. 5.
(21) Descartes, t. VI, p. 20.
(22) Leonardo Pisano, *Practica Geometriae*, pp. 153–5.
(23) Bombelli, p. 19. There, one may realize that, even in the operations with segments, Bombelli precedes Descartes.

areas or volumes, as the ancients would have done, but still as measures of segments (24).

Descartes had to make all his algebraic expressions by means of lines, since he had not yet the principle of continuity, without which the existence of a magnitude had to be guaranteed by a construction, while for us the mathematical existence of a magnitude can be assured by a view of continuity (25): think, for instance, of the third part of an angle, whose existence we admit, even though we cannot effectively construct it (with ruler and compasses).

On this basis Descartes builds his geometric world. To represent by an equation the properties that characterize a curve l (see fig. 1) he considers a straight line r on which he

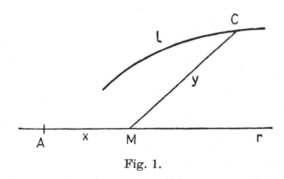

Fig. 1.

takes a point A as the origin of the length measured on the line itself. We draw through a generic point C of the curve l a second straight line in a fixed direction; this meets the line r at a point M. If we indicate the measure of AM by x and the measure of MC by y, the curve will be characterized by an equation in the variables x and y.

In *Géométrie* we do not deal systematically with the equation of the straight line, but when we meet a 1st degree equation in x and y we are clearly dealing with the equation of a straight line.

The straight line AM which functions as the axis of the abscissae, is usually chosen for reasons connected with the curve to be represented.

(24) Descartes, t. VI, pp. 369–71, and 442–4.
(25) Milhaud, pp. 7–8.

When he had dealt with straight lines and conics represented respectively by linear and quadratic equations, Descartes continued by defining the curves of a higher order, classifying them on the basis of the degree of their equations. Thus he built up a geometric world which, with the simple application of the algebraic mechanism, became limitless. Geometry was no longer, as it had been for the ancients, a reality which could be contemplated; it was a world to be built by the spirit, starting from the simplest elements. What is sought is not so much a collection of results as the building up of a method (26).

Descartes accepts the construction of curves by mechanical instruments; he says that as the straight line and the circle are drawn with instruments, the ruler and compasses, there is no reason to reject the use of instruments in order to draw a curve of a higher order (27).

All the same Descartes does not push his generalization beyond all limits, since he excludes from his geometrical world certain curves like the spiral, the quadratrix of Hippias and others, 'a cause qu'on les imagine descrites par deux mouvements separés et qui n'ont entre eux aucun rapport qu'on puisse mesurer exactement' (28). The curves we have mentioned, which he does not accept in his geometry, are transcendental. Thus the transcendental functions—trigonometrical, exponential, logarithmic, etc.—are not included in Descartes' analysis, which kept him only on the threshold of infinitesimal calculus (29).

In Book III of his *Géométrie* 3rd, 4th, 5th and 6th degree equations are solved by means of the intersection of curves, and there he deals too with the classic problems of the duplication of the cube and the trisection of an angle, to which all 3rd degree problems can be reduced (30).

Descartes guessed that it was impossible to solve with ruler and compasses problems that led to a cubic equation irreducible in the field of rationality of the coefficients, and other similar

---

(26) *Géométrie*, liv. II (Descartes, t. VI, chiefly pp. 392-3); Boutroux, pp. 104-9; Bompiani, pp. 313-25.
(27) Descartes, t. VI, pp. 389-90.
(28) Descartes, t. VI, p. 390.
(29) Vacca, **10**.
(30) Descartes, t. VI, pp. 469-73; see also Conti, pp. 353-6, and 385-6.

impossibilities of the kind: his findings were then demonstrated in the spirit of Cartesian geometry (31).

In his *Géométrie* he successfully tackles a problem which had been posed but not solved in all generality by the ancients. This is the so-called problem of Pappus, which can be set out in the following way:

We are given $2n$ straight lines (see fig. 2). Through a point P we draw another $2n$ straight lines which form angles which are equal to one another with the given straight lines. Let

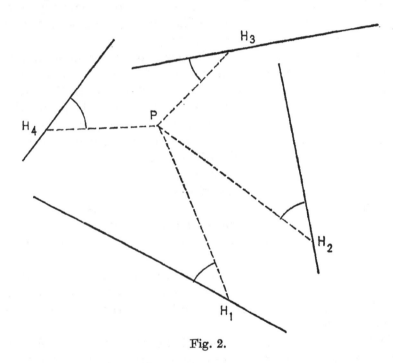

Fig. 2.

$H_1, H_2 \ldots H_{2n}$ be the vertices of these angles. We must find the geometric locus of the point P chosen in such a way that we have equality of the products

$$\overline{P}\overline{H}_1 \ldots \overline{P}\overline{H}_n = \overline{P}\overline{H}_{n+1} \ldots \overline{P}\overline{H}_{2n}$$

In the case in which the number of given straight lines is odd $(2n-1)$ we substitute a given segment for one of the segments $PH_1 \ldots PH_{2n}$.

(31) Descartes, t. VI, pp. 475–6; see Bompiani.

In the case of 3 and 4 straight lines, the problem which, according to Apollonius and Pappus, was presumably posed and partly solved by Aristaeus, was dealt with by Euclid and, more completely, by Apollonius. Descartes proves the power of his method by giving the general solution of the problem, recognizing the algebraic nature of the geometric locus of the point P, and establishing the degree of the equation in question, though in a different way from that which is used at present (32).

So far we have considered Cartesian plane geometry, but Descartes also mentioned the method for treating analytically the problems of geometry in space (33).

Descartes projects a twisted curve drawn from a mobile point in space, on to two planes of reference which are perpendicular to one another. The twisted curve is thus represented by two plane curves which can be studied with the methods we have already mentioned. Here we find the fundamental concept of the method of orthogonal projection in descriptive geometry, called Monge's.

It was on the basis of this Cartesian hint that Clairaut was to build the analytical geometry of space.

In his *Géométrie* Descartes carried out the task he had set himself, according to the conception of mathematics that had guided him; and with his method he unified a huge domain of the world of thought.

'Pour la méthode dont ie me sers tout ce qui tombe sous la consideration des Geometres se reduit a un mesme genre de problemes, qui est de chercher la valeur des racines de quelque Equation' (34).

With his vision of the mathematical world, which he felt he could, at least potentially, dominate entirely, Descartes thus concluded his *Géométrie*:

'Et i'espère que nos neveux me sçauront gré, non seulement des choses que i'ay icy expliquées, mais aussy de celles que i'ay

(32) Descartes, t. VI, pp. 377–87, pp. 396–411. See also the note by P. Tannery in Descartes, t. VI, 721–25. Descartes calls "Curves of order n" those curves which we should define as of order $2n$ or $2n-1$.

(33) Descartes, t. VI, pp. 440–1, near the end of the second book of the *Géométrie*. Descartes tries only one problem on skew curves: that of finding the normal to a curve, at a given point. Unfortunately, he makes a mistake. (See p. 441 of the volume cited.)

(34) Descartes, t. VI, p. 475.

omises volontairement affin de leur laisser le plaisir de les inventer' (35).

§5. *Relations between the work of Descartes and the later developments of mathematics*

Descartes deserves the gratitude of his mathematical descendants: he gave them a lucid and powerful language in which to express algebraic reasoning, and opened up wide fields for them to work in (36). Yet those who came after him were to push on in a way he had not really foreseen; in particular the infinitesimal calculus, which, as we have seen, Descartes approached, was developed in a way that could be called non-Cartesian; so also was the theory of infinite aggregates.

But Leibniz's lucid symbolism for the infinitesimal calculus harmonized perfectly with Descartes' algebraic symbolism; while the symbolism of infinitesimal calculus was to form a part, according to Leibniz, of a more general science, 'characteristica universalis', to which Descartes also seems sometimes to have aspired, as when he said that his method should contain and express 'prima rationis humanae rudimenta' ... 'ad veritates ex quovis subjecto eliciendas' (37).

Yet there is an essential difference between the views of Descartes and Leibniz with regard to logic (38). For Descartes the necessity of the fundamental truth depends on the will of God who has established them from all eternity (39). The coherence of a system would therefore not be enough to guarantee its truth, and its functional formalism would not guarantee the truth of a science. But according to Leibniz, the necessary truths subsist by their own nature, not through a choice of God. So, from these truths, which would be identical, a system could be constructed the coherence of which would guarantee its truth. In any case, for Descartes as for Leibniz,

(35) Descartes, t. VI, p. 485.
(36) Descartes, t. VIII, pp. 327–8, t. IX, 2nd part, pp. 323–7, art CCV.
(37) And also 'hanc methodum omni alia nobis humanitus tradita cognitione potiorem, utpote aliarum omnium fontem esse' (Descartes, t. XI, p. 374).
(38) On this topic, see Schrecker, pp. 336–67.
(39) Descartes, t. IX, p. 236, and also: 'Les vérités mathématiques, lesquelles vous nommez éternelles, ont esté establies de Dieu et en dependent entièrement, aussy bien que tout le reste des créatures. C'est en effet parler de Dieu comme d'un Juppiter ou Saturne, et l'assujétir au Stix et aux destinées, que de dire que ces vérités sont indépendantes de lui.' (Letter to F. Mersenne, April 15th, 1630, t. I, p. 145.)

the methodology was based on metaphysical hypotheses: while for Descartes the reality of science is based on 'cogito', for Leibniz the positiveness of the results of a symbolic science is based on a conception of the rationality of the real.

Mathematical thought was to move on in other non-Cartesian directions: as we have said, Descartes' mathematics is essentially the science of the ratios between magnitudes, a metrical science. Well, as we know, branches of modern mathematics exist in which the metrical concepts do not apply; among them, for instance, projective geometry which studies not the metrical but the graphic properties of figures, invariant for projection and section, in the spirit of the *Brouillon project* of Desargues, whom Descartes, following his metric ideal, vainly advised to use algebra to simplify his demonstrations.

But while on the one hand projective geometry was built up puristically, without metrical ideas, on the other the relations between analytical and projective geometry were enormously fertile, and made important contributions to these two branches of mathematics.

Still further from metrical ideas is another branch of modern mathematics: the 'analysis situs', according to the term introduced by Leibniz, or topology, which concerns the properties which figures maintain while they are deformed continuously without duplication or laceration.

But Descartes made an important discovery from which we can deduce the so-called Eulerian relation, which links the numbers f, v, s of faces, vertices and edges of a polyhedron (40):

$$f + v = s + 2$$

Of this relation we know the topological significance, which transcends the theory of the polyhedra.

But the non-Euclidean geometries were to develop still further from these theories, and from the mathematical world of Descartes. And yet it was their analytical interpretation, according to the spirit of Cartesian geometry, that first demonstrated their legitimacy. Lobachevsky writes in his *Pangeometry*: 'A mere glance at the equations which express the relations between the sides and angles of rectilinear triangles

(40) Descartes, t. X, pp. 257–76 and table, chiefly p. 258. See Natucci, pp. 287–8.

will show that, starting from there, Pangeometry becomes an analytical method which acts as a substitute for, and generalizes, the methods of geometry' (41).

So, even the branches of modern mathematics which theoretically are outside the boundaries of Cartesian geometry are vitally linked with Descartes' work, which remains an essential part of the mathematical thought of our time, in its straining towards clear and distinct ideas, in the symbolism and language of algebra, in its synthesis of algebra with geometry, in its constant searchings, in its application of general methods, and in its reflections on them: methodology, in fact.

§6. *The logic of Pascal and of Port Royal* (42)

The Cartesian views on logic are reflected in the writings of Pascal, who thus very clearly expresses his ideas on the subject (43):

'Cette véritable méthode, que formerait les démonstrations dans la plus haute excellence, s'il était possible d'y arriver, consisterait en deux choses principales:...définir tous les termes et prouver toutes les propositions'; but 'cette méthode ...est absolument impossible: car il est évident que les premiers termes qu'on voudrait définir en supposeraint des précédents pour servir à leur esplication, et que de même les premières propositions qu'on voudrait prouver en supposeraint d'autres que les précédassent, et ainsi il est clair qu'on n'arriverait jamais aux premières. Aussi, en poussant les recherches de plus en plus on arrive nécessairement à des mots primitifs qu'on ne peut plus définir, et à des principes si clairs, qu'on n'en trouve plus davantage pour servir à leur preuve.'

Having thus set down what is the most perfect logical order realizable by man in the construction of a rational science, Pascal gives the rules for reaching this ideal (44): '*Règles pour les définitions.*—1) N'entreprendre de definir aucune des choses tellement connues d'elles-même, qu'on n'ait point de termes plus clairs pour les expliquer.

'2) N'omettre aucun des termes un peu obscurs ou équivoques sans définition.

(41) Bonola, 1, p. 83.
(42) Enriques, 2, pp. 72–5.
(43) Pascal, pp. 59–90.
(44) Pascal, pp. 82–3.

'3) N'employer dans la définition des termes que des mots parfaitement connus ou déjà expliqués.

'*Règles pour les axiomes.*—1) N'omettre aucun des principes nécessaires sans avoir demandé si on l'accorde, quelque clair et évident qu'il puisse être.

'2) Ne demander en axiomes que des choses parfaitement évidentes d'elles-même.

'*Règles pour les démonstrations.*—1) N'entreprendre de démontrer aucune des choses qui sont tellement évidents d'elles-même qu'on n'ait rien de plus clair pour la démontrer et prouver.

'2) Prouver toutes les propositions un peu obscures, et n'employer à leur preuve que des axiomes très-évidents, ou des propositions deja accordées ou demontrées.

'3) Substituer toujours mentalement les définitions à la place des définis, pour ne pas être trompés par l'équivoque des termes que les définitions ont restreint.'

With these rules we approach the principles established from the criticism of contemporary logic, and with them reach the conception of the hypothetical-deductive system, with which we shall deal later (Chapter XIX, §1).

Cartesian thought appears again in *Logique ou art de penser*, called of Port Royal, attributed to the contemporaries of Pascal, Arnauld and Nicole.

§7. *Giambattista Vico*

Giambattista Vico (1668–1774), although clearly he admired Descartes (45), expressed the first form of his gnoseology (46) as a criticism of Cartesian thought, in his *De nostri temporis studiorum ratione* (1708), *De antiquissima Italorum sapientia* (1710), and in his discussion on the subject during the four years 1708–12. On the Cartesian principle '*cogito ergo sum*', the foundation of the certainty of thought and of being, Vico observed that a sceptic might object that this principle gives us consciousness (of the particular or of the certainty), not science (of the universal or of the true) (47).

Now, Vico wonders, what is the criterion that makes science

(45) See *De Mente heroica*, in Vico, **2**, p. 925; Croce, **2**, p. 268.
(46) Croce, **2**, chiefly at pp. 3–5 and 8–9.
(47) *De antiquissima Italorum sapientia*, lib. I, c. I, III, in Vico, **1**, pp. 138–140.

possible? He replies that the condition of knowing a thing is to make it, the truth is the fact itself: *'verum esse ipsum factum'* (48), the knowledge and the operation must be converted into one another, identified with one another, as happens in God through the intellect and the will.

For man, according to the first form of Vico's gnoseology, science is impossible, except mathematics, or more particularly geometry, since mathematics is the only knowledge which man possesses in a way identical with that of divine knowledge, not through its own evidence, but because mathematics is an operative science, not only in the solving of problems but in the demonstration of theorems too: *'geometrica demonstramus quia facimus'* (49).

In the second and definitive form of his gnoseology (50), which matured through his historical and juridical studies in the decade following 1712, Vico, in his *Principi di scienza nuova* (in the three elaborations of 1725, 1730 and 1744) extended the principle *'verum et factum . . . convertuntur'* to the whole human world, and to the world of history, to which he attributes a reality greater than that of the world of geometry (51).

Vico's thought elaborates an aspect of the modern conception of mathematics understood as a construction of the human mind, and therefore true, as Vailati observed, even if experience does not confirm it (52).

In its second form Vico's gnoseology is also in harmony with the present view of the history of science (53).

(48) Vico, **1**, p. 131.
(49) *De nostri temporis studiorum ratione*, IV, in Vico, **1**, p. 85.
(50) Croce, **2**, chiefly pp. 28–30.
(51) *Principi di scienza nuova*, lib. 1; Del metodo, in Vico, **2**, p. 489.
(52) Vailati, **1**, pp. 201 and 434; Brusotti, p. 895.
(53) See ch. I, §1 of this book.

CHAPTER XIII

Modern Infinitesimal Analysis and the Philosophical Thought of its Constructors

§1. *Commentators and continuers of the work of Archimedes*

The problems of infinitesimal character, as we have seen, were presented under many aspects in classical antiquity; and the ways of solving them, after a long critical elaboration, were systematized rigorously by the method of exhaustion attributed to Eudoxus of Cnido and applied by Euclid in Book XII of the *Elements*, which reached its highest point and its most illuminating discoveries in the work of Archimedes. After Archimedes, who already complained of the indifference with which his discoveries were received by the mathematicians of his day, decadence set in, and for centuries there was no progress worth noting.

In the Renaissance infinitesimal analysis (1) began to develop again, starting with the translation and study of, and the comment on, Archimedes' then known work on the subject, which used the method of exhaustion. The first translations of his work into Latin were those of N. Tartaglia, Maurolico (1494–1575), and Commandino (1509–75), who followed his work in the field of mechanics : Tartaglia in his book *Della travagliata inventione* dealt with the way of recovering a sunken ship, and in his *Scienza nuova* left (not without moral preoccupations) the first tables of ballistics; Maurolico and Comandino determined the mass-centre of various solids.

Maurolico, as Vacca has established, was the first to recognize explicitly the principle of complete induction (2), of which we

(1) For the topics dealt with in this chapter, see: Castelnuovo, **5**; Enriques, **12**, pp. 45–67; Enriques, **2**, pp. 51–109; Geymonat, **2**, pp. 57–183.
(2) Zariski, **3**, pp. 170–2.

find traces in Euclid's *Elements* (propositions 8 and 9 of Book IX), to which we shall return in dealing with the systematization of rational arithmetic according to Peano. Later Guldin (1577–1643) worked in the spirit of Archimedes' methods of exhaustion; he is remembered mostly for the famous theorem, which he demonstrated, on the area and the volume of a solid of rotation (3), a theorem which we have already met, without demonstration, in the work of Pappus.

§2. *Luca Valerio* (4)

Luca Valerio (1552–1618) transformed the method of exhaustion into a principle that could be applied to curves and surfaces in general. He demonstrated a theorem which, according to Chisini, in substance expresses Riemann's condition of integrability (5):

'Omni figurae circa diametrum in alteram partem deficienti figura quaedam ex parallelogrammis inscribi potest, et altera circumscribi, ita ut circumscripta superet inscriptam minori spacio quantacumque magnitudine proposita.' That is (see fig. 1): To every plane figure decreasing on both sides from the diameter we can inscribe a figure made of parallelograms and circumscribe another, in such a way that the latter figure is greater than the inscribed one by any given amount (as small as one pleases).

The expression '*in alteram partem deficienti*', which we have translated as 'decreasing on both sides', means that the two arcs of the curve considered BA and AC on opposed bands to the line AD, perpendicular to BC since D is contained in the segment BC, form the diagram of a function increasing in the interval BD, and decreasing in the interval DC.

---

(3) For a modern proof of Guldin's Theorem, see Levi-Civita and Amaldi, vol. 1, pp. 440–1.

(4) Chisini, pp. 100–4.

(5) The comparison between Luca Valerio's theorem and Riemann's condition for integrability is left to the reader. As is known from analysis, Riemann's condition may be stated as follows: The necessary and sufficient condition that a bounded function should be integrable is that $S_n - s_n$ should tend to zero when the number of intervals $(x_{r-1} - x_r)$ tends to infinity in such a way that the length of the longest tends to zero.

Here
$$S_n = U_1(x_1 - a) + U_2(x_2 - x_1) + \ldots + U_n(b - x_{n-1}),$$
$$s_n = L_1(x_1 - a) + L_2(x_2 - x_1) + \ldots + L_n(b - x_{n-1}),$$

where U_i and L_i are respectively the upper and lower bounds of the function in the interval $x_i - x_{i-1}$.

193

The segment DA was divided by Valerio into n equal parts, and through the points of division were drawn to DA perpendicular chords which formed the bases of rectangles inscribed in or circumscribed to the figures having the common height $1/nAD$.

It is clear in the figure that the difference between the sum of the circumscribed rectangles and the sum of the inscribed rectangles is equivalent to the rectangle with a base BC and a height $1/nAD$, a rectangle which can be made as small as one pleases.

A similar theorem was demonstrated by Valerio for the surfaces: 'Omni solido circa axim in alteram partem deficienti,

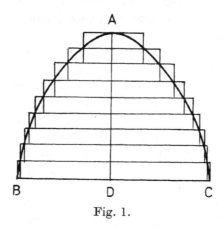

Fig. 1.

cuius basis sit circulus vel ellipsis, figura quaedam ex cylindris, vel cylindri portionibus aequalium altitudinum inscribi potest, et altera circumscribi, ita ut circumscripta superet inscriptam minori excessu quacumque magnitudine proposita.'

In another lemma Valerio changes the indirect reasoning, which is always used in the method of exhaustion, into a passage to the limit: 'Si maior vel minor prima ad una maiorem vel minorem secunda, minore excessu vel defectu quantacumque magnitudine proposita nominatam habuerit proportionem, prima ad secundam eandem nominatam habuerit proportionem.'

This obscure wording is interpreted by Chisini with the help of the following demonstration:

'If a magnitude A_n, greater or less than a former magnitude A, differing from this A by an excess or defect less than any

194

given quantity, has a given ratio k to a magnitude B_n, this also greater or less, (varying) together with A_n, than a second magnitude B, differing from this B by an excess or defect less than any given quantity: then A bears to B the same ratio k.'
In modern language:
If by hypothesis

$$\lim_{n \to \infty} A_n = A, \quad \lim_{n \to \infty} B_n = B, \quad \text{and} \quad \frac{A_n}{B_n} = k$$

then we shall have:

$$\frac{\lim_{n \to \infty} A_n}{\lim_{n \to \infty} B_n} = k$$

Valerio, using these principles, gives us new determinations of volumes and mass-centres of solids limited by parallel planes and surfaces of the 2nd order.

Galileo, in spite of differences of opinion on the Copernican question between himself and Valerio, called him 'the new Archimedes of our day'.

§3. *The direction of science in the sixteenth and seventeenth centuries and the methods of indivisibility*

The method of exhaustion, which was excellent when it wished to demonstrate results that had already been obtained, was not so useful for purposes of discovery, and so it could not satisfy the sixteenth and seventeenth century mathematicians who wished to do research, either in the world of pure thought, or in the physical world.

A process which was far more useful in the search for new theorems was found instead in Archimedes' *Method*, where he showed how to establish the equivalence of two plane or solid figures or their ratio by comparing their infinitesimal elements, and weighing them on ideal scales.

Torricelli guessed that a process of the kind, which could be revived in a simpler and more powerful form in the geometry of indivisibles, must have existed among the ancients; as is shown in a passage of his, here translated from the Latin (6):

'Really I should not dare to state that this geometry of

(6) See Torricelli, **2**, vol. I, p. I, pp. 139–40.

indivisibles is actually a modern discovery. I should think that the ancient geometers used this method in the discovery of their most difficult theorems, although in their demonstrations they preferred another means, either to hide the secret of their art, or else so as not to give their jealous detractors any reason for criticism. But however it is, it is certain that this geometry means a wonderful saving of work in the demonstrations, and that it establishes innumerable almost inscrutable theorems by means of short, direct, affirmative demonstrations, which could not be done through the doctrine of the ancients. This geometry of indivisibles, which is really a royal road through the mathematical bushes, was first seen and shown for the public good by that wonderful inventor, Cavalieri.'

Among sixteenth and seventeenth century thinkers there is a remarkable predilection for the mathematical infinite, not found in the writings of the Alexandrine mathematicians. I think there were two reasons for this. One is the Renaissance development of perspective in painting, which considered vanishing points corresponding to points at infinity. The other is the conception of the infinite in divine thought, found in the philosophy inspired by Christianity, especially in the writings of St. Augustine (see Chapter IX, §1).

The decomposition of a figure into infinitesimal elements, so as to enable one to measure the size of the figure itself, appears in the thought of Leonardo da Vinci, expressed in the Atlantic Codex (7): 'This proof is persuasive if we imagine the circle divided by very narrow parallels, like very fine hairs continually in contact with one another ...' Thus, through images, are foreshadowed the indivisibles of the school of Galileo.

In this period infinitesimal analysis received an impulse from the astronomical and physical researches of Galileo (1564–1642) and of Kepler (1571–1630), to whom we owe some applications of the method of indivisibles.

§4. *The indivisibles and infinite aggregates in the thought of Galileo* (8)

Galileo's thoughts on the infinite and the infinitesimal are

(7) Severi, **5**.
(8) Carruccio, **8**; Capone Braga.

linked on the one hand to the origins of infinitesimal analysis in the modern age by their connection with Cavalieri and Torricelli's methods of indivisibles, but on the other hand anticipate the positions taken by Bolzano and Cantor on the subject. We will here set down exactly the meaning and basis of what he said.

The word *indivisibles*, which can be linked with the first of Euclidean terms (the point is that which has no parts), makes us think of the geometry of Bonaventura Cavalieri (9); who on 15th December 1621 (10) in a letter to Galileo expressed doubts on the fundamentals of his own method of indivisibles which he was then constructing.

Galileo did not answer that letter, but Cavalieri continued to write, showing his growing faith in the method he had built up, and urging him to 'do something about the doctrine of indivisibles, as he had thought of doing some years before'(11).

Obviously Galileo, as his disciples and friends knew, had been meditating on indivisibles for some time, but the chance of writing on the subject was given him by a book, published in Venice in 1633, by Antonio Rocco, a peripatetic philosopher, in which Galileo's position in his *Dialogue on the two greatest systems*, which opposed Aristotle's theories, was attacked. The interesting notes which Galileo added to his adversary's mediocre work were published only in the national edition (12).

Now Rocco, among his criticisms of the *Dialogue on the two greatest systems*, observes that he has always 'found difficult and unintelligible, not to say false', the often-used expression *Sphaera tangit planum in puncto*; because 'if a sphere were dragged along a perfect plane it would trace out a curve, and always touch the plane at a point; this means that the parts of the curves would be points, and would come to be composed of them: which in mathematics and philosophy is considered quite untrue, since every quantity is made up of ever divisible parts.'

After refuting these and other observations of Rocco's in

(9) Cavalieri, **1**.
(10) Galilei, **1**, vol. XIII, p. 81.
(11) See a letter written by Cavalieri to Galileo, Bologna, January 10th, 1634. (Galilei, **1**, vol. XVI, p. 15.) Some remarks on infinity may be found in Galilei, **2**, pp. 243–61.
(12) Galilei, **1**, vol. VII, and chiefly pp. 682–3.

short notes, Galileo examines the matter at greater length and concedes that the saying '*Sphaera tangit planum in puncto* has been, until then, almost unintelligible, but never false'. He says that both truths exist: the continuum consists of parts which are always divisible, the continuum consists of indivisibles; in fact as truth is one, these two affirmations must express the same reality. Then he says in the text (13):

'Open your eyes to the light which perhaps has been veiled until now, and see clearly that the continuum is divisible into parts which are always divisible only because it consists of indivisibles; if division and subdivision are to continue for ever, the multitude of parts must necessarily be such that it can never be overcome; this means that they must be infinite parts, otherwise a division could end; and if they are infinite, they must not be extended, because infinite amounts of extended quantities make up an infinite quantity, whereas we are speaking of a finite amount; but the highest and last, indeed the first components of the continuum, are infinite indivisibles.

'. . . Here the fake philosophers turn up with their acts and their powers, saying that the parts of the continuum are infinite in power but always finite in act: something which they may understand and which keeps them quiet, but I can see no sense in; maybe Signor Rocco can. So I wonder how in a line which is four palms long are contained four parts, that is four lines each one palm long; are they contained in act, or only in power? If I am told that they are contained only in power, while they are not divided and marked, and in act therefore when they are cut, I will prove that extended parts neither in act nor in power can be infinite in the line. And I wonder again if when I actualize the four parts dividing into four parts the line four palms long it grows or diminishes or does not change size. I think I shall be told that it stays the same size, either it contains its extended parts in act, or has them in power, and not being able to hold them in act, it will not contain them in power: and thus parts which are infinite neither in act nor in power can be on the finite line.' Galileo's mistrust and hatred of the potential infinite, which he was to show again in his *Mathematical discourses and demonstrations on two*

(13) Galilei, 1, vol. VII, pp. 745–50.

new sciences, is remarkably like Cantor's attitude on the same subject (14).

Galileo says he is ready to break a line up into its infinite points; in his *Mathematical discourses and demonstrations on two new sciences* he used a very expressive phrase: 'resolving the whole of infinity at one go' (15). As the perimeter of a regular polygon of n sides is divided into n parts, so in a circle, considered as a polygon of infinitely many sides, the infinite number of points of the circumference will be come into being; '... which circle' says Galileo (16), 'will have all the requisites of all other polygons and others, even more marvellous.

'The polygon of a hundred sides built on a plane touches it with one of its sides, that is with a hundredth part of its perimeter; the circle placed in the same way touches it equally with one of its infinite sides, that is with a point. That polygon, in turning, presses on the plane as it turns a continuous straight line, composed of a hundred parts of its ambit; the circle as it turns traces a straight line composed of the infinite and equal points of its circumference. Other results and other wonderful things you will hear another time (17), when I hope to show that the road we generally take in trying to understand the ways of nature is well trodden by the philosophers towards the end they wish for, banishing from their minds the infinites, indivisibles, and vacua, as vain and pernicious concepts, hateful to nature, just as if a painter or a smith were to tell his pupil to banish paints, brushes, anvil, hammer, file, as materials or tools that are useless, indeed harmful to these pursuits.'

Galileo carries on with the discussion, noting among other things that 'the man who carefully draws a line made up of points, does not take up one, or two or a thousand or a million, but an infinite number; since to give divisibility and quantity is a quality of infinity, which is something very far from the attributes and conditions to which the numbers and sizes contained by our intellect are subject; there, such concepts as greater, less, equality have no place, nor have equal or unequal;

(14) Cantor, *Über die verschiedenen Standpunkte in Bezug auf das aktuelle Unendliche*, p. 373.
(15) Galilei, 1, vol. VIII, p. 93; see Bortolotti, 16, p. 359.
(16) Galilei, 1, vol. VII, pp. 747–9.
(17) Galileo deals amply with this matter in his *Discorsi e dimostrazioni matematiche intorno a due nuove scienze*: see Galilei, 1, vol. VIII, pp. 80–2 and 91–3.

every part (if you can call it a part) of the infinite is infinite; and if a line a hundred palms long is greater than a line which is one palm long, the points of the first are not greater in number than the points of the second, for both are infinite.'

There seems an echo of the thought of Anaxagoras, whom we have already mentioned: 'in the great as well as in the small there is an equal number of parts. . . .' Roger Bacon, too, had observed that the points of two different segments could be placed in a one-to-one correspondence one to the other, but from this observation he thought he could deduce the logical impossibility of the mathematical infinite.

Galileo, as we shall see more clearly in what follows, was not stopped by such difficulties; but it seems he did not suspect the existence of infinite aggregates of different powers; except that when asked if the extended parts in the finite continuum (for instance, a segment) are finite or infinite, Salviati (in the *Mathematical discourses and demonstrations on two new sciences*) answers that they are 'neither finite nor infinite': between the finite and the infinite there is 'a third middling term': 'the answer to every given number' (18).

Galileo's views on indivisibles and infinites which appeared in his arguments with Rocco are usefully amplified in his *Mathematical discourses and demonstrations on two new sciences*, published at Leiden in 1638 (19).

It was physical considerations that urged him on to it.

Galileo, like Cantor later, linked his mathematical theories on actual infinitesimals to his conception of the structure of matter (20). The three speakers in the dialogue of the new sciences are discussing the reasons for the cohesion of bodies: Salviati suggests that the cohesion itself is due to the space between the particles of the bodies; he uses the word 'paradoxes' here about something which concerns the infinite and the infinitesimal, and says: 'Well then, now that we've got to paradoxes, let's see if we can show in some way that the possibility of finding infinite vacua does not contradict the idea of continuous extension' (21).

(18) Galilei, **1**, vol. VIII, p. 81.
(19) Mieli.
(20) Lassvitz; this work is quoted in a paper by Goldbeck. See also Cantor, p. 373.
(21) Galilei, **1**, vol. VIII, p. 68.

To demonstrate this Salviati uses the ingenious mechanical device of two regular polygons with n concentric and similar sides which are joined to one another (see fig. 2).

Let us make the larger polygon roll round on the extension of one of its sides : these sides settle down on the extension filling a segment *AS* the length of which is equal to the perimeter of the polygon. Meantime the sides of the lesser polygon would also settle on the extension of one side of the said polygon, but they would leave $n-1$ empty spaces, *IO, PU*, etc.

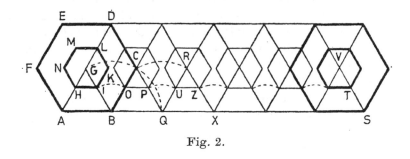

Fig. 2.

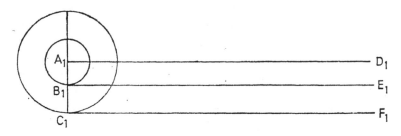

Fig. 3.

Salviati therefore proceeds to the limit (as we should say) considering, instead of the two polygons, two concentric circumferences (see fig. 3) similar to those which were considered by Aristotle (*Mechanics*, Chapter XXV) (22). While the larger rolls without sliding, touching successively the points of a segment C_1F_1, equal in length to the same circumference,

(22) Aristotle, **1**, vol. IV, pp. 68–70. Brunet and Mieli, p. 495. We recall that Galileo's teaching at Padua dealt also with Aristotle's mechanics. See Gliozzi.

the second passes through points of a segment B_1E_1 equal to C_1F_1, greater than the length of the smaller circumference: it cannot be said that the lesser circumference slides, nor that it rolls without sliding in the ordinary sense of the word (23). According to Galileo this small circumference as it moved would touch B_1E_1 at infinitely many points between which would be placed infinitely many other points not touched by the same circumference, which would give us the image of the infinite vacua.

His reasoning on the subject of these two regular concentric and homothetic polygons was to be used to explain not just the rarefaction (obtained in the preceding considerations), but also the condensation by rolling the lesser, inside polygon ($HIK \ldots$) and therefore rolling the circumference of the lesser radius without a pause (24).

Simplicio's (25) objection is examined: 'Making a line up of points, the divisible of indivisibles, the extension of the non-extended, seems to me pretty hard to get round.'

Salviati (26) answers that a finite number of indivisibles would never give a divisible; but that infinite indivisibles could give a divisible. Simplicio then observes that as with two lines, one of which is greater than the other, both are made up of infinite points, we should have 'an infinite greater than an infinite' which 'seemed a concept that could in no way be understood'.

As we have already seen, this difficulty, which the Aristotelian philosophers brought up, had already been settled by St. Thomas Aquinas. Salviati replies: 'These are the sort of difficulties which arise from talking, which we with our finite intellect find with regard to the infinite, by giving it the attributes we give to finite and finished things; which I think is not a good idea, because I believe that these attributes of larger and smaller and equality do not relate to infinites, which one cannot call greater or less or equal to another.' There follows the famous reasoning on the aggregate of the integers and on that of their squares; it ends thus: '. . . it must be said that the squared numbers are as many as all numbers, as there are as

(23) Galilei, 1, vol. VIII, p. 71.
(24) Galilei, 1, vol. VIII, pp. 93–6.
(25) Galilei, 1, vol. VIII p. 72.
(26) Galilei, 1, vol. VIII, pp. 76–9.

many of them as there are roots, and all numbers are roots: and yet from the beginning we say that all numbers are many more than all squares, the greater part of them not being squared. . . . But when Simplicio gives me several unequal lines and asks me how it can be that there are not more points in the greater than in the less, I answer that there are not more nor less nor an equal number, but an infinite number in both; or really, if I were to answer that the points of one are as many as there are squared numbers, of another more than all numbers, of the little one, as many as there are cube numbers, would I not have given him the satisfaction of putting more in one than in the other and an infinite number in both?' We should note that Galileo does not consider the difference between the power of the continuum and that of the enumerable. Here Sagredo observes that squares grow ever rarer as numbers increase 'moving towards ever larger numbers means moving away from the infinite'. Later (27) Salviati takes up the subject again, saying that 'if any number can be said to be infinite this number is unity. It is really there that you find conditions and requirements for the infinite number, since it contains in itself all squares, all cubes and all numbers.' Galileo had already considered this analogy between the One and the Infinite under its philosophical and theological aspect, in the first day of the *Dialogue on the two greatest systems* (28), where we read: 'As for the truth, which mathematical demonstrations show us, it is the same as divine knowledge knows; but I admit that the way in which God knows the infinite propositions, of which we know only a few, is wholly better than ours, which proceeds through speech, passing from conclusion to conclusion, whereas his is a simple intuition: and where we, for instance, to gain knowledge of some of the attributes of the circle, of which there are an infinite number, starting from one of the simplest and taking hold of it by its definition, move on, by talking about it, to another attribute, and from this to a third, and then to a fourth, etc., the divine intellect, with the simple apprehension of its essence, understands, without temporal speech, all the infinity of those attributes, which are also virtually understood in the definition of all things, and finally, by being infinite,

(27) Galilei, **1**, vol. VIII, pp. 82–3.
(28) Galilei, **1**, vol. VII, p. 129; Carruccio, 25.

perhaps are one thing only in their essence and in the divine mind.'

These views of Galileo are linked with the thought of St. Augustine, who considers 'the divine mind perfectly immutable, capable of understanding any infinity and numbering all innumerable things without succession of thoughts' (29). He is also linked with Cantor, through the interest which they both showed in the philosophical and theological questions connected with the mathematical infinite (30).

While Cavalieri appears to regard indivisibles with a certain reserve, Galileo's faith in the actual mathematical infinitesimal and infinite seems firm.

§5. *Fra' Bonaventura Cavalieri*

The method of indivisibles, some aspects of which we have illustrated, was applied systematically to many important problems, and to questions fundamental to the construction of infinitesimal analysis, above all by Cavalieri and Torricelli (31).

The theory of indivisibles of Cavalieri, who mentions Kepler among his forerunners, starts from the consideration of a flat surface as composed of infinite chords intercepted across the surface by a system of parallels; each of these chords, considered as a rectangle of infinitesimal thickness, is the indivisible element. Cavalieri says he does not use the actual infinite, but on this point his language is obscure. In any case the applications which he makes of the principle which bears his name are correct: 'If two plane areas are cut by a system of straight parallels, and above each one of these two equal chords are intercepted, the two areas are equal; if the corresponding chords have a constant ratio, the areas will have the same ratio' (32).

Cavalieri makes a similar principle the basis of his researches into the measurement of solids: if the sections of two solids, obtained by two systems of parallel planes, are equivalent, the two solids are equivalent too; or, put more generally:

(29) St. Augustine, 5, bk. XII, ch. 17; Enriques–De Santillana, 1, p. 640.
(30) Cantor, 1, *Mitteilungen zur Lehre vom Transfiniten* (Ges. Abh. pp. 378 ff.).
(31) Chisini, 1, p. 104 ff.; Bortolotti, 9; Conforto.
(32) Castelnuovo, 5, p. 21.

if the corresponding sections have a certain constant ratio, the ratio between the two solids is the same (33).

Here we will give some examples of the application of Cavalieri's method of indivisibles, starting from the determination of the area of the ellipse (34) (which was already known to Archimedes). Let us consider (see fig. 4) an ellipse with major axis $2a$ and minor axis $2b$, and a circle with the major axis of the ellipse as its diameter. We observe that the circle and the ellipse intercept on each ordinate segments whose ratio is a/b.

By Cavalieri's principle, the areas of the circle and the ellipse should be in the ratio a/b.

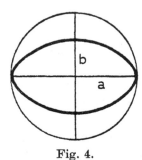

Fig. 4.

If the area of the ellipse is A, we therefore have:

$$\pi a^2 : A = a : b$$

therefore

$$A = \pi ab$$

The method of determining the volume of the sphere through indivisibles (35) is the following:

Let us consider (see fig. 5) a rectangle $ABCD$ in which $BC = 2AB$. Let O and P be respectively the mid-points of BC and DA. Let us join O to A and D and draw a circle with a centre O and radius OC.

Let us draw through a generic point Q of the segment OP a perpendicular to the said segment, which meets AB and CD respectively at the points R and S, the circumference at the

(33) A statement of Cavalieri's postulate on plane and solid figures may be found in Cavalieri, **2**, p. 25.
(34) Cavalieri, **1**, p. 211.
(35) Cavalieri, **1**, p. 259.

points T and U, and OD and OA respectively at the points V and Z. Let us rotate the figure so obtained on the axis OP: the rectangle $ABCD$ will generate a cylinder, the triangle OAD a cone, the semicircle BPC a hemisphere, the straight line RS a plane. If we take the hemisphere from the cylinder we obtain a solid which reminds one of a saucepan and which with Galileo we will call the 'round razor' (36).

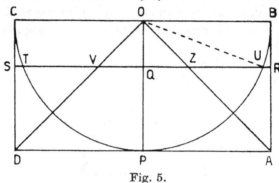

Fig. 5.

Let us show that the plane through Q perpendicular to OP intersects the cone and the round razor in two figures, a circle and a circular crown, of equal area. In fact let us put $\overline{QZ} = x$, $\overline{OU} = r$, observing that $OQ = QZ$, and we shall have that the area of the circle considered is πx^2, while the area of the circular crown is given by

$$\pi r^2 - \pi(r^2 - x^2) = \pi x^2.$$

The volumes of the cone and the round razor will therefore, by Cavalieri's principle, be equal. We shall have:

volume of hemisphere
$$= \text{volume of cylinder} - \text{volume of round razor}$$
$$= \text{volume of cylinder} - \text{volume of cone}$$
$$= \pi r^3 - \tfrac{1}{3}\pi r^3$$
$$= \tfrac{2}{3}\pi r^3$$

The volume of the sphere is therefore given by $\tfrac{4}{3}\pi r^3$.

Cavalieri managed to solve the problem which in modern language is called the calculation of the integral

$$\int_0^r x\,dx$$

(36) Galilei, **1**, vol. VIII, p. 74.

This problem is presented in a slightly different form from Galileo's in the determination of the spaces traversed by falling weights with a speed proportional to their time of fall. As similar triangles are proportional to the squares of their corresponding sides, the area of the right-angled triangle of short sides x and y with $x=y$, will be Kx^2, in which from a particular case (for instance, $x = 1$) we get $K=\frac{1}{2}$.

We therefore pass to the case of $\int_0^x x^2\, dx$. Instead of referring to the area of the parabola $y=x^2$, Cavalieri prefers to build, corresponding to every value of x, a variable square, always on a plane perpendicular to a fixed axis. We thus obtain by varying x an infinity of pyramids whose volumes are proportional to the cubes of the corresponding edges. So

$$\int_0^x x^2\, dx = Kx^3$$

and we get then that $K=\frac{1}{3}$.

At this point a modern geometer would extend the results obtained to space of n dimensions and would find

$$\int_0^x x^n\, dx = \frac{x^{n+1}}{n+1}$$

Cavalieri, although he does not conceive of the logical possibility of space of more than three dimensions, realizes that there is a similarity in virtue of which this preceding formula can also be used for $n > 2$.

§6. *Evangelista Torricelli* (37)

To E. Torricelli (1608–47) we owe an extension of Cavalieri's theory through the use of curved indivisibles (38), the fundamental concept of which can be expressed in the following way: to compare two plane figures, for example, we cut the first with a system of curves and the second with a system of parallels; if every curved indivisible of the first figure is equal to the corresponding indivisible segment of the second figure, that is,

(37) Bortolotti has clarified Torricelli's contribution to analysis: see Carruccio, **18**.
(38) Castelnuovo, **5**, p. 28–9.

207

according to present-day language, if every infinitesimal quadrangle between the two infinitely close curves of the second figure is equivalent to the trapezoid of the other, the two figures have the same area. The process consists substantially in comparing an integral in curvilinear co-ordinates with an integral in Cartesian co-ordinates. Similar methods can be used for the calculation of volumes.

Let us give some examples of the application of Torricelli's method of curved indivisibles (39).

Partly translating Torricelli's thought into modern language,

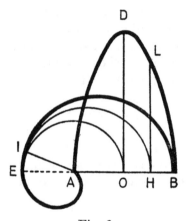

Fig. 6.

for reasons of simplicity, let us consider (see fig. 6) the parabola ADB with an equation

$$y = kx(a-x)$$

and the spiral of Archimedes $BIEA$ with an equation

$$\vartheta = k(a-\rho).$$

Through every point H of the segment AB we have, putting $x = h$,

$$\overline{HL} = kh(a-h) \tag{I}$$

and indicating with I the point at which the circumference with the centre A and the radius H meets the spiral, giving us a

$$\overset{\frown}{HAI} = k(a-h)$$

(39) Torricelli, **2**, vol. 1, p. II, pp. 429–31; see Bortolotti, **9**.

208

value for the arc of the circle

$$\overset{\frown}{HI} = kh(a-h) \tag{II}$$

Comparing (I) and (II) we may conclude that all the ordinates of the parabola are equal in length respectively to all the arcs $\overset{\frown}{HI}$ of the circle, where I is a point of the spiral.

We conclude that the area of the parabolic segment is equal to that contained by the spiral and the segment AB.

Let us give an example relating to solids: determination of a volume of a paraboloid of rotation (the result already known to Archimedes).

Let us consider (see fig. 7) the solid (parabolic conoid)

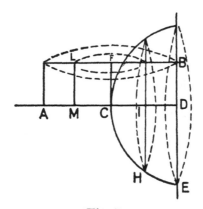

Fig. 7.

generated by the rotation of a parabola EC round an axis CD and a cylinder which has as the radius of the base the portion of the axis CD contained in the segment of the conoid to be measured, and as generator the segment DB equal to the parameter p which appears in the equation of the parabola $y^2 = 2px$, referred to the axis of the abscissa CD and to the axis of the ordinate tangent at C to the parabola. For every point I of the axis of the conoid we take a plane perpendicular to it, and consider the cylinder of which a generator is $ML = BD$, and the radius of the base is CI, and IH an ordinate of the parabola.

Since $y^2 = 2px$, we shall have

$$\overline{IH}^2 = 2p.\overline{CI} = 2\overline{BD}.\overline{CI} = 2\overline{CI}.\overline{ML};$$

14   209

therefore

$$\pi\overline{IH}^2 = 2\pi\overline{CI}.\overline{ML}$$

That is, all the circular sections of the conoid considered are equivalent to the curved surfaces of cylinders with the radius of the base CI and height BD. The volume of the conoid will therefore be given us by the formula:

$$V = \pi\overline{CD}^2.\overline{BD} = \tfrac{1}{2}\pi 2\overline{BD}.\overline{CD}.\overline{CD} = \tfrac{1}{2}\pi\overline{DE}^2.\overline{CD}$$

We conclude that the volume of the parabolic conoid is half the volume of the reduced cylinder.

It is easy to see that the clumsy use of indivisibles could lead to absurdities. Those who opposed Cavalieri's and Torricelli's

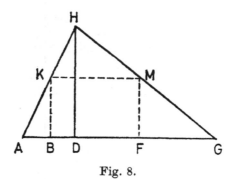

Fig. 8.

use of indivisibles suggested examples of its mistaken use, and the two mathematicians themselves thought up similar examples to forestall objections, or to go deeper into the matter.

Of the many paradoxes they foresaw and discussed we will, for reasons of space, consider only one (40).

Let us consider the triangle AGH (see fig. 8) of height HD and let $3AD = DG$.

From a point K of the side AH let us drop the perpendicular KB on to the base AG and at every segment KB of the triangle AHD let us make correspond a segment MF of the triangle GHD where M is a point HG, on the parallel through K to AG, and F is the foot of the perpendicular from M on to DG. The two triangles ADH and GDH should be equal (41).

(40) Bortolotti, **9**.
(41) Cavalieri's letter to Torricelli (April 5th, 1644): Torricelli, **2**, vol. III, pp. 170–1.

Cavalieri overcomes the paradox expressed and other similar ones by basing himself on the concept which he had formed of plane figures, considered as woven material with infinitely many parallel threads, and of the solid figures as books composed of infinite parallel pages : if we think of the triangle *AHD* as composed of 100 threads perpendicular to *AD*, in the triangle *DHG* with a base three times *DG* we should therefore be able to find 300 perpendiculars to *DG*; while with the process we have mentioned we would take only a hundred, 'and so in its way to the infinite', said Cavalieri.

Torricelli gives us a confutation of the paradoxes of the type which are considered, which is of more general character, based on considerations of the ratio of two infinitesimals.

In Torricelli's *Delle tangenti delle parabole per lineas supplementares* (42) we read 'That indivisibles are all equal to one another, that is the points to the points, the lines in length to the lines, and the surfaces in depth to the surfaces, is something that in my opinion is not only hard to prove but actually false. If there are two concentric circles, and from the centre all the lines are drawn to all the points of the greater periphery, there is no doubt that as many points will cross the lines on the lesser periphery, and each of these will be so much less than any of those, according to how much smaller the diameter is than the other diameter.

'If there are two parallelograms on the same base *AB*, and from all the points of *AB* parallels are drawn to the sides, in both the parallelograms *AC* and *AD* all the *AC*'s taken together are equal to all the *AD*'s taken together, and are also equal in number (because in the two cases there are as many lines as there are points on *AB*); therefore one is equal to the other, but they are unequal in length, and therefore although they are indivisibles, they are of unequal width and reciprocal to the lengths.'

Replying to criticisms of the method of indivisibles, Pascal wrote in 1658: 'Tout ce qui est démontré par les véritables règles des indivisibles se démontrera aussi à la rigueur et à la manière des anciens. Et c'est pourquoi je ne ferai aucune difficulté dans la suite d'user ce langage' (43).

In his mathematical works Torricelli used both the ancients'

(42) Torricelli, **2**, vol. I, p. II, pp. 320–1.
(43) See Enriques, **12**, p. 52.

method of exhaustion and the method of indivisibles. All the same, even in the works undertaken '*secundum methodum antiquorum*' his attitude with regard to the mathematical infinite differed from that of the Alexandrine mathematicians. This is shown, for instance, in the following passage of his *De infinitis spiralibus*, of which we give a translation: 'A mobile point on a finite arc of a hyperbola always approaches an asymptote of the same hyperbola and the distance of the said point from that of the asymptote can be made as small as you wish. This is an obvious sign that this same hyperbola at last comes into contact with the asymptote, and the contrary has never been shown. In fact Apollonius demonstrates that the hyperbola and its asymptote have no points in common at a finite distance; but we, on the other hand, say that the two lines have a point of contact after a distance greater than any finite distance' (44). It is clear from this that Torricelli considers points at infinity as existing in the world of thought, an idea the Alexandrian geometers would not admit.

For reasons of space we will not pause over Torricelli's demonstration of the result which, according to present-day mathematical language, would be expressed in the following terms:

If
$$y^n = kx^m$$

with n and m any positive integers

$$\frac{y \, dx}{x \, dy} = \frac{n}{m}$$

With the method of exhaustion Torricelli demonstrates this result, which can be verified at once with the rules of derivation now in use; and besides he observes, although with a different language from our present one, that the ratio of the two elements of area $y \, dx / x \, dy$ is equal to that of the sub-tangents and of the abscissa x

$$\frac{y \, dx}{x \, dy} = \frac{s}{x}$$

$$\left(\text{we should write:} \quad \frac{dy}{dx} = \frac{y}{s} \right)$$

(44) See Torricelli, **3**, p. 24.

This result of a general character applied to the curves $y^n = kx^m$ for which the ratio $y\,dx/x\,dy$ has been calculated allows us to find the tangents of all the curves of the type which we have considered (45).

From this, Torricelli implicitly demonstrates the well-known fundamental formula:

$$\int_{x_1}^{x} x^a\,dx = \frac{1}{a+1}\left(x^{a+1} - x_1^{a+1}\right)$$

for a any positive or negative rational different from -1.

Cavalieri demonstrated this formula for some integral values of a, Fermat extended it to the case of any rational positive a, and Torricelli was the first to establish it for negative values of a (46).

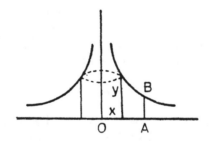

Fig. 9.

Torricelli made other very important discoveries in integral calculus with regard to the extension of the concept of the integral to functions which have infinities in the field of integration, or a field of integration which is infinitely extended (47).

Let us consider, with Torricelli (changing the language for reasons of simplicity), a hyperbola which has the equation $xy = 2k^2$ (see fig. 9) and a surface S limited by a branch of the hyperbola itself, by Cartesian axes of origin O, and by the segment AB parallel to the axis y, where B is a point of the hyperbola and A is the projection of B on the axis x.

(45) Bortolotti, 9.
(46) Bortolotti, 15, pp. 463–4.
(47) Torricelli, 2, vol. I, p. I: *De solido hyperbolico acuto*, pp. 173–221. Vol. I, p. II: *De infinitis hyperbolis*, pp. 231–74; *De infinitis parabolis*, p. 321; Bortolotti, 6 and 7.

Let us rotate the surface S round the axis y. It is shown that the volume of the infinitely long solid thus obtained is equal to a cylinder of radius $2k$ and height OA. According to the principles of curved indivisibles we consider this solid as a totality of the enveloping cylinders having as an axis the axis of the ordinates, radius of base x and height y. The cylinders have a surface $2\pi xy = \pi(4k^2)$; each is therefore of area equal to that of a circle of radius $2k$.

According to the principle of curved indivisibles the volume of the solid considered is therefore equal to that of the cylinder with radius of base $2k$ and height OA.

Encouraged by this result Torricelli went on to examine the curves $x^m y^n = c^n$ and established in which cases the surfaces and volumes of rotation of solids of infinite length were finite, thus determining the conditions of existence of the relative integrals extended over infinite intervals.

Another contribution of his to the development of analysis was the introduction of exponential and logarithmic curves; the curve $y = ce^x$, the logarithmic spiral

$$\rho = ae^{-b\theta}$$

This latter curve leads us to look at Torricelli's *De infinitis spiralibus* where, for the first time in the history of mathematics, a curved arc is rectified with ruler and compasses (48). The object of the book is to deal systematically with the fundamental properties of the logarithmic spiral (called by Torricelli geometrical), the drawing of any one of its arcs, and the quadrature of the surface bounded by the curve itself.

In the rest of Torricelli's work, which we have still to look at, his research into infinitesimal analysis is interwoven with his work on mechanics.

The calculation of areas and volumes on a basis of the method of indivisibles leads to the calculation of definite integrals, while the consideration of the ratio of the speed to the space traversed leads to the concept of indefinite integrals (functions of

(48) Torricelli, **2**; vol. I, p. II, pp. 349–73; Loria, 1. Before the publication of this essay historians of mathematics believed that the first curve to be rectified was the semi-cubical parabola, rectified in 1659 by Neil, Von Heuraeth and Fermat. See also Bortolotti, **10**, Agostini, **1**. For Descartes' opinion on this point, see §8 of this chapter.

the upper limit of integration), and Torricelli established substantially the formula

$$s(t) = \int_0^t v(t)\, dt.$$

Given the velocity of movement we obtain the space by squaring (49).

To reconstruct Torricelli's thought on the theorem that establishes the inverse character of the operations of integration and derivation we should remember the third day of the *Mathematical discourses and demonstrations on two new sciences*, composed by Galileo in Arcetri towards the end of

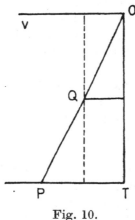

Fig. 10.

his life. There Galileo represents the uniformly varying movement by means of a diagram (see fig. 10), where the time t is assumed as the abscissa and the velocity $\bar{v}$ as the ordinate: $v = gt$.

If O is the origin, P the point of the diagram relating to the instant t, and T the projection on the t-axis, the space traversed from the instant zero to the instant t is measured by the area of the triangle OTP, that is by $\frac{1}{2}gt^2$. If we wished to apply a general formula given later to this particular case we should write:

$$s = \int_0^t gt\, dt.$$

(49) Bortolotti, 15, pp. 472–4.

Galileo's researches were continued by Torricelli in his work *De motu gravium*, published in 1644, and then in the unpublished notes which were put in order and interpreted by Ettore Bortolotti.

Considering the two diagrams of space and speed as functions of time, Torricelli stated that the ordinates of the space-curve are proportional to the areas enclosed by the speed-curve, while the ordinates of points on the speed-curve are the angular coefficients of the tangents of the space-curve.

This confirms, from a mechanical point of view, the reciprocal character of the operations of integration and derivation known under the name of Barrow's theorem, published by Newton's master in the *Lectiones opticae et geometriae* in 1670. Barrow, who was profoundly influenced by the Italian school, mentions Galileo and Torricelli as his forerunners. But the importance of this result could be evaluated fully only after the discovery of general, quick and easy rules for the calculus of derivatives.

This kinematic interpretation of the principles of calculus brings to mind a method Torricelli proposed for drawing a tangent to a curve considered as described by a point, the motion of which results from the composition of two movements : the resultant of the velocities of the two motions lies on the tangent to the curve at this point. The method was used to draw the tangent to the spiral of Archimedes, as Torricelli told Galileo in a letter of 29th June 1641 (50). Galileo answered approving. Torricelli published the method in a scholium to his treatise *De Motu* and applied it to the cycloid (51).

One of the theories which Torricelli notably advanced is that of the mass-centres of figures, as Bortolotti clearly showed; he showed too the way Torricelli's thought took to reach the fundamental relation which characterizes the position of the mass-centre of a figure possessing a diameter.

Torricelli told Michaelangelo Ricci about this in a letter on 12th August 1645, and Cavalieri in one on 7th April 1646. This is how he demonstrated it in the letter to Cavalieri (see fig. 11).

'Let $ABCD$ be any plane figure you like with a diameter $CA[a]$ and a mass-centre E. I say that $CE[X]$ is to $EA[a-X]$,

(50) Torricelli, **2**, vol. III, pp. 55–6; Agostini, **2**.
(51) Torricelli, **1**, p. 121.

as all the rectangles of the ordinates and of the abscissae taken from the vertex C, of which one is $DI.IC[\int_0^a x.f(x)\ dx]$, stand in relation to all the rectangles of the ordinates and of the parts remaining of the diameter after having subtracted the abscissae, of which one is $DI.IA[\int_0^a (a-x)f(x)dx]$.

'In fact let us have a figure $CFGH$ similar, equal, similarly placed and co-axial with $ABCD$, and let O be the mass-centre of $CFGH$. If we suspend the system of $ABCD$, and $CFGH$ as a balance at the point C, shall we have or not have equilibrium. We shall have that: the moment DI:moment $HL = DI.IC:HL.LC$. And it is always thus:

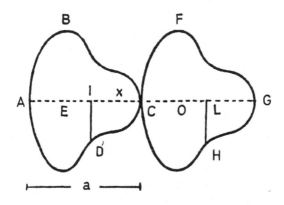

Fig. 11.

and so in its complexity the whole moment of figure $ABCD$ is to the whole moment of figure $CFGH$ as all the rectangles of which $DI.IC$ is one stand in relation to all the rectangles of which $HL.LC$ is one. But CE and CO are also to one another like moments of the figures; therefore CE will stay to CO, that is at the equal segment EA, as all the rectangles of which $DI.IC$ is one are to all the rectangles of which $HL.LC$ is one, that is $DI.IA$.'

This means:

$$X:(a-X) = \int_0^a x f(x)\ dx:\int_0^a (a-x)f(x)\ dx.$$

217

From which we get

$$X = \frac{\displaystyle\int_0^a x\, f(x)\, dx}{\displaystyle\int_0^a f(x)\, dx}$$

Note that the demonstration is at once applicable to figures which have no diameter, when the distance from an axis or from a plane, suitably chosen, is substituted for the portions of the diameter (52).

We will end this rapid and incomplete review of Torricelli's most important work in the field of infinitesimal analysis by recalling that we owe him the introduction of the concept of the envelope of a family of curves (53). Torricelli thought of it when he was working on the problem of the determination of the envelope to the parabolas described by projectiles issuing from the same gun with constant speed but with varying inclination. The envelope found by Torricelli is, as we know, a parabola.

Torricelli meant to set down his latest very fruitful discoveries in a work which was to be called *De lineis novis*, but before he could do so he died, after a short illness, at the age of 39. The fate of his manuscripts, too, helped to diminish the influence of his important discoveries. Yet on the development of infinitesimal analysis his influence was still very important. Leibniz wrote thus to Manfredi: 'Cavalieri and Torricelli were the initiators and promoters of the sublimest geometry . . . and others pushed ahead with their help' (54).

In his book *The origins of infinitesimal calculus in the modern age* (page 107) Castelnuovo concludes: 'I have already said that there was no such person as the founder of infinitesimal calculus, but that various mathematicians contributed to its discovery. The reader may ask who, before Newton and Leibniz, came closest to it. Although I must give warning that such a judgment is necessarily subjective, I should like to mention the names of P. Fermat and E. Torricelli.'

(52) Bortolotti, **4** and **15**, pp. 479–86.
(53) Marcolongo; Bortolotti, **3**.
(54) Bortolotti, **2**: Vacca, **11**.

§7. *Pietro Mengoli* (55)

Pietro Mengoli (1626–86), follower of Cavalieri and his successor in the university chair at Bologna, took a critical stand with regard to the construction of calculus, aiming to set the discoveries on a rigorous basis, without making use of the actual infinite and infinitesimal.

He studied infinite series systematically, particular cases of which had appeared in the work of Oresme, of P. A. Cataldi (56), the conceiver (1625) of continuous fractions, Torricelli, and Gregorio da San Vincenzo (1584–1667). With regard to the early research on infinite algorithms, we should remember J. Wallis, who, through brilliant inductive processes (extending to the case of a fractional *n* formulae previously established for integral values only), gave us the value of π in the form of an infinite product:

$$\frac{\pi}{2} = \frac{2}{1} \cdot \frac{2}{3} \cdot \frac{4}{3} \cdot \frac{4}{5} \cdot \frac{6}{5} \cdot \frac{6}{7} \cdot \frac{8}{7} \cdots$$

Mengoli studied the convergence and divergence of series, establishing some fundamental results about them: a condition necessary for the convergence of a series of which a_n is the generic term is $\lim_{n \to \infty} a_n = 0$; if the partial sums of a series with positive terms are bounded, the series is convergent. As Oresme had already done, he established the divergence of the harmonic series forty years before Jacques Bernoulli (1664–1705), to whom the discovery is sometimes mistakenly attributed; he gave a development in series of logarithms almost ten years before N. Mercator (1620–87) gave the development of $\log (1 + x)$, and he summed many other series.

Mengoli also gave a definition of limit with the relative theory (57), from which he was led to a definition of the definite integral which is not substantially different from that of A. Cauchy (1789–1857). In relation to this Mengoli calls *form* the trapezoid made up of all the ordinates of a curve

(55) Agostini, **3**: In this book there is an ample treatment of Mengoli's work, and a bibliography. Eneström, Vacca, and Agostini made clear the importance of Mengoli's work.

(56) Bortolotti, **19**, pp. 78–91.

(57) Cassina, **6**, pp. 89–101; in this paper Mengoli's concept of the limit is studied thoroughly.

$y=f(x)$. With the base of the trapezoid divided into n equal parts (on the axis of the abscissae) let x_i and x_{i+1} be the abscissae of the ends of one of these intervals, and m_i and M_i respectively the smallest and the greatest of the ordinates in the interval considered. We have, according to Mengoli, three figures formed from rectangles, the measurements of which tend to the size of the form.

We shall have:

inscribed figure $\quad s_n = \sum_{i=0}^{n} m_i(x_{i+1}-x_i)$

circumscribed figure $\quad S_n = \sum_{i=0}^{n} M_i(x_{i+1}-x_i)$

ascribed figure $\quad \sigma_n = \sum_{i=0}^{n} f(x_i)(x_{i+1}-x_i)$

or else

$$\sigma_n = \sum_{i=0}^{n} f(x_{i+1})(x_{i+1}-x_i)$$

Mengoli shows that

$$\lim_{n \to \infty} S_n = \lim_{n \to \infty} s_n;$$

and since

$$s_n \leqq \sigma_n \leqq S_n$$

σ_n also tends to the common limit of s_n and S_n.

But the form is also contained between the *inscribed* and *circumscribed* figures. Translating Mengoli's thought into modern terms we should say that there exists the measure of the *form*, which is given by the limits of the measures of the ascribed figures.

§8. *R. Descartes and P. Fermat*

We will now examine Descartes' contribution to the development of infinitesimal calculus. In the *Géométrie* the possibility is mentioned of determining the area enclosed by a curve with a given equation (58), but the rectification of a curve is mistakenly called a superhuman task (59). We owe the first success in this field, as we have seen, to Torricelli.

(58) Descartes, vol. VI, p. 413.
(59) Descartes, vol. VI, p. 412.

Facing the question of the angle between two plane curves Descartes goes on to consider their tangents and therefore their normals; he then concentrates his attention on this problem. He solves the problem of the construction of the normal in the case of algebraic curves, and from this solution goes on at once to the construction of the tangent. From the geometrical point of view his procedure is as follows:

Let us consider (see fig. 12) a given curve of which C is a generic point. We propose to find, on the axis of the abscissae, the point P, such that the line CP will be normal to the given curve. This is verified when the circle with the centre P and

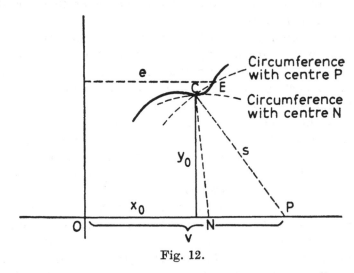

Fig. 12.

radius CP has two intersections with the curve coincident in C. The problem is solved analytically by means of the equation (I) $f(x, y) = 0$ of the given curve (60). Let x_0, y_0 be the co-ordinates of C. O being the origin, we say that $OP = v$, and $CP = s$. We shall have:

(II) $$(x - v)^2 + y^2 = s^2$$

In equation (I) the terms of even degree in y are separated from those of odd degree, being taken over to the other side of the equation; then the two members of the equation are squared and we obtain an equation (III). Then we eliminate

(60) See Bompiani.

y between (II) and (III), substituting in (III) the value of y^2 obtained from (II). Thus we obtain an equation (IV) $F_{2n}(x) = 0$. Let E be another point in which the circle of centre P and radius CP meets the given curve and let e be the abscissa of E. If CP is normal to the curve in C, E must coincide with C. That is $F_{2n}(x)$ must be divisible by $(x-e)^2$, and therefore we have

(V) $$(x-e)^2 F_{2n-2}(x) = 0$$

After developing and ordering equation (V) the coefficients thus obtained are equated to those respectively of the same degree of (IV). Thus we obtain equations which determine s and v, whence the normal CP.

Descartes applies his method to the determination of the tangents to the parabola, to curves of the third order, and to his ovals (61), which are important in his researches on lenses. In this connection, Descartes solved one of those problems in which one must determine a curve, given the properties of its tangents, problems which lead to a differential equation.

The process indicated for the construction of the normal to a curve led to the solving of the problem of the determination of the tangent to an algebraic curve at one of its points, but does not apply to transcendental curves.

P. Fermat, contemporary with Descartes, thought of a different method of determining the tangents to the curves, which substantially consists in determining the angular coefficient of the tangent to the curve $y = f(x)$ as derived from the function $f(x)$ calculated at the point of contact (62). A discussion arose between the two mathematicians, each maintaining his method was the better one; and it proved fruitful because it led to the rival methods being applied in ways that contributed to the construction of the infinitesimal calculus.

Fermat's *Methodus ad disquirendan maximam et minimam* of 1638 should be remembered too (63).

Suppose we are seeking the maximum or minimum of the

(61) Descartes' oval is the locus of the points P which satisfy the condition; $\mu FP + \mu' F'P = $ const., where F and F' are two fixed points. It is a generalization of the ellipse. Descartes discovered another curve, the folium, the equation of which is: $x^3 + y^3 - 3axy = 0$.

(62) Castelnuovo, 5, pp. 56–60.

(63) Fermat, t. I, p. 133.

expression $f(x)$. There we substitute $x+e$ for x, and the new expression thus obtained must be equal to the original one. Actually the author, using the language of Diophantus, uses the verb '*adaequare*' which means '*almost make equal*'. So the equal terms of the two members are suppressed, everything is divided by e, and the terms still containing e are suppressed. In the end the equation thus obtained is solved. It is clear that this process is equivalent to annulling the derivative of $f(x)$.

The method can be applied, for instance, to find the maximum rectangle, which has as its base and height the two parts into which a segment of constant length can be divided. If a is the segment and x is one of its parts, we are looking for the maximum of $x(a-x)$. We say:

$$(x+e)[a-(x+e)] = x(a-x).$$

Developing this, suppressing the equal terms of the two members and dividing by e, we get

$$-x+a-x-e = 0$$

and saying $e=0$,

$$x = \frac{a}{2}$$

With regard to the Cartesian diagram of the function $y=f(x)$, we should note that Fermat recognized as a necessary condition for the existence of the maximum or minimum the existence, where this is verified, of a tangent parallel to the axis of x, although recognizing that in the case of the inflexion we can have a tangent parallel to the axis of x without a maximum or a minimum.

It is a remarkable fact that Descartes successfully solved a problem of typically infinitesimal character like that of the normal to a curve, without using potential or actual infinitesimals, or passing to the limit, properly so-called, but with a purely algebraic process. This way of thinking of his should be seen in relation to his ideas on the mathematical infinite.

From various passages of Descartes on the subject the following, from *Principia Philosophiae* (64), clearly expresses his attitude to the physical and mathematical infinite.

(64) Pars I, art. XXVI and XXVII (Descartes, t. VIII, pp. 14–15).

'Thus'—I translate from his Latin—'we shall never weary ourselves with discussing the infinite. The fact is that, as we are finite, it would be absurd for us to say anything on the subject and thus try to make it finite and grasp it. We should not bother either to answer those who ask if, given an infinite line, half of it would be infinite too, or else if an infinite number is even or odd, or other questions of the kind: since only those with infinite minds can ponder on such questions. We, on the other hand, do not say that everything we can find no end for is infinite, but we regard it as indefinite. But since we cannot imagine an extension so great that we cannot understand that there can be a greater one, we say that the size of possible things is indefinite. And since we cannot imagine a number of stars so great that God could not have been able to create more, we suppose the number of the stars is indefinite too; and so on in other similar cases.

'And we call these things indefinite rather than infinite either to reserve the title of infinite for God alone, because in Him alone in every way, not only do we not know any limits, but we positively know that there are none; or else because we do not positively understand in the same way that other things are without limits of any kind, but only negatively declare that their limits, if they have them, cannot be found by us.'

Descartes' speculative position, which in this passage is decidedly agnostic in regard to the physical and mathematical infinite, explains his negative attitude to Galileo's views on the actual mathematical infinite and infinitesimal, and on the geometry of indivisibles of Bonaventura Cavalieri.

But I do not want to end this glance at Descartes' views on the mathematical infinite on this negative note. In a letter to Mersenne on 26th April 1643, Descartes considers inexpressible infinite magnitudes, and takes up more clearly a question which the ancient mathematicians, bewildered by the discovery of incommensurable magnitudes, in which they had found an inexpressible element (see Chapter III, §3), had already faced.

Descartes' mention of the subject recalls a fertile discovery of Cantor's, clarified by Richard and by Borel: There exist in the geometrical continuum (if this is not an abuse of the word to use the verb 'exist') elements which cannot be defined (65).

(65) See Geymonat, **2**, pp. 292–4, and this book, ch. XVII, §1.

In the passage of Descartes in question we also find affirmed the existence of transcendental numbers, and the way in which, according to Cantor, it can be demonstrated is almost given. We read (66):

'Je ne sçay pas ce que me demande M. de Vitry la Ville, touchant les grandeurs inexplicables; car il est certain que toutes celles qui sont comprises dans les equations, s'expliquent par quelques signes, puisque l'equation mesme qui les contient est une façon de les esprimer. Mais, outre celles la, il y en a une infinité d'autres qui ne peuvent pas mesme estre comprises en aucune equation.'

§9. Isaac Newton and G. W. Leibniz

What was needed for a system of infinitesimal calculus, after the researches we have already examined, was to have the theorem relating to the inverse character of the operations of derivation and integration placed at the basis of the calculus itself, and to establish fixed and simple rules for deriving the functions commonly found in analysis.

This was done by Isaac Newton (1642–1727) and G. W. Leibniz (1646–1716). Newton's infinitesimal analysis is found within the framework of his researches into the construction of the system of the world, given us in the *Philosophiae naturalis principia mathematica* (London, 1687). Yet in this work Newton still uses laborious infinitesimal processes, inspired by the ancients' rigorous method of exhaustion, to anticipate any objections to his fundamental discoveries in the field of mechanics.

The difference between his conception of calculus and that of the method of the indivisibles is clearly shown from a glance at the chapter of the *Principia* on the movement of bodies (67):

'For demonstrations are shorter by the method of indivisibles; but because the hypothesis of indivisibles seems somewhat harsh, and therefore that method is reckoned less geometrical, I chose rather to reduce the demonstration of the following Proposition to the first and last ratios of nascent and evanescent quantities, that is, to the limits of those ratios. . . . Therefore if hereafter I should happen to consider quantities made up of

(66) Descartes, vol. III, p. 658.
(67) See Newton, 2.

particles, or should use little curved lines for right ones, I would not be understood to mean indivisibles, but evanescent divisible quantities; not the sums and ratios of determinate parts, but always the limits of sums and ratios. . . .

'It may . . . be objected, that if the ultimate ratios are given, their ultimate magnitudes will also be given: and so all quantities will consist of indivisibles, which is contrary to what Euclid has demonstrated concerning incommensurables, in the tenth Book of his *Elements*. But this objection is founded on a false supposition. For those ultimate ratios with which quantities vanish are not truly the ratios of ultimate quantities, but limits towards which the ratios of quantities decreasing without limit do always converge; and to which they approach nearer than by any given difference, but never go beyond, nor in effect attain to, till the quantities are diminished *ad infinitum.*'

A language nearer to that of our infinitesimal calculus is found in the *Tractatus de quadratura curvarum*, published in London in 1704 (68). There appear the fundamental concepts of a *fluent* and a *fluxion*.

The fluents are magnitudes, functions of time, and the fluxions are derivatives with respect to time. The variable time t, however, can be assumed arbitrarily, substituting, as we should say today, a new variable t', a function of t, for t.

Let us now read some particularly significant passages of the *Tractatus de quadratura curvarum*, translated from the Latin.

'*Introduction*. In this work I deal with mathematical magnitudes, not as made up of small parts but as generated by a continuous movement.

'Lines are drawn, not by the addition of parts, but by the continuous movement of points; surfaces, by the movement of lines; solids, by the movement of surfaces; angles, through the rotation of their sides; times, by continuous flux, and so on in similar cases.

'These generations really take place in nature, and are to be seen every day in the movement of bodies. . . . And so, considering that generated quantities, growing for an equal time, become greater or less according to the greater or less speed at

(68) Carruccio, **5**. A translation of this work of Newton's has been published in the appendix to Castelnuovo, **5**.

which they grow, I have looked for a way of determining magnitudes from the speed of movements or from the increments with which they are generated; calling these speeds of growth *fluxions*, and the quantities generated *fluents*, I gradually, during the years 1665 and 1666 reached the method of fluxions, which I am using here in the quadrature of curves.'

. . . .

'3. The quantity x flows uniformly and we have to find the fluxion of the quantity x^n [where n can be whole or fractional, positive or negative].

'In the time in which the flowing quantity x becomes $x + h$, the quantity x^n becomes $(x + h)^n$, that is, according to the method of infinite series:

$$x^n + nhx^{n-1} + \frac{n(n-1)}{2}h^2x^{n-2} + \ldots$$

And the increments

$$h \text{ and } nhx^{n-1} + \frac{n(n-1)}{2}h^2x^{n-2} + \ldots$$

stand in relation to one another as 1 is to

$$nx^{n-1} + \frac{n(n-1)}{2}hx^{n-2} + \ldots$$

Now, if the increment h disappears their ultimate ratio will be

$$1 : nx^{n-1}$$

'With similar reasoning, through the method of the first and last ratios, we can calculate the fluxions of straight or curved lines in any case whatever, and the fluxions of surfaces, angles, and other quantities. . . .

'To find the fluents, having been given the fluxions, is a more difficult problem, and the first step in its solution is equivalent to the squaring of the curves. . . .'

To indicate the fluxions Newton uses a symbolism which has remained in use in rational mechanics to indicate the derivatives in regard to time.

'In what follows I consider indeterminate quantities growing or decreasing as if by a continuous movement, that is fluent or defluent, and I indicate them with the letters z, y, x and u and their fluxions or the velocity with which they grow by the same letters with dots on them $\dot{z}$, $\dot{y}$, $\dot{x}$, and $\dot{u}$.

'There are also the fluxions of these more or less fast fluxions or mutations, which we may call the second fluxions of the same z, y, x and u and indicated by $\ddot{z}$, $\ddot{y}$, $\ddot{x}$ and $\ddot{u}$.

'Thus these quantities z, y, x and u can be considered as fluxions of others which I will indicate thus: $\dot{z}$, $\dot{y}$, $\dot{x}$, $\dot{u}$.'

Besides this, Newton indicates the fluxions of $\sqrt{az-z^2}$ by $\sqrt{az\overset{\cdot}{-}z^2}$ and the fluxion of

$$\frac{az+z^2}{a-z} \quad \text{by} \quad \frac{az+z^2}{a\overset{\cdot}{-}z}$$

'*Proposition I:*

'*Problem I.* Given an equation that contains any number of fluent quantities, to find the fluxions.

'*Solution.* Multiply each term of the [rational integral] equation by the exponent of one of the fluent quantities contained in that term, and in the single products obtained substitute the relative fluxion for one of the factors of the power. [Do the same for the other fluents]; and the sum of all these products with their signs will be the new equation.

'*Explanation*: Let $a, b, c, d \ldots$ be constant determinate quantities and let us be given an equation containing any fluent quantities $z, y, x \ldots$ such as:

(1) $$x^3 - xy^2 + a^2z - b^3 = 0$$

'First multiply the terms by the exponents of the powers of x, and in the single products, instead of a factor x, the base of the power, or of x to the first power, write $\dot{x}$; the sum will be

$$3\dot{x}x^2 - \dot{x}y^2$$

Do the same for y and we shall have

$$-2xy\dot{y}$$

Do the same for z and we shall have

$$a^2\dot{z}$$

Say the sum is zero and we shall have the equation

$$3\dot{x}x^2 - \dot{x}y^2 - 2xy\dot{y} + a^2\dot{z} = 0$$

'I say that with this equation we define the relations between the fluxions of the sizes z, y and x.

'*Demonstration.* Let h be a very small quantity, and let

$$h\dot{z},\ h\dot{y},\ h\dot{x}$$

be moments of the quantities z, y and x, that is synchronized momentary increments. If the fluent quantities are already z, y and x, these, after a moment, increased by their increments $h\dot{z}$, $h\dot{y}$, $h\dot{x}$ become:

$$z+h\dot{z},\ y+h\dot{y},\ x+h\dot{x}$$

'These expressions, written in equation (1) in place of z, y, x give place to the equation:

$$x^3 + 3x^2h\dot{x} + 3xh^2\dot{x}^2 + h^3\dot{x}^3 - xy^2 - h\dot{x}y^2 - 2xh\dot{y}y$$
$$- 2\dot{x}h^2\dot{y}y - xh^2\dot{y}^2 - xh^3\dot{y}^2 + a^2z + a^2h\dot{z} - b^3 = 0$$

'Subtract from this member by member equation (1) and divide the result by h; we shall get:

$$3\dot{x}x^2 + 3\dot{x}^2hx + \dot{x}^3h^2 - \dot{x}y^2 - 2xy\dot{y}$$
$$- 2\dot{x}h\dot{y}y - xh\dot{y}^2 - \dot{x}h^2\dot{y}^2 + a^2\dot{z} = 0$$

'Now make the quantity h tend to zero, suppress the evanescent terms, and there will remain:

$$3\dot{x}x^2 - \dot{x}y^2 - 2xy\dot{y} + a^2\dot{z} = 0 \qquad \text{Q.E.D.}$$

'*Complementary explanation.* In the same way, if the equation was

$$x^3 - xy^2 + a^2\sqrt{ax - y^2} - b^3 = 0$$

we should obtain

$$3x^2\dot{x} - \dot{x}y^2 - 2xy\dot{y} + a^2\overline{\sqrt{ax - y^2}}\,\dot{} = 0$$

If we wish to make the fluxion $\overline{\sqrt{ax - y^2}}\,\dot{}$ disappear, write

$$\sqrt{ax - y^2} = z$$

and then

$$ax - y^2 = z^2$$

From which, from what we have said:

$$a\dot{x} - 2\dot{y}y = 2\dot{z}z$$

from which

$$\frac{a\dot{x} - 2\dot{y}y}{2z} = \dot{z}$$

That is

$$\frac{a\dot{x} - 2\dot{y}y}{2\sqrt{ax - y^2}} = \sqrt{ax - y^2}$$

and therefore

$$3x^2\dot{x} - \dot{x}y^2 - 2xy\dot{y} + \frac{a^3\dot{x} - 2a^2\dot{y}y}{2\sqrt{ax - y^2}} = 0$$

'Through repeated operations we reach second, third, and following fluxions.'

Newton thus determines various curves which can be squared by determining the functions of which the integral is already known. He manages this by deriving a known function.

Let us now consider the first work on differential calculus, Leibniz's *Nova methodus pro maximis et minimis, itemque tangentibus, quae nec fractas nec irrationales quantitates moratur et singulare pro illis calculi genus* (69).

In this work, published in 1684 at Leipzig in the *Acta eruditorum*, the differential calculus was set down for the first time, in substantially the same form and with the same notation as is used today.

The methods of derivation known before 1684, like Newton's process which we have just given, required to have the equation reduced to a polynomial form; and this reduction was sometimes long and complicated. Leibniz's method did away with all this, by giving for the first time the rules of differentiation for fractions and radicals; and in fact this was expressed in the title, which can be translated thus: 'A new method for maxima and minima, and for tangents, which is not held up by fractional or irrational quantities; and a singular way of solving these problems.'

Although Leibniz uses the expressive symbolism of differentials, he still gives no explicit definition of differentials themselves. It seems as if these, like Newton's fluxions, are defined by the rules given for working with them. Leibniz says that dx

(69) Carruccio, **1**. A translation of this work of Leibniz has also been published in Castelnuovo, **5**.

and dy are proportional to the momentary increments of x and of y. But from what he writes it is not clear whether he considered differentials as potential or actual infinitesimals (70).

It would appear that he thought he could construct the calculus without recourse to the actual infinitesimal, although drawn by philosophical considerations to affirm the actual infinite and infinitesimal (71). 'Je suis tellement pour l'infini actuel qu'au lieu d'admettre, que la nature l'abhorre, comme l'on dit vulgairement, je tiens qu'elle l'affecte partout pour mieux marquer la perfection de son Créateur.'

a being a constant, and v, w, y and z ordinates of curves corresponding to the abscissa x, in Leibniz's work of 1684 we find the following rules of differentiation:

$$da = 0 \qquad dax = a\,dx$$
$$d(z - y + w + x) = dz - dy + dw + dx$$
$$dxv = x\,dv + v\,dx$$
$$d\frac{v}{y} = \frac{\pm vdy \mp ydv}{y^2}$$

(In this latter formula, since he has not fixed any exact conventions for the signs in geometry, Leibniz is forced to complicate the rule with ambiguous signs.)

$$dx^a = ax^{a-1}\,dx$$
$$d\sqrt[b]{x^a} = \frac{a}{b}\,dx\sqrt[b]{x^{a-b}}$$
$$d\frac{1}{\sqrt[b]{x^a}} = \frac{-a\,dx}{b\sqrt[b]{x^{a+b}}}$$

'The rule of integral powers,' Leibniz continues, 'would have sufficed to determine the differentials of the fractions and of the roots; and in fact the power becomes a fraction when the exponent is negative and changes into a root when the exponent is fractional: but I have preferred to deduce these results myself, rather than leave them to others to deduce, because they are fairly general and are often met with, and in a subject which is in itself complex, it is preferable to search for easiness.

(70) Vivanti.
(71) Lettre à Foucher: Leibniz, **2**, t. I, p. 416.

'With the knowledge of this particular algorithm, or of this calculus which I call differential, all other differential equations can be solved by means of the common calculus, and maxima and minima obtained, and tangents too, in such a way that the fractions or the irrationals need not be made to disappear, or other difficulties, as had to be done according to the methods published until now.

'The demonstration of all the rules expressed will be easy for anyone familiar with these studies. . . .'

For reasons of space we will not dwell on Leibniz's interesting examples of the application of the differential calculus in this work; among them is the deduction of the law of the refraction of light.

Later he introduced the integral symbol in the form which is used at present.

Leibniz's symbolism of the differential and integral calculus, which wonderfully influenced the development of infinitesimal analysis, should be placed in relation to the 'characteristica universalis' symbolism which was one of the main motives that inspired his philosophy.

He hoped to turn the complex concepts into simple ones by expressing them with a few characteristic symbols, from the combination of which, with suitable rules, not only the known truths would be found but new undiscovered ones too.

In Leibniz's philosophy the small differences are especially important: 'petites perceptions', which establish the continuity of the life of the individual, 'little hints and differences' in nature. These small differences, which could not be adequately treated in the analysis of finite quantities inspired by Descartes, are expressed and systemized in Leibniz's infinitesimal analysis: the symbolism of which, Leibniz considered, formed part of its 'characteristica universalis'. We shall return to this subject in Chapter XVII, §1; for the moment we will deal with Leibniz's thought on the value of a suitable symbolism: 'What one asks of symbols is that they shall be suitable for research; this is so chiefly when they express in a concise way and as it were paint the intimate nature of things, because then they wonderfully spare one the effort of thought'. The symbolism of Leibniz's calculus satisfies this.

Leibniz's philosophical spirit leads him to generalities,

suggesting to him that concept of function as an 'expression de calcul', formed by the operative symbols of analysis—a concept which foreshadows Dirichlet's function of real variables, whereas Newton starts from concrete examples, generally in the field of algebra. Consider therefore the functions $y(x)$ obtained from the algebraic equation $f(x, y) = 0$. Thus we have cases which are outside the framework of Leibniz's definition of functions, and open the way to the study of the *singularities* of curves.

Newton and Leibniz quarrelled at length about which of them first invented the infinitesimal calculus, and their supporters continued the quarrel. But, as we have seen, it had no single inventor, but was built up gradually from ancient times. Newton's and Leibniz's contributions to the wonderful construction of analysis, however, bear the obvious mark of their genius.

§10. *A glance at the later developments of infinitesimal analysis*

Although we cannot linger over this subject, we wish however to deal with some of the more important aspects. To sum up and integrate the history of the development of the infinitesimal calculus from Archimedes to Lagrange we give a table which, according to Vacca, represents the influences which the greatest mathematicians who worked on the subject had on one another (see page 234).

The infinitesimal calculus developed in such a way that it led to problems which no longer sought to determine unknown quantities, but to find the form of functions satisfying given conditions: as for instance problems which can be solved by means of differential equations, the historical precedent of which can be found in the spiral of Archimedes, in the loxodrome of Pedro de Nuñes (1492–1577), a curve which cuts the meridians of a spherical surface at a constant angle, and the problems which led Napier to the invention of logarithms.

This invention answered the need to simplify numerical calculation in the application of mathematics to astronomy, mechanics, physics, the calculation of compound interest, etc.

The calculus with fractional exponents of N. Chuquet (1484), the comparison between arithmetical and geometrical progressions of M. Stiefel (1487), with sums which correspond

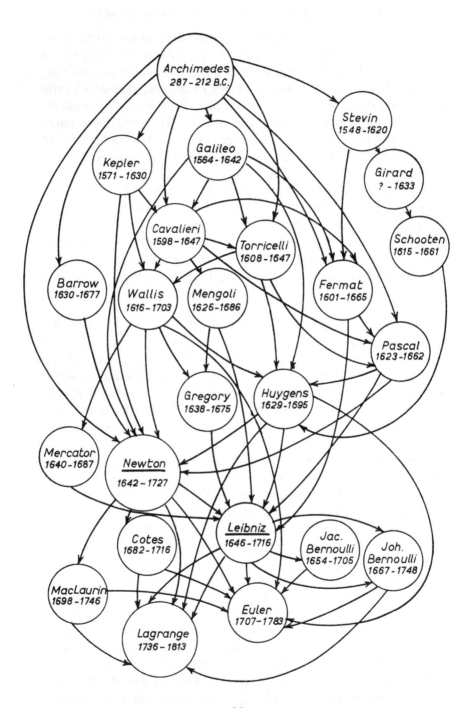

to products, the tables for the calculation of compound interest of S. Stevin (1548–1620), who studied statics, and of Jobst Bürgis, are steps towards the invention of logarithms (72).

John Napier (1550–1617) studied the logarithmic function basing himself on a mechanical consideration which, using modern language, can be expressed in the following way (see fig. 13).

Two points X and Y move respectively on two parallels OX and UA and x and y are their abscissae. X moves uniformly: $x = vt$ and Y describes the segment UA in such a way that with suitable conventions of the signs we have:

$$\frac{-dy}{dt} : v = y : \overline{UA}$$

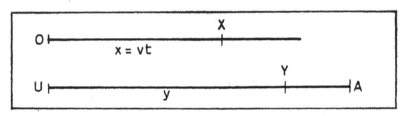

Fig. 13.

Putting $\overline{UA} = 1$ and remembering that $t = x/v$ we shall have $dy/dx = -y$

$$x = -\log y$$

Similar problems were solved by Galileo (determination of the movement of a body the constant acceleration of which is known), and by his school, and, as we have seen, by Descartes too and those who came after him.

We do not dwell on equations with partial derivatives, or on integral and integro-differential equations, explored by V. Volterra (73), but will glance rapidly at the problems of the calculus of variations (74), which are concerned with the

(72) Vacca, 5.
(73) See Volterra: *L'evoluzione delle idee fondamentali del calcolo infinitesimale*, pp. 159–88; *L'applicazione del calcolo a fenomeni di ereditarietà*, pp. 189–218.
(74) See Enriques, 11, pp. 468–71. There is also a good bibliography. On p. 440, the fundamental property of geodesics is dealt with.

determination of lines, surfaces, etc., which minimize a given function.

A problem of this kind which Newton solved is that of determining the solid of revolution of minimum resistance. Another is that of the brachistochrone, that is the trajectory which is described in the minimum time by a weight which falls under gravity alone from one point to another not situated on the same vertical; we obtain a cycloid with a horizontal base. The problem was proposed and solved by Jean Bernoulli in 1687.

The researches of Jacques Bernoulli on the same subject were the point of departure for a general analytical study of the isoperimetric problems, in which the work of the ancients (Pappus and Zenodorus) comes up again in the particular case of the curves of given length that enclose a maximum area.

Another problem of this type is that of 'finding a plane curve of a given length, which, turning round on an axis in its plane, generates a surface of maximum or minimum volume'. In 1744 Euler found that the curve was the catenary.

The fundamental property of geodesics, or curves of minimum length on a surface, limited by two fixed points, was discovered by Jean Bernoulli: the osculating plane to the curve is everywhere normal to the surface.

The most general and complete solution of the isoperimetric problem is found in Euler's work: *Methodus inveniendi lineas curvas maximi minimive proprietate gaudentes* . . . (1744).

With the researches of Lagrange (*Memoirs of the Academy of Science of Turin*, volume 2, 1762) and the studies of Gauss in 1833 on multiple integrals with variable limits, the algorithm of the calculus of variations reached its highest applicability and perfection as far as the discovery of curves and surfaces corresponding to conditions of maximum and minimum was concerned. But the question of deciding whether the extremals which integrate the corresponding differential equations effectively give rise to the maximum or minimum proposed, in spite of Lagrange's and Jacobi's important work on the subject, remained obscure until the end of the last century. A rigorous solution of the problem was given by Weierstrass, and independently by Darboux.

These problems could be considered from a new angle with

the development of a general theory of the functions of lines through the work of Volterra, Ascoli, Arzelà, Hilbert, Lebesgue, Caratheodory, Hadamard, and Tonelli (75). For a glance at the development of the calculus which furnished the algorithmic instruments for the general theory of relativity, see the last part of §4, Chapter XV.

From the time of Galileo, who saw the book of the universe in mathematical characters (76), through the work of Newton and the other great scientists up to our own day, the infinitesimal calculus has appeared as a powerful instrument in the field of mechanics and mathematical physics.

§11. *Criticisms of the principles of infinitesimal analysis* (77)

In spite of decades of wonderful success with the calculus, following Newton's and Leibniz's systemizations, the foundations of the calculus itself still seemed so insecure that when a young man expressed doubts on it d'Alembert would answer: 'Allez de l'avant: la foi vous viendra'.

This pragmatic view was useful from the point of view of research in some phases of the development of the calculus, but a more vigilant critical attitude swept it away.

The Berlin Academy, presided over by Lagrange, in 1784 held a competition on the concept of the mathematical infinite, asking for 'a clear and precise theory of what is called the infinite in mathematics'. The prize was given to Simon L'Huilier, whose work, published in 1795, shows an atmosphere of reaction against the methods of Leibniz, and a return to those of the ancients.

The modern methods were championed by L. Carnot in his *Reflexions sur la métaphysique du calcul infinitésimal*. But a more definitive answer to the criticisms on the calculus was given by A. L. Cauchy in his *Analyse Algébrique* of 1821. Cauchy's analysis was based on the concept of limits. 'Leibniz and Newton,' writes H. Weyl, 'saw clearly that the infinitesimal

(75) Tonelli, 1. We recall also Fantappiè's work on analytic functionals.
(76) 'Philosophy is written in this great book, which is open before our eyes (I say the universe), but it can be understood only by those who know the language in which it is written. It is written in mathematical language.' (Galileo, 1, vol. VI, p. 232.) For Galileo's opinion on the importance of infinitesimal concepts for understanding natural phenomena, see §4 of this chapter.
(77) On this matter, see Enriques, 12, pp. 66–7. Geymonat, 2, pp. 153–94.

calculus is only a way to the limit; but they did not see clearly that this way must not only determine the value of the limit, but must before all else *guarantee its existence*.'

Cauchy's philosophical position, which must be seen strictly in relation to the Aristotelean tradition, definitely rejects the actual mathematical infinite and infinitesimal, basing the calculus exclusively on concepts of potential infinite and infinitesimal. In fact, the famous paradoxes already considered by Galileo, in which the elements of an infinite aggregate are placed in biunivocal correspondence with the elements of an actual part of the same aggregate, prove, according to Cauchy, the impossibility of thinking of an infinity of objects as co-existent and constituting a single whole.

With fighting zeal H. N. Abel (78) struggled to 'establish without any doubt the principle of complete rigour'. He particularly opposed the incautious use of divergent series: 'It is shameful that anyone should dare to base any demonstration on divergent series. Through them you can demonstrate whatever you like: they are founts of disappointment and of error'. Yet he was not blind to the scientific interest of the series he fought against, which more recently have been used in a new way (by Borel), with suitable caution.

B. Bolzano (of whom we shall speak later with regard to the theory of aggregates) made an important contribution to the critical systemization of the calculus, as did K. Weierstrass, B. Riemann, G. Cantor, H. E. Heine, R. Dedekind, B. H. Méray, G. Darboux, C. Jordan, U. Dini, D. Arzelà, and G. Peano.

With the systemization of the calculus, carried out according to Cauchy's ideas, it seemed that the actual mathematical infinite was for ever barred from the world of modern mathematics. But with Cantor's theory of sets and the non-Archimedean geometry of G. Veronese (1891) the actual mathematical infinite reappeared in the mathematical world, as we shall see in Chapter XVII.

(78) Enriques, **2**, pp. 151–4; Carruccio, 24.

CHAPTER XIV

From the Origins of Projective Geometry to the Development of the Erlanger Programme

§1. *Origins of projective and descriptive geometry* (1)

We have met with processes relating to perspective in ancient times, and properties invariant for the operations of projection and section (Anaxagoras, Democritus, Vitruvius, and Pappus). But it was the artists of the Renaissance who felt especially the need to represent the figures of three-dimensional space in perspective on a surface. Rules of perspective were applied by Ambrogio Lorenzetti of Siena (1344), and by Masaccio (1401–1428); geometrical ideas on the subject were developed by Brunelleschi (1377–1446), and by Paolo Uccello (1397–1475).

These rules were studied in a theoretical way by Leon Battista Alberti (1404–72), Piero della Francesca (1410–92), Leonardo da Vinci (1452–1519), and Albrecht Dürer (1471–1528), who was a pupil at Bologna of Scipione del Ferro.

A rational systemization of the principles of perspective is found in the works of Commandino (1588), of J. Barozzi called il Vignola (1507–73), and of Guidubaldo dal Monte (1545–1607), who sets down and demonstrates the law of the point of flight (*punctus concursus*): '... all straight lines parallel to one another and to the horizon, however far inclined to the plane of the square, always converge towards a point on the horizontal line, and that point is the one where the line is met by the straight line which is led parallel from the eye to the original straight lines'.

As we have already seen, with regard to the development of infinitesimal analysis in the Renaissance, the consideration of a

(1) Enriques, 12, pp. 30–1 and 255–6; Amodeo, pp. 6–45; Frajese, 2; Viola.

point which can be represented graphically, corresponding to a point at infinity, made the mathematicians of the time familiar with the concept of the infinite.

As for the ideas from which projective geometry was to develop, Kepler introduced the point at infinity of the straight line and the focus at infinity of the parabola and, with his theory of continuity, passed from properties of conics of one kind to those of another.

Projective geometry, properly so-called, began with Desargues and Pascal, to whom we owe the famous theorems on conics known by their names, theorems particular cases of which we have met with in Pappus.

The development of analytical geometry, which progressed in the mathematical world chiefly through the influence of Cartesian thought, for a long time (from Desargues and Pascal until Monge) overshadowed the synthetic methods of projective and descriptive geometry. Yet the projective processes were not entirely abandoned: we owe to Newton a projective classification of plane curves of the third order: '*genesis curvarum per umbras*' in his *Enumeratio linearum tertii ordinis* (London, 1706).

A number of minor geometers kept the tradition of synthetic geometry alive at this time, until there was a reaction against the analytical processes, which were recognized as being sometimes longer and more complicated than the graphic processes. This reaction had its origins in the geometric tradition of the artists and in technical necessity (stone cutting, design of machines, etc.), and Monge showed how special figures could be represented systematically by means of their projections on two planes, perpendicular to one another.

The *Traité des propriétés des figures* by Poncelet (1788–1867) is very important from the conceptual point of view; there we find a distinction made between the metrical and graphic properties of figures, and an effort to find invariant metrical expressions for the operations of projection and section (cross-ratio). Poncelet used the method by which, through a vision of continuity, we move from the metrical to the projective properties, considered as a more general expression of the former.

His principle of continuity was criticized by Cauchy, who

observed that in certain cases it could lead to errors; but the examples Cauchy used to criticize the principle of continuity were outside the algebraic field, to which Poncelet always referred.

'And yet,' observes Enriques, 'the theory of functions of a complex variable founded by Cauchy is itself the most complete justification of Poncelet's views.'

§2. *The principle of duality on the sphere* (2)

With Poncelet and Gergonne we acquire full understanding of the principle of duality, the discovery of which was to have an important influence on the development of logic.

But the first laws of duality known to mathematicians were, as we know, those used in the geometry of pencils of straight lines and planes, and spherical surfaces.

It is natural that the discovery of the law of duality on the sphere should have preceded the discovery of the laws of duality in the plane, because the first is valuable for its graphic properties as much as for its metrical properties, does not need the distinction between these two types of properties, and can be established by means of elementary geometry; whereas the second, which is valid only for graphic properties, needs the distinction to be established.

Historically we start by considering two supplementary trihedra T, T', such that the faces of T are supplementary in regard to the normal sections of the dihedra of T', and the normal sections of the dihedra of T are supplementary to the faces of T'. These trihedra were studied by the Arab mathematician Nasir Eddin (3).

If two polar trihedra of the same vertex are cut through a spherical surface with the centre at this vertex we obtain two triangles called supplementary; a triangle is called supplementary to a given one when it has as sides the supplements of the angles of the given triangle, and as angles the supplements of the sides of the given triangle.

Perhaps the supplementary triangle was already known to Tycho Brahe: the Danish astronomer knew the relation

$$\cos A = \sin B \sin C \cos a - \cos B \cos C$$

(2) Carruccio, **2**.
(3) Frajese, **1**.

which is derived from the theorem of the cosine, from which Lagrange deduced the whole of spherical trigonometry (4):

$$\cos a = \cos b.\cos c + \sin b.\sin c.\cos A,$$

applied to the supplementary triangle, where a, b, and c are the sides and A, B, C are the angles of the triangle to which the latter relation is applied. In a work by Viète we find a rather involved statement which can be interpreted as referring to a couple of supplementary triangles, an example which was found among other less interesting matters.

The concept of the supplementary triangle was clearly defined in a posthumous work of Snell, who, it seemed, wished to isolate and clarify a pre-existing discovery.

As it evolved through the centuries the geometry of the sphere and spherical trigonometry reached the simplicity and harmony which we now admire in it.

The law of duality for the sphere was illumined by being incorporated in Gergonne's general considerations on duality (5).

An enunciation of this law, in which ideas on the theory of isoperimeters also appear, was given without a demonstration by Steiner (1796–1863), who made a useful contribution to the theory from the synthetic point of view (6). Thus, a theorem T relating to a spherical polygon, containing a relation between sides, angles, perimeter, area, maximum and minimum, was transformed, according to this law, into another theorem T', in which the terms angles, sides, areas, perimeter, minimum and maximum, were respectively substituted for the preceding terms.

The following is helpful for an understanding of the law: consider the pencil of straight lines and planes which has its centre at the centre of the sphere. To every line of that pencil we make correspond the perpendicular plane passing through the centre, and to every plane the perpendicular straight line through the centre. From this correspondence it turns out that a law of duality referring not only to the relations

(4) Comparing this theorem with Tycho Brahe's formula, Lagrange shows that it is possible to deduce a new theorem from every theorem of spherical trigonometry, by considering the supplementary triangle: see Lagrange, t. VII, pp. 345–6.
(5) 'Annales de Mathématique', t. XV, 1825, p. 302.
(6) Steiner.

of incidence, but also to the metrical relations, is valid, making planes, straight lines, dihedra, and angles correspond respectively to straight lines, planes, angles, and dihedra.

Now let us consider the intersections of a pencil with the concentric sphere, and make the intersections of corresponding elements correspond to one another.

Thus to every point P on the sphere corresponds the great circle made up of points equidistant from P, but inversely, to every such circle two points correspond which one might call the antipodes.

But it is not difficult to make a single point correspond to a great circle. It is enough, for instance, to consider a spherical polygon as the common region to n hemispheres and to assume as a dual figure the polygon determined by the poles of the respectively opposite hemispheres.

To every *arc* of the great circle, determined by two points, corresponds an *angle* determined by two great circles, and conversely.

If we have the polygon P with the n sides $\lambda_1, \lambda_2 \ldots \lambda_n$, its perimeter will be:

$$p = \lambda_1 + \lambda_2 + \ldots \lambda_n. \tag{I}$$

The angles of the dual figure will be

$$\pi - \lambda_1, \pi - \lambda_2, \ldots \pi - \lambda_n$$

(This is obvious in a triangle and the result can be extended to any polygon.)

Remembering that the spherical polygon of n sides having angles $\alpha_1, \alpha_2 \ldots \alpha_n$ has an area

$$A = [\alpha_1 + \alpha_2 + \ldots + \alpha_n - \pi(n-2)]r^2,$$

where r is the radius of the sphere, the area of the dual polygon, with $r = 1$, will be given by:

$$\begin{aligned}
A &= (\pi - \lambda_1) + (\pi - \lambda_2) + \ldots + (\pi - \lambda_n) - \pi(n-2) \\
&= \pi_n - (\lambda_1 + \lambda_2 + \ldots + \lambda_n) - \pi(n-2) \\
&= 2\pi - (\lambda_1 + \lambda_2 + \ldots + \lambda_n) = 2\pi - p
\end{aligned} \tag{II}$$

Therefore the perimeter of a spherical polygon, and the area of its dual figure, differ only by a change of sign and in an

243

additive constant: to equal perimeters correspond equal areas of the dual figures; and so it is allowable to interchange these *perimeters* with the *areas* in the enunciations of theorems considered.

From the equalities (II) we deduce that while the perimeter of a spherical polygon increases, the area of the dual figure decreases. Thus it is established that the terms *maximum* and *minimum* can also be interchanged.

§3. *Laws of duality in projective geometry* (7)

Poncelet deals with the polar theory with regard to a conic: given a conic in a plane, to every point *P* of this corresponds a straight line *p* which joins the two points of contact of the tangents drawn from *P* to the conic. This theory, foreshadowed by various ancient and modern authors (8), allows us to transform a theorem in which appears a graphic relation between points and lines into another in which the logical structure of the enunciation and the relation of incidence is maintained, but in which the terms *point* and *straight line* are interchanged. In 1806 Brianchon had deduced the theorem which bears his name on the hexagon circumscribed to a conic, from Pascal's theorem on the inscribed hexagon, using a polar transformation of the figure.

The general method of reciprocal polars, which virtually contains the law of duality of projective geometry, was formulated by Poncelet in 1824.

A more complete view of the law of duality, especially from the logical point of view, was given by Gergonne (1771–1859) in his *Considérations philosophiques sur les éléments de la science de l'étendue* of 1826 (9). Here, after distinguishing between the graphic and metrical properties of figures, Gergonne observes that the theorems relating to the graphic properties (which are not dual in themselves) are presented in pairs: in plane geometry the terms 'point' and 'straight line' are interchangeable, and so in spatial geometry are the words 'point' and 'plane', leaving unaltered the term 'straight line'

(7) Enriques, **2**, pp. 132–8.
(8) Kötter, pp. 45–52; Enriques and Chisini, vol. II, pp. 3–9.
(9) Gergonne, p. 125.

and in each case keeping the logical structure of the enuncia-tion, starting from the valid theorem T we obtain a valid theorem T'. T and T' are called dual.

This symmetry or duality constitutes an *a priori* principle for Gergonne, who exhibits the parallel development of the first theorems of projective geometry, starting from the simplest principles. Yet he does not give a complete analysis of the postulates on which his theory is based, to guarantee its logical symmetry. To be valid his reasoning needs a system of pro-jective geometry with purist aims (exempt from metrical ideas), which Staudt gave in 1847.

Since projective geometry G was founded on the basis of exclusively graphic ideas, the law of duality in space is demon-strated synthetically by proving the fact that, if in the postulates of G we interchange the terms 'point' and 'plane', leaving unaltered, of course, the term 'straight line' with the logical structure of the propositions considered, we obtain a proposition which is still valid in the system G. From this we get the fact that every demonstration D of a theorem T of G, obtained by interchanging the terms 'point' and 'plane', is transformed, proposition for proposition, logical passage for logical passage, into a new demonstration D' which is valid in G, which has as its conclusion the theorem T' dual of T.

The law of duality in the plane can be demonstrated by a similar process, or it can be found as a result of the law of duality in space (10).

Poncelet and Gergonne argued over which of them first discovered the law of duality, but we shall not go into the dispute.

Gergonne also made contributions to logic in his *Essai de dialectique rationnelle*, in which he classified the syllogisms on the basis of the rules of the conversion of propositions, starting from the so-called circles of Euler.

Gergonne's theory of definition is also valuable, and he extended the meaning of the theory: 'Si une phrase contient un seul mot dont la signification nous est inconnue, l'énoncé de cette phrase pourra suffire à nous en révéler la valeur. . . . Ces sortes de phrase qui donnent ainsi l'intelligence de l'un des

(10) For the proof of the laws of duality (in the plane, in space, . . .) see Enriques, **5**, chiefly pp. 31–4, 39–41.

mots dont elles se composent au moyen de la signification connue des autres, pourraient être appelées *définitions implicites*, par opposition aux définitions ordinaires qu'on appellerait *définitions explicites* . . .' (11).

'The theory of implicit definition of a system of concepts by means of a system of propositions,' observes Enriques, 'has become essential for contemporary logic.'

§4. *A glance at the later developments of projective geometry* (12)

We will mention briefly the later developments of projective geometry. J. Plücker (1801–68) introduced the co-ordinates which bear his name, which were eminently suited to specify various geometrical entities (straight line in the plane, plane in space, etc.) and also fully justified the law of duality of projective geometry from an analytical point of view. A. F. Moebius (1790–1868) studied homographic and correlative correspondences between two planes and two spaces. Steiner gave the projective generation of various geometrical figures, among them the conic.

The purist need to construct projective geometry without metrical ideas was satisfied by K. G. C. von Staudt (1798–1867).

A gap in the demonstration of the fundamental theorem of Staudt, who sought to apply the postulate of continuity, was filled first by G. Darboux (1842–1917) (13), who reduced the question to the study of the functional equation $f(x+y) = f(x) + f(y)$, and then by Enriques from the purist point of view (14).

The relations between the graphic and metric properties of the figures were studied more deeply, after it had become possible to establish that the metrical properties can be considered as graphic properties relative to particular geometrical entities, which thus constitute the so-called 'absolute'.

This work must be placed in relation to the observation of E. Laguerre (1834–86) that the angle between two straight lines is proportional to the logarithm of the cross-ratio which they

(11) Vacca, **1**, p. 186.
(12) Enriques, **12**, pp. 268–77.
(13) Vitali, pp. 226–30.
(14) Enriques, **5**.

form with the straight lines projecting the two circular points of the plane (1853) (15).

Cayley showed that all the metric properties of figures can be considered as projective or graphic properties with reference in the plane to the circular points or the absolute involution, and, in space, with reference to the absolute circle or the orthogonal polarity in a star of straight lines (16).

Later developments of this idea through the interpretation of non-Euclidean geometry will be examined in the next chapter.

§5. *The Erlanger programme and topology* (17)

Projective geometry studies the invariant properties of figures when these are subjected to the group of transformations formed from the operations of projection and section, while elementary (metric) geometry studies the invariant properties of figures subjected to groups of transformations formed from similitudes, and so on.

(We must remember that by definition a set of transformations constitutes a group, when, if a transformation belongs to the set, the inverse transformation also belongs to it, and if two transformations belong to the set, their product belongs to it too.)

Generalizing the preceding examples we reach the conception of geometry according to Klein, enunciated in the Erlanger programme (1872). There geometry is considered as the study of the invariant properties of figures in respect of an assigned group of transformations.

This general conception of geometry includes *Analysis situs* or *Topology*, which, according to Klein, is the study of the invariant properties of geometric figures with respect to the group of continuous biunivocal transformations (18).

The term '*Analysis situs*' was introduced by Leibniz. A problem of topology was studied by Euler: the problem of the seven bridges of Koenigsberg (to determine a road which would

(15) See Castelnuovo, **1**, pp. 277–8; Amodeo, p. 45.
(16) Enriques, **5**, pp. 180 and 421–2.
(17) Klein.
(18) For a historical and critical account of topology, see e.g., Enriques–Chisini, vol. I (with reference to algebraic geometry); Chisini, **3**. For Leibniz's interpretation of *analysis situs*, see Politano.

cross them all, passing over each once and only once) (19). As we have already mentioned in speaking of Descartes, the Euler formula concerning the elements of a polyhedron has a topological character.

J. B. Listing (1808–82) first dealt systematically with topology (20).

We must remember that topology goes beyond the limits of mathematics understood as a science of quantity, and is nearer to the fusion of mathematics and logic.

§6. *Hyperspace and algebraic geometry* (21)

As long ago as 1533 M. Stiefel noticed that one could go beyond the surface and the solid in considering manifolds with more than three dimensions. Lagrange interpreted mechanics as a four-dimensional geometry. Argand, with regard to the representation of complex numbers, and Jacobi, studying the quadratic form, felt the need to use a many-dimensional language.

The concept of hyperspace was expressed by the philosopher Herbart (1776–1841), who influenced Grassmann (1809–77) and Riemann (1826–66), who promoted research into the manifolds having n dimensions (22).

Hyperspatial ideas were usefully applied in the field of algebraic geometry, understood as the study of algebraic manifolds in a space of n dimensions.

The concept of a space with a number of dimensions greater than three encountered difficulties. There were the philosophical objections: in his *Critique of Pure Reason* Kant affirms that space must necessarily have only three dimensions (23); and besides, as we know, it had often been observed that it was impossible to understand a space of more than three dimensions.

As far as the logical possibility of a space of n dimensions, with $n > 3$, is concerned, this possibility is founded on a study of the n-ples of numbers (real or complex), $x_1, x_2 \ldots x_n$ which

---

(19) Euler.

(20) Listing.

(21) See Enriques, **15**; Terracini; Enriques, **12**, pp. 287–92.

(22) On this argument, see the treatise by Enriques–Chisini; for a general survey of the historical development of algebraic geometry, see Severi, **2**; see also Benedicty. (In English see Hodge and Pedoe; *Methods of Algebraic Geometry*; Semple and Roth: *Introduction to Alg. Geometry*.)

(23) See Kant: 'Transzendentale Aesthetik', §3, p. 67.

determine a point of the space of n dimensions. This logical possibility is based on analysis.

As far as the possibility of understanding hyperspace is concerned, we should remember what Einstein wrote concerning the question of spherical space (24): 'Representing a space means nothing more than representing a compendium of spatial experience'. In this sense there is nothing to prevent us from representing the ideal experience in hyperspace, for instance, in four-dimensional space (going out from the inside of a closed surface without meeting the surface itself, a movement which transforms a three-dimensional object into its image with respect to a plane, etc.).

The physical problem of the structure of physical space, a problem which must be treated with the help of experience, is very difficult.

It is possible to interpret spaces of more than three dimensions in ordinary three-dimensional space, for instance the straight lines of ordinary space constitute a manifold of four dimensions: it is enough to think of the equations of a straight line:

$$z = lx + p$$
$$y = mx + q$$

where l, p, m and q can be considered as the co-ordinates of the variable straight line in a quadridimensional manifold.

Besides, in mechanics time can be interpreted as a fourth dimension t that determines the point-event $P(x, y, z, t)$, as Lagrange had already seen in the field of classical mechanics and as Minkowski conceived in relativistic mechanics.

§7. *Abstract geometry* (25)

So far we have examined rational theories of equal logical structure, but with varying intuitive content. Examples of this kind are made up from a system of propositions of projective geometry and from the system of their duals; from a theory of analytical geometry in which the co-ordinates can be interpreted as co-ordinates of points, or else of other geometric

---

(24) Einstein, **2**, p. 99; on the four-dimensional space of Minkowski, see p. 47. Poincaré (**1**, pp. 88–90) also examines the possibility of representing a four-dimensional space.

(25) Enriques, **2**, pp. 138–41.

entities (e.g. Plücker's idea). Other examples occur to us in the interpretation of non-Euclidean geometry, which we shall deal with in the next chapter. A rational theory which is capable of various geometric interpretations is called *abstract geometry*.

'Klein and Lie,' writes Enriques, 'greatly developed the concept of abstract geometry, which later (after Segre) became an ordinary working tool in the hands of contemporary Italian geometers. In fact nothing is more fertile than the way our intuitive powers are multiplied by this principle: it seems as if mortal eyes, with which we can examine a figure under a certain relationship, have added to them a thousand spiritual eyes capable of seeing so many other transfigurations: while the unity of the object shines in our enriched reason, which allows us to pass easily from one form to another.

'But to use such a principle really fruitfully means using our logical faculties with sureness.'

CHAPTER XV

Criticism of Euclid's Postulate V; Non-Euclidean Geometries

§1. *Efforts to demonstrate Euclid's postulate V*

To the mathematicians who came after Euclid, his fifth postulate seemed less obvious than the rest. This led to efforts to demonstrate it from the other postulates and axioms, taking as a basis the first 28 propositions of Book I of the *Elements*, in which, as we have said, the famous postulate is not used.

These efforts, which began in Euclid's time and continued for a period of about twenty centuries, led to the conclusion that it was impossible to demonstrate the postulate; they also led to the conclusion that geometries in which the same postulate was not valid were logically possible (1).

We have already mentioned the work of Posidonius and of Geminus, based on the introduction of the definition of parallel straight lines as equidistant straight lines, in which a new postulate is hidden. But this, like the work that followed, was all reduced to the explicit or implicit substitution of postulate V by another which was substantially equivalent to it. For this reason the efforts themselves are interesting, among other things because they exhibit many postulates equivalent to Euclid's fifth. Thus Ptolemy (an astronomer of the second century A.D.) showed that a theory of parallels can be built up by admitting the following postulate: if any two parallels are cut by a transversal so that the sum of the interior angles on one side is either greater or less than two right angles, then the analogous property holds for any other pair of parallels.

Proclus, who refers us to Posidonius and Ptolemy's views on the theory of parallels, gives us a postulate which can be

---

(1) When nothing is said to the contrary, we refer always to Bonola, **1** and **2**. See also Enriques, **1**; Fano.

substituted for Euclid's fifth: the distance of two intersecting straight lines increases beyond limit while the distance between two parallels remains finite.

One point Proclus made in a way foreshadows the thought of Lobachevsky. Proclus observes that since the sum of two angles of a triangle is less than two right angles, there *exist* straight lines which when cut by a transversal give internal conjugate angles whose sum is less than two right angles, and which meet on that side of the transversal on which the sum is less than two right angles. 'But if for some pair of straight lines with a transversal a pair of internal angles on the same side is such that their sum is less than two right angles, there exists a point of *incidence*, it remains to be seen if this will be true for *all* such pairs. One might observe that there was a certain deficiency [from two right angles] for which they [the straight lines] do not meet, while all the others for which this deficiency is greater do meet.'

Some Greek contributions to the theory of parallels have reached us through the Arab mathematician an-Nairizi, Latinized into Anaritius (ninth century), in whose commentary a certain Aganis is named (identified by some as Geminus), in whom we meet the concept of Posidonius again.

Among Arabic mathematicians Nasir Eddin (1201–74) dealt with parallels, admitting as evident a rather complicated proposition which we will not stop to examine.

In the Renaissance the interest in critical problems with regard to Euclid's postulate V arose after 1550, when Proclus's commentary became known.

Commandino, Clavio (1537–1612), Cataldi, the inventor of continued fractions and author of the first work on parallels, B. A. Borrelli, author of a book on the movement of animals studied on the basis of mechanical principles, and Giordano Vitale (1633–1711), all dealt with parallels in the way we have already met with among the ancients. Vitale considered a figure which was to acquire singular importance in the work of G. Saccheri: the quadrilateral $ABCD$ in which by hypothesis the angles $\hat{A}$ and $\hat{B}$ are right angles and $AD = BC$.

A new idea appeared with the mathematician J. Wallis (whom we have already encountered in the history of infinitesimal calculus). He demonstrated postulate V by assuming the principle: 'For every figure there exists a similar one of arbitrary size'. But to demonstrate Euclid's postulate, according to Wallis, it is enough to admit that given a triangle one can construct a similar triangle on an arbitrary scale (2).

Let us in fact consider two straight lines a and b cut by a transversal c respectively at the points A and B (see figure 1). Let α and β be the internal angles formed by the transversal on the side of it where $\alpha + \beta < 2$ right angles. Let us construct the straight line b' through the point A in such a way that b and b' form equal corresponding angles with the straight line c. b' will fall in the angle adjacent to α, since by hypothesis $\alpha + \beta < 2$ right angles. Now if we extend continuously the straight line b in such a way that B goes through the segment

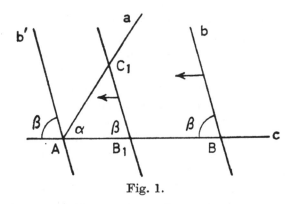

Fig. 1.

AB and the angle which it forms with c is held equal to β, the straight line b, before reaching its final position b', will necessarily have to meet a at the point C_1 since there exist a pair of points situated on the same trajectory of the translation, one of which is found on b in movement, and the other on b', situated on opposing sides in respect of a. Thus we obtain a triangle AB_1C_1 with the angles at A and B_1 respectively equal to α and β. But by Wallis's hypothesis on the possibility of constructing a triangle similar to a given one on an arbitrary scale, we can construct a triangle ABC similar to the triangle AB_1C_1, that is the straight lines a and b must meet at a point which is the third vertex of the triangle ABC. Thus, if

(2) For a thorough study of Wallis's reasoning, see Cassina, **10**.

Wallis's hypothesis is admitted, Euclid's postulate V is demonstrated.

We should remember that Wallis has a place in the history of logic with his *Institutio logicae* (Oxford, 1687) (3). He recognizes that mathematical definitions are only nominal, stressing the mathematician's freedom to give one name rather than another to the objects of his thought (for instance the triangle which to Euclid is merely plane can be spherical to Theodosius); he admits, however, the real definition in the case of the natural sciences.

§2. *Forerunners of non-Euclidean geometry*

The best attempt to demonstrate Euclid's postulate V, that of P. G. Saccheri (1667–1733), was actually, although he did not wish it to be, the prelude to non-Euclidean geometry.

Saccheri, mathematician, philosopher and theologian, studied logic as well (4). In his *Logica demonstrativa* (1679) the observations on the fallacy of the complex definition (where the existence of an entity satisfying conditions the compatibility of which is not proved is presupposed) seem inspired by direct efforts to eliminate Euclid's postulate V, by introducing the definition of parallels as equidistant straight lines. Saccheri sees that this complex definition presupposes that the locus of the points equidistant from a straight line is still a straight line. Besides he notes that the real definitions are nominal definitions to which a postulate or a demonstration of existence is added.

But his most interesting work on the subject is the *Euclides ab omni naevo vindicatus* (Milan 1733) translated into English by G. B. Halsted (Chicago–London, 1920). One of the 'warts' in Euclid's *Elements*, Saccheri says, is postulate V, and Saccheri wishes to demonstrate it himself to free Euclid from it. He proposes to construct his demonstration indirectly; he presupposes the first 28 Euclidian propositions true, in which, as we have seen, postulate V is not applied, and starts by considering a quadrilateral $ABCD$ in which by hypothesis $\hat{A}$ and $\hat{B}$ are right angles, and $AD = BC$.

(3) Enriques, **2**, p. 81.
(4) Enriques, **2**, pp. 81–2; Vailati, **1**, pp. 477–84.

It is easily demonstrated (substantially for reasons of symmetry) that $\hat{C} = \hat{D}$.

Three hypotheses are possible:

1. the two angles are both right angles (as happens in Euclidian geometry);
2. the two angles are both obtuse;
3. the two angles are both acute.

If one of these hypotheses is verified in a particular case it will always be verified (prop. V, VI, VII).

On the hypothesis of the right angle, Saccheri demonstrates in prop. IX the following result:

The sum of the angles of a triangle is equal to two right angles.

On the hypothesis of the obtuse angle the sum of the angles of a triangle is greater than two right angles.

On the hypothesis of the acute angle the sum of the angles of a triangle is less than two right angles.

(In actual fact Saccheri demonstrates this theorem for right-angled triangles, but it can be extended to any triangle almost immediately, since any triangle can always be divided into two right-angled triangles.)

The propositions enunciated are inverted in XV: 'Given any triangle *ABC*, of which the sum of the angles is equal {greater} {[less] than} two right angles, we establish respectively the hypothesis of the right angle {of the obtuse} {[acute] angle}.'

From this we deduce immediately the theorem, once known as Legendre's (published in 1833), which is now more justly called the Theorem of Saccheri–Legendre:

If in a particular triangle the sum of the angles is equal to two right angles, it is equal to two right angles in every triangle.

This theorem, the demonstration of which we shall give in §3 of this chapter, was not explicitly enunciated by Saccheri, though. Saccheri's hypotheses correspond to three geometric systems which were later all recognized as logically possible: the geometries of Euclid, of Riemann, and of Lobachevsky.

Saccheri had the satisfaction of destroying the hypothesis of the obtuse angle (prop. XIII) by presupposing the unlimited prolongability of the straight line, and he had the satisfaction of concluding:

'Igitur hypothesis anguli obtusi est absolute falsa quia se ipsam destruit.'

There remained 'the inimical hypothesis of the acute angle'. To destroy it he deduced various consequences from it, with the object of making it absurd, and thus, not wishing to do so, made the first discoveries in non-Euclidean geometry: he proved the existence of straight lines perpendicular and oblique to the same straight line which do not meet; he recognised (prop. XXIII) the asymptotic behaviour of two coplanar straight lines which are non-incident and without a common perpendicular.

On the basis of this latter result Saccheri hoped to destroy the hypothesis of the acute angle. Here is the (translated) conclusion of prop. XXXIII:

'We must recognize the hypothesis of the acute angle, according to which two straight lines AX and BX must be perpendicular at a same common point X [at infinity] to a third straight line coplanar with them, as absolutely false, because contrary to the nature of straight lines.'

Saccheri does not realize that he is thus introducing a new intuitive fact, extending to infinity the various properties of figures at a finite distance (impossibility of the existence of two perpendiculars through a point to a given straight line).

In the second part of his book Saccheri tries to demonstrate that the hypothesis of the acute angle is absurd in another way, using infinitesimal means, and taking up the concept of equidistance again; but he falls into error. What prevented him from reaping the fruits of his subtle and brilliant researches was his preconception of the unconditional validity of Euclid's geometry.

It is hard to establish the extent of his influence on the mathematicians who developed the theory of parallels after him, and later came to non-Euclidean geometry. Saccheri's work was mentioned in some books in the eighteenth century and the beginning of the nineteenth, but was finally forgotten until the mathematician Eugenio Beltrami called the attention of geometers to him again, and set him up as the forerunner of non-Euclidean geometry.

J. H. Lambert (1728–77), philosopher and mathematician, who first demonstrated the irrationality of π, made interesting contributions to the theory of parallels. It is possible that he knew Saccheri's *Euclides*, which was minutely analysed in the works of G. S. Klügel (5). There, from the examination of about thirty efforts to demonstrate postulate V, emerges the possibility that the postulate is indemonstrable on the basis of the first 28 propositions of Euclid.

Starting with a quadrilateral which by hypothesis has three right angles, Lambert formulates for the fourth the three possible hypotheses: right angles, obtuse, or acute, and finds not only results similar to those of Saccheri, but new theorems as well.

On the hypothesis of the acute angle, let us call the deficiency of a polygon of n sides the difference between the $2(n-2)$ right angles and the sum of the angles of the considered polygon. Lambert observes that in this case the area of a polygon is proportional to its deficiency; for the triangle of the angles $\alpha \beta \gamma$ we shall have:

$$\triangle = \rho^2(\pi - \alpha - \beta - \gamma)$$

While the formula which expresses the area of the spherical triangle is given by

$$\triangle = r^2(\alpha + \beta + \gamma - \pi)$$

where r is the radius of the sphere. Putting in this latter formula $r = i\rho$ we find the preceding result. Probably it was these considerations that made Lambert observe: 'I should almost reach the conclusion that the third hypothesis can be verified on a sphere of imaginary radius'. We will return to this subject with the relative demonstrations in §3 of this chapter.

Lambert notes besides that on the hypothesis of the acute angle it is possible to attribute an absolute significance to the size of segments in the same sense in which, in Euclidean geometry, one can attribute an absolute significance to the measure of angles.

This makes him reject the hypothesis of the acute angle, yet the fact that his work was published only posthumously makes us suppose that he caught a glimpse of some new horizon.

The important mathematicians who flourished between the

(5) Klügel.

end of the eighteenth and the beginning of the nineteenth century had interesting thoughts on the theory of parallels, which d'Alembert called 'the stumbling block or, if you like, the scandal of the whole of geometry'.

Near the end of his life, Lagrange wrote a memoir on parallels, but while he was reading his work to the Academy he interrupted himself saying: 'Il faut que j'y songe encore.' He was convinced that spherical trigonometry was independent of postulate V and had reflected on important relationships between the theory of parallels and the fundamentals of mechanics, inspiring an interesting work on the subject by De Foncenex (1760–1).

The concept of similitude which we have met in Wallis reappeared in Carnot, and in Laplace, who linked his ideas on the subject to the astronomical consequences of the law of universal gravitation.

Fourier too proposed a new systemization of geometry in which he took as a primitive concept that of distance, and discussed it with Monge.

A. M. Legendre aroused interest in problems relating to the theory of parallels among the mathematicians of his day. The most important results contained in his writings on parallels are drawn from Saccheri, but the elegance of his demonstrations and the wide diffusion of his work favoured the progress of the theory.

We must just recall Gauss's friend Wolfgang Bolyai (1775–1856), father of Janos Bolyai, one of the founders of non-Euclidean geometry.

After an effort (1804) to demonstrate the existence of equidistant straight lines, confuted by Gauss, he ended by persuading himself that the negation of postulate V could not lead to a contradiction.

An interesting postulate to which Wolfgang reduces that of Euclid is the following: 'Three points not in a straight line always lie on a circle'.

This idea was taken up again by a pupil of Gauss, L. Wachter (1792–1817), who established that on the hypothesis of the acute angle, Euclid's geometry holds on a sphere of infinite radius.

Wolfgang's important achievement was to make his son

recognize the difficulties and gaps in the theory of parallels; however, he warned him against efforts to demonstrate Euclid's postulate V, because, as he said, it would 'poison his existence'.

§3. *The sum of the angles and the area of a polygon*

We propose to establish some results on the sum of the angles and on the area of a polygon, which we have mentioned with regard to Saccheri and Lambert; and so give examples of the application of the very elementary way in which the simplest questions of geometry independently of Euclid's postulate V or on the hypothesis of its negation were treated.

We will suppose that Euclid's 28 propositions which are independent of postulate V support our views, or the chapters of modern books of elementary geometry in which this postulate or its equivalent have not yet been applied. And so in plane geometry (to which we will limit ourselves for the moment) the fundamental properties of congruent figures, the properties of perpendicular and oblique lines, of the circle (chords, tangents), etc., remain valid.

The results which we will now establish on the sum of the angles of a triangle are taken substantially from Saccheri, but we will follow Legendre's demonstrations, which are simple and rapid.

We start from proposition XVII of Euclid's Book I which we demonstrated in its place:

In every triangle the sum of two angles is always less than two right angles.

On this basis it is possible to establish that:

The sum of the angles of a triangle is less than or equal to two right angles.

In fact (see fig. 2) we consider a triangle ABC and demonstrate *per absurdum* that the sum of its angles cannot be $2\hat{R} + \alpha$ ($\hat{R}$ is the right angle, $\alpha > 0$).

Through the mid-point E of the segment AC draw a straight line to a point F such that $BE = EF$, and join C to F.

The two triangles ABE and FEC are congruent by the first criterion, therefore the two triangles ABC and BCF have an equal sum of the internal angles $(2\hat{R} + \alpha)$.

But at least one of the angles of the triangle BCF has an angle less than or equal to half of the angle $\widehat{ABC}$.

Applying the same process again to the triangle BCF and so on n times, we shall finally get a triangle in which the sum of the internal angles is always $2\hat{R}+\alpha$, but one angle is less than an angle however small, for instance it is less than α.

But then the sum of the two angles remaining must be greater than $2R$, against Euclid's proposition XVII.

The reasoning developed is valid if we suppose the straight line infinite, because we admit that we can always double a

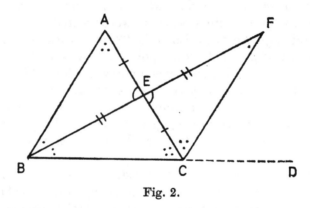

Fig. 2.

segment; it becomes defective in spherical geometry, in which the straight line is not infinite and the sum of the angles of a triangle is greater than two right angles.

With the usual reasoning in which the given polygon is divided into triangles, it is demonstrated that *in a convex polygon of n sides the sum of the angles is less than or equal to $(2n-4)\hat{R}$.*

We call deficiency the (non-negative) difference $(2n-4)\hat{R}-s$, where s is the sum of the angles of the polygon.

If the sum of the angles of a triangle is equal to two right angles, the sum of the angles of any triangle obtained from the first by subdivision by straight lines through a vertex is also equal to two right angles.

In fact (see fig. 3) we are given a triangle ABC in which the sum of the internal angles is equal to two right angles; we divide it into two triangles ADC and DBC. If the sum relating to *ADC* was less than two right angles, given that the complete sum of the angles of the two triangles must be equal to $4\hat{R}$, the sum relating to the others must be greater than $2\hat{R}$, which is absurd.

Any triangle can be divided into two right-angled triangles by means of a straight line passing through a suitably chosen vertex.

We observe in fact that in a triangle (if the straight line is

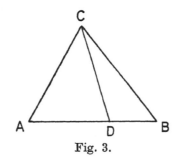

Fig. 3.

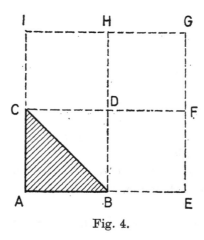

Fig. 4.

unlimited) at least two angles are acute. Drawing the height from the third vertex, this height falls within the opposite side because otherwise a triangle would be formed with a right angle and an obtuse angle, which is absurd.

If in a triangle the sum of the three angles is equal to two right angles, we can always construct a (square) quadrilateral with four right angles and four sides equal to or greater than any given segment.

In fact, let us consider a triangle for which the sum of the internal angles is equal to two right angles. If this is not isosceles right-angled we can reduce it in this case by means of subdivision by straight lines through a vertex.

Joining two triangles in this way (see fig. 4) we can obtain a square. Joining four of these squares we can obtain a new square of double the side and so on.

Thus we obtain a square with sides as large as we please.

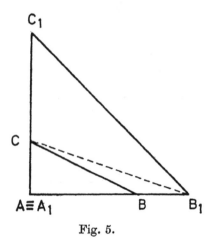

Fig. 5.

A quadrilateral of the type considered is divided by a triangle into two right-angled isosceles triangles of zero deficiency. Therefore we conclude:

If there exists a single triangle of zero deficiency, we can construct a right-angled isosceles triangle of zero deficiency with arbitrarily large sides.

We can now demonstrate *the first part of the theorem of Saccheri–Legendre*:

If the sum of the angles of a particular triangle is equal to two right angles, the same is true for any triangle.

We will first prove the theorem for a right-angled triangle ABC. If by hypothesis there exists a triangle of zero deficiency, we can construct a right-angled isosceles triangle $A_1B_1C_1$ (see fig. 5) of zero deficiency, in which the short sides are greater than those of the given triangle. Now we make the right angles $\widehat{CAB}$ and $\widehat{C_1A_1B_1}$ of the two triangles considered coincide. There being no deficiency in AB_1C_1, the same thing will be true for AB_1C and in the end for ABC.

If the given triangle is not right-angled we can divide it into right-angled triangles but, as we have shown for any of these triangles, the deficiency is zero, and therefore it will be zero for the triangle given, Q.E.D.

Bearing in mind the fact that the sum of the angles of a triangle is less than or equal to two right angles, and also the preceding theorem, we obtain the second part of the theorem of Saccheri–Legendre:

If the sum of the angles of a particular triangle is less than two right angles, the same is true for any triangle.

In fact if it were not so there would exist some triangles in which there is no deficiency, and the same thing would happen in every triangle, in particular the given one, contrary to hypothesis.

We will now examine the relation between the deficiency and the area of a polygon.

The deficiency of a polygon P of n sides, of which the sum of the angles is s, will be indicated by:

$$\delta(P) = (n-2)\pi - s$$

Similarly for a polygon P'

$$\delta(P') = (n'-2)\pi - s'$$

We propose to demonstrate that when a polygon is composed of several others, its deficiency is equal to the sum of the deficiencies of the component polygons.

That is, we wish to demonstrate that

$$\delta(P+P') = \delta(P) + \delta(P')$$

First of all we consider the case in which the straight line that divides a certain polygon Q into two polygons P and P' passes through the vertices of Q. In this case we shall have:

$$\delta(P+P') = (n+n'-2-2)\pi - (s+s') = \delta(P) + \delta(P')$$

If instead the straight line which makes the subdivision

cuts a side of the whole polygon, we must exclude a straight angle from both the expressions which with their difference give $\delta(P+P')$; similarly if the straight line cuts the sides; in every case the deficiency is

$$\delta(P+P') = \delta(P)+\delta(P')$$

We must now note that:

1. Equal or equally composed polygons have equal deficiency.
2. If a polygon is the sum of several others its deficiency is the sum of the deficiencies of the polygons of which it is composed.

Therefore we can apply the general criterion of proportionality and conclude that the polygons considered by geometry in which the deficiency is positive are proportional to their deficiency.

We write:

$$\text{Area of } P = \rho^2\delta(P)$$

In particular in a triangle of angles α, β, γ the area

$$\triangle = \rho^2(\pi-\alpha-\beta-\gamma) \tag{I}$$

The area of a triangle therefore cannot be more than

$$\rho^2\pi$$

It will tend to this maximum when α, β, γ tend to zero.

In this geometry there cannot exist triangles with equal angles and diverse areas because the angles determine the areas: similar triangles with different dimensions do not exist.

We will confront the formula obtained from the area of a triangle in the new geometry with the area of a spherical triangle.

$$A = r^2(\alpha+\beta+\gamma-\pi) \tag{II}$$

If we substitute $i\rho$ for r, the formula II transforms into (I).

We have already recalled this in dealing with Lambert.

§4. *The founders of non-Euclidean geometry*

Karl Friedrick Gauss (1777–1855) is considered the first mathematician to have reached a clear conception of a geometry independent of Euclid's postulate V; but he did not publish the results he reached on the subject before Lobachevsky and Bolyai, to avoid, as he said, 'the shrieks of the dullards'.

Gauss's life was devoted to various branches of mathematics, physics, astronomy and geodesy (the division of the circle into n equal parts and conditions to ensure that the problem can be solved with ruler and compasses; the method of least squares for estimating the errors of observations in astronomy and geodesy; the '*Disquisitiones generales circa superficies curvas*': fundamentals of differential geometry, concept of total curvature which is invariant on a flexible and inextensible surface; questions relating to mechanics: the principle of least action; researches on magnetism . . .). He also studied practical problems of optics: he constructed achromatic lenses and established principles which even today are fundamental in this field. His motto was '*pauca sed matura*'. Yet his work extends to twelve volumes.

On the relationships between science and technique he left this thought: science should be the friend of technique but not its slave, it should bestow gifts, but not serve.

Before coming to non-Euclidean geometry Gauss seems to have tried, like Saccheri and Lambert, to demonstrate Euclid's postulate V by reasoning *per absurdum*. A letter of his to Wolfgang Bolyai in 1799 is interesting:

'As far as I am concerned,' Gauss writes to his friend, 'my works are already well ahead, but the road which I have taken does not lead to the desired end which you say you have reached, but makes me rather doubt the correctness of that geometry.

'I have, it is true, established several things in ways which most men would consider valid, but which to my eyes seem to prove, you might say, *nothing*; for example, if I could demonstrate the possible existence of a rectilinear triangle in which the area is greater than any given area, then I would be in a position to establish the whole geometry with perfect rigour.

'Almost everyone, it is true, would like to give that the title of an axiom, but I would not; in fact it could be that however far from one another the vertices of a triangle were in space, its area would nevertheless be always inferior to an assigned limit.' (Cf. the results established towards the end of §3 of this chapter.)

But Gauss became certain of the logical possibility of non-Euclidean geometry only after long meditations.

Up to 1813 we have evidence of his hesitations in the face of the stumbling blocks that barred the way to the new geometrical world. In 1816 he developed a system of theorems of the new geometry which he called anti-Euclidean, but perhaps he was not yet certain of its logical possibility. In 1824 in a letter to the mathematician Taurinus (of whom we shall speak) he uses the name non-Euclidean geometry and says that all efforts to find a contradiction in it have been vain. In 1831 he says definitely that non-Euclidean geometry has nothing contradictory in itself, in spite of the paradoxical aspects of many of its results.

We shall now see how Gauss defines parallels.

Let us consider (see fig. 6) a straight line AM in the same

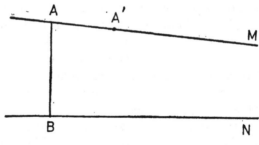

Fig. 6.

plane as BN and not cutting it. AM is called parallel to BN if every straight line through A contained in the angle $B\widehat{A}M$ meets BN. Note the difference between Euclid and Gauss's definitions of parallels.

Gauss then demonstrates that having fixed a certain direction of parallelism the parallel drawn through a point A' of AM coincides with AM: *conservation* of parallelism. Besides this, he demonstrates the *reciprocity* and *transitivity* of the relation of parallelism.

Gauss made many other important discoveries in non-Euclidean geometry: similar figures of different size do not exist, for instance the angles of an equilateral triangle vary with the side and tend to zero as the side tends to infinity.

For the measuring of segments there exists an absolute unity that appears in the formulae as a special constant k. For instance, in a letter Gauss wrote to Schumacher in 1831 we find the expression for the length of a circle of radius r in the form

$$\pi k(e^{r/k} - e^{-r/k})$$

which as $k \to \infty$ tends to the limit $2\pi r$ (as is easily verified by applying the rule of De L'Hospital).

In general, to obtain Euclidean geometry as a limiting case, we have merely to let the constant k tend to infinity in the formulae of non-Euclidean geometry.

From the physical point of view, if non-Euclidean geometry were valid but k were very large it would not be possible to decide through experiment which of the two geometries could best be applied to physical phenomena.

Two lawyers contemporary with Gauss, Schweikart and Taurinus, made an important contribution to the construction of non-Euclidean geometry. Ferdinand Karl Schweikart (1780–1857) published first a work on parallels from a Euclidean point of view; later he developed a geometry independent of postulate V, and in 1818 told Gauss of his results, for which Gauss praised him in a letter in 1819.

Here are some particularly significant passages in what Schweikart wrote: 'There exist two types of geometry—a geometry in a restricted sense—the Euclidean; and an astral geometry.

'The triangles in this latter geometry have the peculiarity that the sum of their three angles is not equal to two right angles.

'Having established this, one can rigorously demonstrate:

'(a) That the sum of the three angles of a triangle is less than two right angles.

'(b) That this sum is smaller the greater the area of the triangle.

'(c) That the height of a right-angled isosceles triangle, although it increases with the sides, cannot, all the same, be greater than a certain segment which I call constant.

. . . .

'Euclidean geometry is valid on the hypothesis that the constant is infinitely large.

'Only then is it true that the sum of the three angles of every triangle is equal to two right angles, and this allows us to demonstrate it easily only if it is admitted that the constant is infinitely large.'

Francis Adolph Taurinus (1794–1874), Schweikart's nephew, was drawn by him to study the theory of parallels. While Schweikart had a clear idea of the possibility of the new geometry, Taurinus remained convinced of the absolute validity of postulate V (at least from a physical point of view). Yet in his work *Theorie der parallellinien* (1825) and *Geometriae prima elementa* (1826), besides developments similar to those of Saccheri and Lambert, we find Gauss's constant k. As he believed that in space all the geometries corresponding to the infinite values of this parameter should be valid at one and the same time, he felt that he should reject this hypothesis.

But Taurinus's most interesting discovery was the following: if in the formulae of spherical trigonometry we substitute for the radius r of the sphere the expression $r \sqrt{-1}$, we get new relations which assume a real form in terms of the hyperbolic functions.

The formulae thus obtained correspond to the hypothesis of the acute angle.

This useful idea allowed Taurinus to obtain the formulae of non-Euclidean trigonometry in the quickest possible way, the area of the triangle, the length and area of the circle, the area and volume of the sphere. Lobachevsky and Bolyai later reached the same results only after a great deal of work.

Taurinus was the first to have a clear idea of Euclidean geometry's intermediate position between spherical geometry and logarithmical-spherical geometry (obtained on the hypothesis of the acute angle, and so called because in the formulae of that trigonometry Taurinus made use of logarithms).

The sum of the angles of a triangle could thus vary from zero (the case of the triangle of maximum area in logarithmical-spherical geometry) to three straight angles in spherical geometry.

Taurinus clearly shows his perplexity about the value of his own discoveries:

'The question of the real essence of logarithmical-spherical geometry, if it contains something possible or if it is only imaginary, goes beyond the limits of the *Elements* (page 68). But I presume that it will not be without importance for mathematics.'

Taurinus published the *Elements* at his own expense, and distributed copies of it to his friends and to important people of his day, but, disgusted by the indifference with which his work was received (except by Gauss), threw the volumes which were left into the fire.

Nicholas Ivanovitch Lobachevsky (1793–1856) was professor of mathematics at the University of Kazan in Russia, where he studied under the direction of Bartels, a friend and contemporary of Gauss. It is difficult to establish Gauss's influence

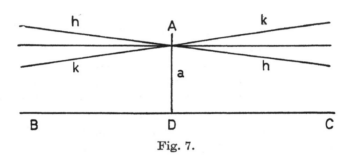

Fig. 7.

on Lobachevsky with any certainty, but the development of Lobachevsky's thought always seems rather independent of the most recent progress of the theory of parallels in his own time, and it seems more likely that he knew of Saccheri and Lambert's discoveries, either directly or indirectly.

Lobachevsky (6), having put forward a group of theorems independent of Euclid's postulate V, and on the sum of the angles of a triangle (compare §§2 and 3) considers a plane pencil of straight lines with a centre *A* and a straight line *BC.* that does not pass through *A* (see fig. 7).

Let *D* be the foot of a perpendicular drawn from *A* on to *BC*. In Euclidean geometry the perpendicular to *AD* through *A* is the only straight line through *A* which does not meet *BC*.

(6) The work of N. I. Lobachevsky, *Geometrische Untersuchungen* . . . was translated into English by G. B. Halsted (1891).

Whereas in Lobachevsky's geometry there exist in the pencil with centre A other straight lines which do not meet BC. The incident straight lines and the non-incident straight lines are separated by two straight lines h and k, which are also non-secant (as is easily established).

These straight lines h and k are called parallels by Lobachevsky and they have each a determinate direction of parallelism (in the figure h towards the right, and in k towards the left).

One of these parallels forms a certain angle with AD (this angle is 90° in Euclidean geometry), called the angle of parallelism, which is a function of $\overline{AD} = a$, indicated by the symbol $\Pi(a)$.

$$\lim_{a \to 0} \Pi(a) = 90°$$
$$\lim_{a \to \infty} \Pi(a) = 0°$$

Lobachevsky demonstrated the main properties of the new parallels (similar to the properties of the old ones): conservation, reciprocity, and transitivity, which we have already mentioned in relation to Gauss. He also recognized the asymptotic behaviour of the new parallels, which in a way had already been established by Saccheri.

The most important part of the 'imaginary geometry' consists of the formulae of non-Euclidean trigonometry. With regard to this Lobachevsky introduces two new figures: the horocycle (a circle of infinite radius) and the horosphere (a sphere of infinite radius). In Euclidean geometry the circle of infinite radius is the straight line, and the sphere of infinite radius is the plane, but in non-Euclidean geometry this is not so. Lobachevsky established that on the horosphere Euclidean geometry is valid when the horocycles are interpreted as straight lines.

The most important results which Lobachevsky deduced from his formulae are the following:

1. For triangles with infinitesimal sides the formulae of ordinary trigonometry can be substituted for the formulae of 'imaginary' or non-Euclidean trigonometry, to within infinitesimals of a higher order.

2. If for the sides a, b, c of a triangle in non-Euclidean geometry we substitute the imaginary sides ia, ib, ic, in the formulae

of Lobachevsky's trigonometry we get the formulae of spherical trigonometry.

3. If on the plane or in non-Euclidean space we fix the system of co-ordinates similar to the Cartesian system, we can calculate, with the methods of analytical geometry, length, areas, and volumes.

Lobachevsky's philosophical thought on space is shown clearly in the following passage of his *New fundamentals of geometry* of 1835.

'The unfruitfulness of the efforts made since Euclid's time in the space of two millennia, made me suspect that the truth which we wished to demonstrate was not contained in the facts themselves, and that experience, for instance of astronomical observations, might be helpful, as in the case of other natural laws.'

As we have shown in Chapter XII, Lobachevsky realized that his geometry was non-contradictory, by relying on analytical considerations. He also studied the position of his geometry with regard to physical reality.

Now it is known that the formulae of non-Euclidean geometry transform into those of Euclidean geometry when a certain parameter k (contained implicitly in Lobachevsky's formulas) tends to infinity; through his astronomical studies Lobachevsky established that k is either very large or actually infinite; that is to say, in the experimental field the Euclidean hypothesis is valid.

While Lobachevsky tried above all to construct a geometrical system on the negation of Euclid's postulate V, Janos (John) Bolyai (1802–60) established theorems and geometrical constructions independently of this postulate. In 1823 Bolyai discovered the true nature of the problem that he was tackling, by finding the fundamental formula which links the angle of parallelism $\Pi(a)$ (which we mentioned above) to the corresponding segment a:

$$e^{-a/k} = \tan \tfrac{1}{2}\Pi(a)$$

Janos wrote to his father Wolfgang: 'All that has been found until now seems to me a card castle in comparison with this tower.' He mentions the geometry that is independent of postulate V, and now seems to him like the absolute science of space. And he adds: 'I have created a new world from nothing.'

Wolfgang urged his son to publish his discoveries; he wrote: 'There is some truth in the fact that many things have their season, in which they are found at the same time in several places, just like the violets which come up everywhere in spring.'

And in fact, precisely in the year in which Bolyai's work was published, 1829, Lobachevsky's first work on non-Euclidean geometry was published too.

Janos Bolyai's discoveries were published in appendices to the first volume of his father's work, under the title: *Appendix scientiam spatii absolute veram exhibens: a veritate aut falsitate Axiomatis XI Euclidei, a priori unquam decidenda indipendentem: adjecta ad casum falsitatis quadratura circuli geometrica.*

A copy of this work was sent by Wolfgang to his friend Gauss, who in 1832 replied: 'If I start by saying that I cannot praise this work, you certainly will be surprised; but I cannot say anything else; to praise it would be to praise myself; in fact its whole contents, the way your son has elucidated it, the results he has reached, coincide almost entirely with the meditations which have partly occupied me for thirty to thirty-five years.

'... It was my intention in time to write it all down so that at least it would not perish with me. And so it is a pleasant surprise to see that I can now be spared this labour, and I am extremely content that it should be the son of my old friend who has preceded me in such an important way.'

Wolfgang was very pleased with this answer, but Janos was not, for he suspected that Gauss wanted to get the credit for his discoveries, and although later he had to admit that his suspicions were unfounded an unreasonable dislike remained.

In the non-Euclidean geometry of Gauss, Lobachevsky, and Bolyai the hypothesis which we have called, with Saccheri, the hypothesis of the obtuse angle was excluded; this hypothesis was rejected because the straight line was supposed to be infinite. The possibility of a geometry in which the hypothesis of the obtuse angle is valid, the straight line is not infinite, and the sum of the angles of a triangle is greater than two right angles, was established by Bernhard Riemann (1826–66).

Riemann's work, through its originality and the profundity

of its thought, was fundamentally important in the development of modern mathematics.

We will limit ourselves to recalling his theory of functions of a complex variable and his concept of Riemann surfaces; and the theory of definite integrals, which found a new systemization through his work.

But above all we must deal with his research into the principles of geometry.

A work of his on the subject in 1854 is the point of departure for modern differential geometry and prepared the way for the construction of the absolute differential calculus of Ricci-Curbastro (1853–1925), reconstructed by T. Levi-Civita (1873–1941).

As we know, the absolute differential calculus is the algorithmic instrument which is used in the general theory of relativity.

§5. *The differential-metric trend of non-Euclidean geometry*

To understand Riemann's geometry and the interpretation of this geometry and that of Lobachevsky on a surface with a constant curvature, we will consider a surface S, on which we propose to found a geometry.

Given two points A and B of S, we consider the arc $\overset{\frown}{AB}$ (in general determinate) which follows the shortest path from the point A to the point B on S. The arc $\overset{\frown}{AB}$ thus defined is called the geodesic arc.

On a plane the geodesics are straight lines, on a spherical surface they are great circles.

In the geometry of surfaces the geodesics have a place similar to that of straight lines in plane geometry.

Now we will define two equal figures in the geometry of surfaces, figures which are called geodetically equal.

Two figures drawn on a surface are called geodetically equal when they can be made to correspond point for point in such a way that the distance between any two points of the first figure measured on the relative geodesics is equal to the distance between the two corresponding points, of the second figure, measured on the relative geodesic.

We can express this concept of geodetic equality between figures in a more intuitive way: we consider the surface on

18   273

which we have founded our geometry as a flexible and inextensible leaf: two figures drawn on this leaf are geodetically equal if they can be placed on each other through a movement of the surface in which the surface bends while keeping its inextensibility.

For instance, a cylindrical surface can be applied without extension, duplication or tearing to a plane region. In this case the two figures drawn on the cylindrical surface will be called geodetically equal if they can spread over equal figures in the plane.

Two geodetically equal figures are not generally congruent in space in the ordinary sense.

Basing ourselves on these principles we can construct a geometry of surfaces, which we prefer to refer to a conveniently limited region of the same surface.

Two surfaces which can be transformed one into the other with flexion and without extension have the same geometry. For instance on the cylindrical surface a geometry similar to plane geometry is valid. A different geometry is valid for spherical surfaces, since the spherical surface is not applicable to the plane.

Plane geometry and spherical geometry have a common character, though, because the sphere, like the plane, can move freely over itself (this is something which generally does not happen with other surfaces, for instance the ellipsoid). This means that there exists a character which is invariant for all surfaces. This character is curvature, which Gauss defines in the following way:

Let us consider a surface F and a point P on it, and through P draw the normal n to the surface on which we choose a positive sense; the infinite planes passing through n cut F in infinite curves: two of these situated on two orthogonal planes have respectively the greatest and least curvature, $1/r_1$ and $1/r_2$. The total curvature of the surface is given as:

$$K = \frac{1}{r_1} \cdot \frac{1}{r_2}$$

If the sections of the surface have a constant curvature $1/r$, as happens for the sphere, then $K = 1/r^2$.

Now it is possible to construct surfaces with constant curvature; there are three possible cases:

$$K = 0 \qquad K > 0 \qquad K < 0$$

For $K = 0$ we have the plane with all the developable surfaces, that is applicable to the plane.

For $K > 0$ we find the sphere and the surfaces applicable to the spherical surface.

For $K < 0$ we find the pseudosphere and the surfaces applicable to the pseudosphere.

The pseudosphere is a surface of revolution which has as its meridian a curve, with an asymptote, called a tractrix, characterized by the following property: the segment of the tangent to the tractrix, contained between the point of contact and the asymptote, has a constant length (see fig. 8).

Fig. 8.

We should remember, as was explained in Chapter XIV, the various ways in which the same abstract geometry can be interpreted, and apply this to the case of the geometry of a surface with constant curvature, and to a plane geometry using the postulates of Euclidean geometry, except for the fifth.

We can thus obtain the following table:

Geometry on a surface of constant curvature	Plane geometry
(*a*) Surface (conveniently limited)	(*a'*) Region of plane (conveniently limited)
(*b*) Point	(*b'*) Point
(*c*) Geodesic	(*c'*) Straight line
(*d*) Arc of geodesic	(*d'*) Segment
(*e*) Linear properties of the geodesic	(*e'*) Linear properties of the straight line
(*f*) Two points determine a geodesic	(*f'*) Two points determine a straight line
(*g*) Fundamental properties of the equality of the geodesic arcs and of the angles	(*g'*) Postulates of the segmentary and angular congruence
(*h*) If two geodesic triangles have two sides and the included angle equal, the remaining sides and angles are equal	(*h'*) If two rectilinear triangles have two sides and the included angle equal, the remaining sides and angles are equal

It follows from this that those properties which do not depend on postulate V (formulated in an abstract way) and which refer to a limited region of the surface or of the plane, can be considered common to the geometry of surfaces with a constant curvature and to plane Euclidean geometry.

But in the Euclidean plane the sum of the angles of a triangle is equal to two right angles; and it is not asserted that the same is true in the geometry of surfaces.

For example, let us examine the table already considered, particularly in a case in which the surface is a sphere.

This has a very important effect: since Riemann's geometry and the geometry of the sphere (both considered in a limited region) are two interpretations of the same abstract geometry, a contradiction in one of these geometries should be reflected by a contradiction in the other; that is, if a contradiction exists in Riemann's non-Euclidean geometry, it should also exist in the geometry of the sphere, that is in Euclidean geometry.

Spherical geometry	Riemann's plane geometry
(*i*) The sum of the angles of a spherical triangle is greater than 2 right angles	(*i'*) The sum of the angles of a rectilinear triangle is greater than 2 right angles
(*j*) Great circles without points in common do not exist	(*j'*) Coplanar straight lines which do not meet (parallel according to the Euclidean definition) do not exist

One could reason in the same way about the geometry of the pseudospheres and Lobachevsky's geometry.

So this is a way of demonstrating the logical possibilities of the non-Euclidean geometries of Lobachevsky and of Riemann, when we confine ourselves to considering them in a limited region of space.

Is it possible to find surfaces of positive or negative constant curvature in which the geometries of Lobachevsky and of Riemann are integrally valid?

This can be answered as follows:

1. No regular (that is non-singular) analytic surface exists for which the geometry of Lobachevsky–Bolyai is valid in its entirety (theorem of Hilbert).

2. A surface for which Riemann's plane geometry is valid in its entirety must necessarily be closed.

The only regular analytic closed surface with constant positive curvature is the sphere (theorem of Liebmann).

But on the sphere in whose normal (limited) regions Riemann's geometry is valid, two great circles (corresponding to straight lines) meet in two points.

So we may conclude:

In ordinary space there do not exist surfaces on which the geometry of the non-Euclidean plane is integrally valid.

These results had already been foreseen by the mathematician A. Genocchi.

While Riemann realized, from 1854 onwards, the possibility of interpreting the geometry of two-dimensional manifolds on ordinary surfaces, the interpretation of non-Euclidean geometry on surfaces with constant curvature was developed by Eugenio Beltrami (1868).

The logical possibility of Riemann's non-Euclidean geometry extended to the whole plane can be demonstrated by observing that the geometry of the star of straight lines and of planes, and Riemann's non-Euclidean geometry, are two interpretations of the same abstract geometry.

We can now draw up a second table:

Geometry of the star of straight lines and planes	Riemann's geometry
Pencil	Plane
Straight line	Point
Plane (pencil)	Straight line
Angle of two straight lines	Segment
Dihedron	Angle
Trihedron	Triangle
Planes of the star without a straight line in common do not exist	Non-intersecting straight lines do not exist
The sum of the three dihedra of a trihedron is greater than two straight dihedra.	The sum of the three angles of a triangle is greater than two right angles.

The demonstration of the logical possibilities of non-Euclidean geometry will now be given in a more complete form in another way.

§6. *The projective-metrical direction of non-Euclidean geometry*

Developing Cayley's ideas on the metrical properties of figures, considered as graphic relations of these figures with the absolute (see Chapter XIV, §4) Beltrami (1868) and Klein noted that the various non-Euclidean geometries could be interpreted in a projective metrical system in which the absolute is a conic (7).

(7) Enriques, **12**, pp. 303–4.

To make this concept clearer we must use some ideas on the homographies which leave a conic invariant.

A homography between two planes always transforms a conic K into a conic K'. Given that a homography is determined by four pairs of corresponding points, no triad of which belongs to the same straight line, to make K' correspond to K by a homography we make three points of K': A', B' and C', correspond to three points of K: A, B and C; to the point O common to the tangents a and b to K at A and B respectively we make correspond the analogous point O' (see fig. 9).

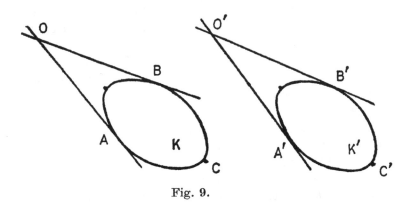

Fig. 9.

The homography that transforms $ABCO$ into $A'B'C'O'$ must also transform K into K', because the correspondent of K must pass through $A'B'C'$ and be the tangent to the straight lines a' and b' (corresponding respectively to a and b of the homography).

The correspondent of K is therefore identified with K', being a conic fixed by five conditions.

In particular we can make K coincide with K': a homography which leaves invariant the conic $K \equiv K'$ will be individualized by making the three points ABC of the conic correspond to three other points chosen arbitrarily on the conic itself.

Now we will give an interpretation of Lobachevsky's geometry by means of a metrical projective system referred to a conic K.

As the complement of this we can give an interpretation of the concepts of distance and of angle in metrics referred to K.

279

Lobachevsky's plane geometry	Metrical-projective system with regard to a real cone K
(a) Plane	(a') Plane region interior to a conic K
(b) Straight line (unlimited)	(b') Chord of K (excluding the ends)
(c) Points following one another on a straight line	(c') Points following one another on the chord
(d) Movement of a plane	(d') Homography that leaves K invariant
(e) Congruent plane figures	(e') Figures that can be transformed by an homography that leaves K invariant
(f) Postulates of incidence, order and continuity relating to points and straight lines	(f') Relation of incidence, order and continuity relating to points internal to K and to chords of K
(g) Postulates of movement: the movements of planes form a group of transformations; there exist two movements of the plane that carry a half-line of the plane into a chosen half-line of the plane; a movement of the plane that leaves three points invariant is the identity	(g') Properties of the homographies that leave K invariant: these homographies form a group of transformations (as is easily seen); there exist two homographies that leave K invariant and transform the segment AN into a segment $A'N'$ (where A and A' are interior to K and N and N' belong to K) (see fig. 10); a homography that leaves K invariant and three points ABC interior to K is the identity (see fig. 11).

The interpretations we are seeking of the distance AB and of the angle ab should be invariant in respect of transformations which leave K invariant. These conditions are satisfied respectively by the cross-ratios $(ABMN)$ and $(abmn)$ where M and N are the intersections of the straight line AB with K, and m and n are the imaginary tangents drawn to K from the vertex of the angle.

But we must bear another condition in mind: distance and angle are additive invariants: if A, B, C are in a line and

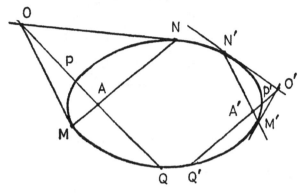

Fig. 10.

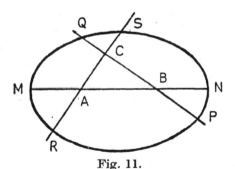

Fig. 11.

a, b, c pass through a point, distance $AC =$ distance $AB +$ distance BC, $\widehat{ac} = \widehat{ab} + \widehat{bc}$.

Instead the cross-ratios considered have, as we can check immediately, a multiplicative character. The condition of additivity is satisfied by putting

$$\text{distance } (AB) = c \log (ABMN)$$
$$\widehat{ab} = c' \log (abmn)$$

Now we ask: in the metrical-projective system considered, is Euclid's postulate valid (which is equivalent to the existence and the uniqueness of parallels through a point to a given straight line) or the hypothesis of Lobachevsky? Considering Fig. 12 we see at once that it is satisfied by Lobachevsky's

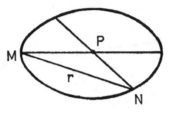

Fig. 12.

hypothesis. More precisely, returning to the table:

Lobachevsky's plane geometry	Metrical-projective system with regard to a real conic
(*h*) Given a straight line *r* and a point *P*, the straight lines of the pencil with centre *P* divide into incident and non-incident straight lines	(*h'*) Given a chord *r* and a point *P* internal to *K*, the chords through *P* divide into incident and non-incident chords
(*i*) The non-incident straight lines which are limits of incident straight lines are called parallels	(*i'*) The chords with an extreme point in common correspond to the parallels
(*j*) There exist two parallels to *r* through *P*	(*j'*) There exist two chords through *P* passing through the extremes of *r*

The fundamental properties: conservation, reciprocity, transitivity of the idea of parallelism, are easily verified in the above representation.

It is also easy to verify the asymptotic behaviour of the parallels in Lobachevsky's geometry. Let *AB* be two points or two chords meeting at a point *Z* of *K*, and let *M* and *N* be the extremes of the chord which joins A and B. We shall have:

$$\text{dist } (AB) = c \log (ABMN) = c \log \frac{AM}{BM} \cdot \frac{AN}{BN}$$

If A, B, M and N all tend to the point Z the distance (AB) tends to $c \log 1 = 0$.

If instead of a real conic K we take an imaginary conic we can get a representation of Riemann's geometry.

To get a representation of the non-Euclidean geometry of space, all we need do is interpret the points of the space as interior points of a quadric, the straight lines as chords of the quadric, and so on.

All this shows us that non-Euclidean geometries can be interpreted within the ambience of Euclidean geometry, and so if a contradiction exists in one of the non-Euclidean geometries, this contradiction has repercussions in Euclidean geometry as well. So if we admit that Euclidean geometry is exempt from intrinsic contradictions, we may say the same of every one of the non-Euclidean geometries.

All this shows too that postulate V cannot be deduced from the other postulates on which Euclid's geometry is based.

§7. *The problem regarding physical space, in philosophy and in science* (8)

In ancient times the modern ideas on space had already appeared: the Eleatics denied the existence of a void and considered movement as purely relative, while Democritus believed in the void and the absolute motion of atoms in respect of space, empty space—which is something.

Descartes takes up the Eleatic position again, while Galileo (and later Newton) finds it necessary, in order to construct dynamics, to consider absolute movement with respect to empty space.

The critics of the Galilean-Newtonian concept could not succeed in the face of the scientific success of the dynamical theory; but their efforts brought us to a profounder conception of space: Leibniz wondered: 'Is space something outside the order of existing things?'

(8) Enriques, **10** and **12**, pp. 304–8; Einstein, **2**, ch. XIV, §6, pp. 95–100.

Kant suggested a new view of it: the subjectivity of space, which is no longer the order of external things but the order conferred by the intellect on sensible data. He tried to save the character of necessity in science, above all in mathematics and in physics; and found necessity in propositions that to us do not have it: for instance: 'space cannot have more than three dimensions'. Now we know that nothing prevents us from considering a space of n dimensions with $n > 3$ (compare Chapter XIV, §6).

In fact it is an experimental problem to decide if there are or are not more than three dimensions in physical space.

Substantially the geometry that Kant wished to found on the solid basis of his *a priori* philosophy was the Euclidean geometry of three-dimensional space.

Kant's physics were the same as Galileo's and Newton's.

Now just when the *Critique of Pure Reason* (first edition, 1781) was being published and upheld in philosophical circles, Gauss, Lobachevsky, Bolyai, and other mathematicians we have mentioned managed to prove the logical possibility of a geometry that was different from the Euclidean.

Besides, we have seen that the geometry which is built on the rejection of Euclid's postulate V depends for its validity on a parameter, which for a particular value ($k = \infty$) gives rise to Euclidean geometry. So our sensible experiences can be explained equally well by Euclidean geometry as by a non-Euclidean geometry in which k is sufficiently large, since physical measurements are only approximate.

As we have already said, for the founders of non-Euclidean geometry geometrical truths were only experimental.

According to H. Poincaré geometrical propositions should be considered as a system of conventions through which physical facts are expressed, as magnitudes refer to a system of measures. And as physical magnitudes can be referred to one system of measurement or another, so the same physical facts can be expressed by one geometry or another.

Let us suppose, for example (Poincaré says substantially) (9), that in a certain space the light rays propagate themselves along lines which do not enjoy the properties of Euclidean

(9) Poincaré, **1**, p. 93.

straight lines. In such a case there are two ways in which we can systemize the observed physical facts :

(*a*) Change the geometry.

(*b*) Change the properties of the light rays.

According to Poincaré, the second solution would be the more advantageous.

Not all scientists were fully persuaded by this way of considering geometry as a system of conventions; Poincaré's point of view was criticized by Enriques (10).

§8. *Possibility of a finite but not limited world*

Let us consider living and thinking beings in a two-dimensional space : for instance, a plane or a spherical surface ; in the second case the space of these beings would be finite because if they drew a circle of radius *r* (taken on a spherical surface) as *r* increased the circle would grow to a maximum value, then it would decrease. But this space would not be limited because in whatever direction these beings moved they would not find a boundary.

Now we ask : can we, with our way of thinking, represent a space that is finite but not limited, as a spherical space would be? If by representing a space we mean imagining experiences we could have in this space, it *is* possible. Einstein (11) writes on the subject : 'From a point we draw so many straight lines (we hold threads in all directions) and on each of them we lay off with the measuring rod the segment *r*. All the free extremities of these segments fall on a free surface the area (*F*) of which can be measured by a model square. If the world is Euclidean $F = 4\pi r^2$; if it is spherical $F < 4\pi r^2$.

'As the radius *r* grows, *F* grows from zero to a maximum determined by the cosmic radius, and then, as *r* grows further, *F* diminishes gradually until it becomes zero. The radial straight lines from the original point first draw ever further apart, and then draw closer to meet at last at the point opposite the one whence they originated, and thus they have run through the whole of the spherical space. We can easily see that spherical space in three dimensions is similar to that in two (spherical surface). It is finite in volume without having limits.'

(10) Enriques, **6**, ch. IV, §6, pp. 154–7.
(11) Einstein, **2**, p. 99.

CHAPTER XVI

Combinatorial Analysis and Calculus of Probability

§1. *Meaning, origins, and applications of combinatorial analysis*

Combinatorial analysis (1) is a branch of arithmetic, the object of which is to count sets of objects and symbols which can be formed under determined conditions; it finds its application in the most diverse fields of mathematics, in particular in algebra (for instance in the theory of determinants), in the theory of the groups of substitutions and in the calculus of probability.

The origins of combinatorial analysis are linked to the developments of the nth power of a binomial (2), which, perhaps for the first time, was calculated by Omar Khayyam, towards the end of the eleventh century A.D., in his *Algebra* (3); it is probably to him too that we owe the discovery of the tables of binomial coefficients determined by successive additions.

This table appears in a scholium to a Greek manuscript of Euclid which goes back to the thirteenth century (4), and in a work of the Chinese mathematician Chu Shih Chieh (who does not say it is his own), in 1303.

In the west, following other mathematicians, N. Tartaglia in his *Treatise on Arithmetic* (Venice, 1556) published the arithmetical triangle in question, which took its name from him. The development of the binomial series for negative and fractional exponents was discovered by Newton about 1666 and communicated in a letter to Leibniz in 1676.

(1) See Severi, 1; Gigli, 2.
(2) Vacca, 9.
(3) Omar Khayyam.
(4) Euclid, t. VIII, p. 290.

§2. *A brief historical glance at the calculus of probability* (5)

In Greek thought, particularly in Aristotle, we already find the first notions on which the calculus of probability is based: chance, fortune, choice, means (6). But a more direct ancestor of the present calculus had its origins in games of chance; and later it was applied in the most varied fields of modern physics, in statistics, and in actuarial mathematics. Subtle researches were undertaken concerning its basis and its rational structure in relation to the most recent trends of modern logic.

The history of the calculus of probability is generally thought to begin with Dante's comment in the *Divine Comedy* in 1447 on a game of dice called 'zara':

> 'Quando si parte il gioco de la zara
> colui che perde si riman dolente,
> repetendo le volte e tristo impara.'
> > (*Purgatorio*, C. VI, v. 1–3)

The commentator explains the rules of the game, and the possible and favourable chances.

Problems of the calculus of probability are met with in some passages of the writings of Cardano and of Galileo, and in the correspondence of Pascal and Fermat.

In 1713 Jacques Bernoulli's *Ars conjectandi* was published posthumously, and here we find the famous theorem with which we will deal later.

It is to S. Laplace (1749–1827), who wrote the *Théorie analytique des probabilités* (1812) and the *Essai philosophique des probabilités* (Paris, 1812), that we owe the classic exposition of the calculus of probability. P. Ruffini took a stand against his philosophical position in his *Critical reflections on the philosophical essay concerning probability of Count La Place* (Modena, 1821).

The first applications of the calculus of probability to physics were those relating to thermodynamics: the work of J. C. Maxwell (1831–79) and L. Boltzmann (1844–1906). Contemporary physics is dominated by probabilistic and statistical ideas.

(5) See Castelnuovo, 4.
(6) Vacca, 7.

§3. *Some aspects of the foundations and development of the calculus of probability according to the classical conception* (7)

The classical concept of *probability* is defined in the following way: 'The probability of an event is the ratio of the number of chances which favour the event to the number of possible chances, provided that all the cases considered are equally possible.'

This probability of an event is evaluated before the event itself occurs (*a priori*) on the basis of the modality according to which it happens: whereas the *frequency* is determined after the verification of the event (*a posteriori*):

'The ratio ν/n of the number ν of cases in which the event has shown itself and the number n of trials made, is called the frequency relating to the event.'

A relationship between the concepts of probability and of frequency thus defined is given us by the *empirical law of chance.*

'In a series of repeated trials performed a great number of times under the same conditions, each of the possible events is shown with a (relative) frequency which is approximately equal to its probability, the approximation improving ordinarily with the growing number of trials.'

In non-technical language this is sometimes expressed by saying: the frequency of a chance event in a series of experiences, made under the same conditions, tends, with the increase in the number of these experiences, towards a limit which is equal to the probability of the event.

But we must make it clear that the expression 'tends towards a limit' has *not* the precise sense that it has in infinitesimal calculus.

Its exact meaning is found in the theorem of Jacques Bernoulli:

'In a series of n trials an event which in each of them has the constant probability p occurs ν times; the probability that the difference $(\nu/n) - p$ is, in arithmetical value, less than a given number, tends towards certainty with the increase in the number n of trials.'

(7) Castelnuovo, **2**, ed. 1947, vol. I; pp. 1–9, 56, 103–7, 205–12. We have taken from this book the statements between quotation marks.

'That is, the probability that

$$\left| \frac{\nu}{n} - p \right| < \varepsilon$$

tends to unity as n tends to infinity'.

But the two enunciations—that of the empirical law of chance and that of Bernoulli's theorem—which are thus confused, are irrevocably different, and it is not possible to demonstrate one of them by using only the other. Yet there is an important relationship between them: by using Bernoulli's theorem, by admitting the empirical law of chance when the event is very likely, when we have a so-called 'practical certainty', we can demonstrate the empirical law of chance in all the other cases.

In fact Bernoulli's theorem relating to the event E of probability p considers the event E' in which

$$\left| \frac{\nu}{n} - p \right| < \varepsilon$$

By Bernoulli's theorem, the probability of E' can be made as near to 1 as we please, that is E' is practically certain when n is very large. This affirmation coincides substantially with the empirical law of chance.

Castelnuovo concludes:

'Thus the real character of mathematical enquiry, which cannot claim to demonstrate a natural law *a priori* although it can lead it to another empirical law which is more easily acceptable, is clearly shown.'

Among the many important later developments of the calculus of probability we will limit ourselves to indicating the exponential law regarding the theory of errors of observation discovered by Gauss and published in his *Theoria motus corporum coelestium* of 1809. Gauss's very subtle and interesting demonstration, although it laid itself open to criticism of the explicit and implicit hypotheses on which he based it, was conducted according to a functional process. We get the following result:

'The probability of committing an error contained between

z and $z+dz$ in the measurement of a magnitude is expressed by the formula

$$\frac{h}{\sqrt{\pi}}e^{-h^2z^2}\,dz$$

where h is a constant depending on the accuracy of measurement.'

§4. *Various conceptions of probability*

The *classical conception* of probability, which appears in the definition quoted in the last paragraph, is open to criticism.

Castelnuovo (8) considers 'unsatisfactory' the way in which the caution needed to assess probability correctly is mentioned in the definition, in the clause : 'so that all cases shall be equally possible'. He observes that 'if the calculus of probability is studied in view of its applications . . . the correct enumeration of possible and favourable cases must depend exclusively on the physical particulars of the concrete problem examined. And when some doubt arises as to the assessment of cases which are equally possible, the experimental verification based on the empirical law of chance can provide a criterion to decide which among various acceptable *a priori* hypotheses should really be accepted. But it is clear that if one must determine, by actual experiment, a frequency for pronouncing judgment on a probability, the calculus would lose one of its objects, which consists in foreseeing the frequency, starting from probability.

'In actual fact the forecast can be made in many cases, but with varying criteria, according to the nature of the problems.'

With these considerations we approach the '*empirical conception* (9) [of probability] based on the concept of repeatable events in which the frequency in a great number of trials (by the 'empirical law of chance') gives us the probability almost exactly and almost certainly'.

An idealization of this is given by the *asymptotic conception*, in which we consider an infinite succession of trials and define probability as the limit of frequency.

Finally we have the '*subjective conception* [supported by

(8) Castelnuovo, **2**, ed. 1926, pp. 3, 5–9.
(9) The four interpretations of probability (classical, empirical, asymptotic, subjective) are drawn from De Finetti, **4**, pp. 10–11.

B. De Finetti] which considers probability as a measurement of the degree of faith of a determinate subject in the happening of an event'.

Probability thus defined should be such as to satisfy a necessary and sufficient condition of intrinsic consistency (10); it is thus defined on the basis of assessments which would be undertaken by a hypothetical bookie (11).

'Assessing the probability of the event E as equal to p means, to the hypothetical bookie, declaring oneself ready to accept any bet with any competitor who is free to fix at his pleasure the bet S (positive or negative) in such a way that the gain G (positive or negative) of the competitor will satisfy

$$G(E) = (1-p)S$$
$$G(-E) = -pS$$

in the two cases E and $-E$, and that is on the hypothesis that the event E will be verified or will not be verified, as the case may be.

'A man is consistent in assessing the probability of certain events if, whatever set of bets $S_1 S_2 \ldots S_n$ a competitor makes on any set of events $E_1, E_2 \ldots E_n$ among those he has considered, it is not possible that the competitor's gain G, in any positive case, should turn out to be positive.'

On the basis of this definition we once more obtain the fundamental theorems of the classical calculus of probability.

The conception of probability chosen must be set in relation to the system of postulates to be placed at the basis of the calculus of probability. For a deeper understanding of the subject I would refer to the writings of De Finetti (12).

Other definitions of probability, connected to one or other of the various ideas we have mentioned, can be found in contemporary neo-positivist thought (13).

We find the concept of probability in an example of non-Archimedean magnitudes (see Chapter XVII, §3) and in the interpretation of some many-valued logics (Chapter XX, §2).

(10) De Finetti, 1, pp. 258–61.
(11) De Finetti, 2, pp. 308.
(12) See De Finetti, 4, and the related bibliography.
(13) Barone, pp. 67–82, 88–93, 377–80, ch. XX, §3 of this book.

CHAPTER XVII

Sets, Functions, Curves, Non-Archimedean Geometries

§1. *The theory of sets* (1)

In the history of mathematics the problems of the actual mathematical infinite have appeared many times under various aspects, as we have already seen.

While, as we have also seen, the rigorous system of the infinitesimal calculus constructed by Cauchy, and by his followers from his work, seems to have banned the actual infinite from mathematics, in a very different sense Bolzano's thought tends to remove the apparent contradictions relating to infinite sets.

But the modern theory of infinite sets is above all the work of Georg Cantor (1845–1918), who with his publications between 1874 and 1883 opened up new paths to the human mind, and built up a system of knowledge which was of the highest importance for the progress of logic, and the relationship between logic and analysis.

We propose to reconstruct rapidly some of the most important aspects of the theory of sets, which deals especially with the possibility or impossibility of establishing a biunivocal correspondence between several aggregates.

With the terms 'entity', 'element', 'object', we indicate a primitive concept which does not allow of definition in terms of simpler concepts; above all we might say of the term 'sets' that it is assumed as a primitive concept and is synonymous with 'class' or 'aggregates' (2).

To indicate that an object e belongs to a set I we write, in Peano's notation:

$$e \in I.$$

With regard to this concept, we must ask a serious critical

(1) See Enriques, **2**, pp. 151–60.
(2) See Sierpinski, p. 1 ff.; Vitali and Sansone, part I, p. 1 ff.; Severi, **1**, pp. 92–7.

question: when can a set be considered as existent (in the world of logical thought)?

On this point mathematicians are divided into two large groups, which we can call roughly 'idealist' and 'empiricist', with various subtle differences within each group. For an exposition of the two points of view, we should go back to the dialogue between the idealist and the empiricist of Paul Du Bois-Reymond in his *Die Allgemeine Functionentheorie* (Tübingen, 1882). Idealists and empiricists are respectively the spiritual heirs of the realists and the nominalists whom we encountered in dealing with the mediaeval controversy on universals.

A matter which is closely linked with this consists in establishing exactly the relations between the existence of a set and the existence of the elements that constitute it. L. Geymonat (3), who has made a profound analysis of the question, gives three mental positions possible on the subject, which lead to various systems (syntaxes) of rules, with which to operate on the verb 'exist'.

'*First solution.* We can call a set existent only when every element in it has a well-determined existence, that is, when an exact definition exists (which can be enunciated with a finite number of words) for every one of its elements. . . .' This is the empiricists' solution.

'*Second solution.* We can speak (under determinate conditions) of the existence of a set, even when we do not have precise rules for constructing effectively every element of the set; we must besides accept the principle . . . to conclude . . . from the existence of a set a certain existence (in a generalized sense) of all its elements.' This is the idealistic solution. An intermediary position can also be taken.

'*Third solution.* We can, as in the preceding solution, speak (under determinate conditions which, for reasons of brevity, we will not stop to analyse) of the existence of a set even if we do not have precise rules for effectively constructing every element of the set; but we cannot . . . deduce, from the existence of the set, the existence of every one of its elements.'

These solutions are particularly important when they are applied to sets of real numbers (in which, as we shall see,

(3) Geymonat, **2**, pp. 291–305, chiefly pp. 298–9.

elements exist which cannot be defined with a finite number of linguistic signs and symbols) and when they are placed in relation to the postulate of Zermelo, with which we shall deal later.

For the moment, leaving aside the questions which we have mentioned in this section, we will deal with some of the most suggestive parts of the classical theory, sets according to the views of G. Cantor.

Two sets are said to be of the same *power* and of equal *infinite cardinal number* if it is possible to establish a biunivocal correspondence between the elements of the sets themselves. As Galileo observed, the elements of an infinite set can be placed in biunivocal correspondence with the elements that constitute a part of the same set (e.g. the case of whole numbers and their squares). Therefore for infinite sets the axioms of inequality do not apply: a set and its parts can have equal power.

What is more: Dedekind assumes as a definition of an infinite set the property of its being placeable in biunivocal correspondence with one of its own parts.

The abstract concept of the power of a set is identified with that of the *infinite cardinal number* (F. Severi).

Let M and N be two sets and m and n their infinite cardinal numbers. We will examine several possible cases:

(1) M and N can be placed in biunivocal correspondence: we then write $m = n$;

(2) M can be placed in biunivocal correspondence with a part of N, and M cannot be placed in biunivocal correspondence with N: we then write $m < n$;

(3) N can be placed in biunivocal correspondence with part of M and M cannot be placed in biunivocal correspondence with N: we then write $m > n$;

(4) M can be placed in biunivocal correspondence with a part of N and N can be placed in biunivocal correspondence with a part of M: in this case it is shown that it is also possible to establish a biunivocal correspondence between M and N, that is $m = n$ (theorem of Cantor-Bernstein) (4).

(4) I reproduce the proof by Geymonat, **2**, pp. 212–13, and add the figure. An interesting proof, in symbolic language, with critical remarks, is in Peano, **6**.

The enunciation and the demonstration are valid for all sets and make no reference to particular sets, but it may help to have a figure in which the sets M and N are represented by segments (sets of points) and the correspondences are realized through projections (see fig. 1). By hypothesis a correspondence which we shall call α makes a part of N, N_1, correspond to M, while a correspondence β makes a part of M, M_1, correspond to N. Starting from α and β we want to construct a correspondence γ between M and N. With this object we should observe that in virtue of the correspondence β, a part of M, M_2, corresponds to N_1. The transformation $\alpha\beta$ allows us to establish a correspondence between M and M_2. Our aim would

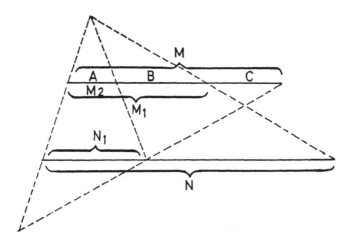

Fig. 1.

be achieved if we knew how to establish a correspondence between M and M_2, since we have already been given the correspondence between M_1 and N. Say:

$$M_2 = A; \qquad M_1 - M_2 = B; \qquad M - M_1 = C$$

And therefore we know the correspondence between

$$A + B + C \text{ and } A$$

We wish to construct the correspondence between

$$A + B + C \text{ and } A + B$$

With this object we observe that in virtue of the correspondence $\alpha\beta$, according to which A corresponds to $A+B+C$ and always in virtue of $\alpha\beta$, three sets, A_1, B_1 and C_1, contained in A, correspond to A, B and C. To A_1 B_1 and C_1 correspond, according to $\alpha\beta$, three sets, A_2, B_2 and C_2 contained in A_1, and so on. We will indicate with D the common part (if any) of all the sets A_i which could also be an empty set. Then we have:

$$M = A+B+C = B+C+B_1+C_1+B_2+C_2+\ldots+D$$
$$M_1 = A+B = B+C_1+B_1+C_2+B_2+C_3+\ldots+D$$

The correspondence we are looking for is achieved in the following way:

$$B \rightleftarrows B$$
$$C \rightleftarrows C_1$$
$$B_1 \rightleftarrows B_1$$
$$C_1 \rightleftarrows C_2$$
$$B_2 \rightleftarrows B_2$$
$$C_2 \rightleftarrows C_3$$

$$\ldots\ldots$$

$$D \rightleftarrows D$$

The theorem is thus demonstrated.

These considerations do not exclude the case of non-comparable sets M and N, for which it is not possible to establish which of the three cases is verified: $m=n$, $m<n$, $m>n$ (trichotomy).

All the same, examples of non-comparable sets have never been given.

A set is called *denumerable* when it can be placed in biunivocal correspondence with the series of natural numbers. An infinite set, contained in a denumerable set, is denumerable. In fact let I be the infinite set contained in the denumerable set I'. The latter can, by hypothesis, be placed in biunivocal correspondence with the natural series; the elements of I' can therefore be represented by $a_1, a_2 \ldots a_3$. Going through this succession we meet a first element of I, a second element of $I \ldots$ and so on to infinity; I is therefore denumerable.

Theorem. *The set of rational numbers is denumerable* (5).

We first demonstrate the result for the positive rational

(5) Calò, pp. 521–32; for Liouville's proof, see pp. 517–24.

numbers. Let us consider all the irreducible fractions p/q (p and q not less than unity). We will call the sum $p+q$ the *height* of the fractional number p/q. We put the rational numbers considered in aggregates of increasing height, where the elements of each aggregate of a given height are finite in number. The rational numbers of equal height will thus be arranged according to increasing values; we then will obtain a succession $a_1, a_2, a_3 \ldots$ of rational positive numbers in correspondence with the positive integers. If we wish to order all rational numbers, including zero, all we need do is consider the succession:

$$0, a_1, -a_1, a_2, -a_2, \ldots.$$

Theorem.   The set of real algebraic numbers is denumerable.

All the algebraic numbers are obtained as roots of algebraic equations with integral coefficients prime to one another, with $a_0 > 0$:

$$a_0 x^n + a_1 x^{n-1} + \ldots + a_n = 0 \qquad \text{(I)}$$

We call the height of an algebraic number, root of equation (I), the positive integer

$$h = (n-1) + |a_0| + |a_1| + \ldots + |a_n|$$

For a given value of h, h being the sum of positive integers, these can be chosen in a finite number of ways, and the distribution of the signs of $a_0 \ldots a_n$ can occur in a finite number of ways. To a given value of h therefore corresponds an aggregate of equations, finite in number; each equation has a finite number of roots. So the algebraic numbers can be placed in such a way that those of a lesser height precede those of a greater height, and the algebraic numbers of given heights (finite in number) will be placed according to absolute increasing values; similarly we can arrange the negative numbers, before these. In this way we can establish a biunivocal correspondence between real algebraic numbers and natural numbers.

Theorem.   Transcendental numbers exist.

We consider the succession of algebraic numbers constructed

according to the preceding theorem, where the numbers have been written in the form of decimals, adopting in the case where there exist two possible forms, one with an infinity of zeros and the other with an infinity of nines, one of these, for instance the first. It is easy to define a new number N which is none of the algebraic numbers of the former succession.

To do so all we need do is define N, taking an arbitrary integral part, the first decimal figure different from 9, and from the first decimal figure of the first algebraic number ... the nth decimal figure different from nine and the nth decimal figure of the nth algebraic number ... (it is easy to make precise the law on the basis of which the figures of N are determined).

The number N will not be the first algebraic number because it differs from it by the first decimal figure ... it will not be the nth because it differs from the nth decimal figure ... it will not be any of the algebraic numbers, it will be a transcendental number. Q.E.D.

Descartes alluded to transcendental numbers in a passage which we have quoted (Chapter XIII, §8). Their existence was demonstrated for the first time by Liouville (*Comptes rendus*, 1844) who used the expression of numbers as continued fractions. From the researches of Cantor on the properties of sets of algebraic numbers comes the demonstration we have given, in which substantially we follow Klein's exposition.

In 1874 Cantor demonstrated the following:

Theorem: The set of real numbers has a power greater than that of the denumerable.

The theorem is valid for the set I of real numbers contained within a certain interval, for instance between 0 and 1.

Let us first of all observe that the set of natural numbers can be placed in biunivocal correspondence with a part of I (for instance, the set of the inverses of the natural numbers). We must still demonstrate that it is not possible to establish a biunivocal correspondence between I and the series of natural numbers. With this object let us reason *per absurdum* : suppose we have managed to establish a biunivocal correspondence between I and the series of natural numbers. In the succession thus obtained of the elements of I, the real numbers are written in decimal form, avoiding the succession of infinite nines in the decimal figures, so that every number will be represented in one way only:

$$0, a_{11}\, a_{12} \ldots a_{1n}$$
$$0, a_{21}\, a_{22} \ldots a_{2n}$$
$$\cdot \quad \cdot \quad \cdot \quad \cdot \quad \cdot \quad \cdot$$
$$0, a_{n1}\, a_{n2} \ldots a_{nn}$$
$$\cdot \quad \cdot \quad \cdot \quad \cdot \quad \cdot \quad \cdot$$

Now let us define a number $0, b_1\, b_2 \ldots b_n \ldots$ which has the first digit b_1 different from a_{11} and from 9, b_2 different from a_{22} and from 9, and so on into the infinite (for instance: if $a_{nn} = 0$, $b_n = 1$, if $a_{nn} > 0$, $b_n = 0$). This number does not coincide with any of the numbers of the succession considered, which therefore does not include all the real numbers contained between 0 and 1, contrary to hypothesis.

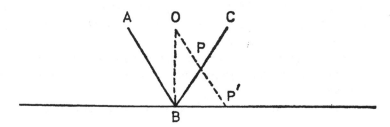

Fig. 2.

The set of real numbers contained between 0 and 1 is not denumerable but has a greater power. Q.E.D.

The power of the set of real numbers contained between 0 and 1 is called the power of the continuum.

The power of the points of a segment is equal to that of the continuum, since it is possible to establish, as we know from elementary geometry, a biunivocal correspondence between real numbers between 0 and 1 and points of a segment (taken as a unit of measure). The points of a straight line have equal power, since it is possible to establish a biunivocal correspondence between the points of a segment (excluding the extremes) and the points of a straight line (without the point at infinity) through a suitable projection (see fig. 2) (6).

(6) Geymonat, **2**, p. 279.

We cut the given segment into two equal segments, not belonging to the same straight lines, AB and BC, and project them from the middle point O of the segment AC on to the straight line through B perpendicular to OB. Through this projection, to every point P of AB or of BC the point P' of the perpendicular corresponds, and vice versa.

To every point of the straight line corresponds a relative real number from which it results that the set of all these has the power of the continuum.

In 1877 Cantor demonstrated that the linear continuum, the surface continuum, and the continuum in three dimensions all have the power of the continuum. In particular we demonstrate the following:

Theorem. The power of the set of points of a square is equal to that of the set of points of one of its sides.

Conducting the demonstration according to Waismann (7) we consider a square $ABCD$, whose points we refer to a system of co-ordinates in which AB is the axis of x, AD the axis of y, and AB the unit of measure.

In representing real numbers in decimal form so as to have a single way of writing the figures, we exclude the infinite succession of zeros.

At a point Q of the square correspond its co-ordinates:

$$x = 0, a_1 a_2 \ldots a_n \ldots$$
$$y = 0, b_1 b_2 \ldots b_n \ldots$$

We can make correspond to Q the point P of the side AB of the abscissa

$$t = 0, a_1 b_1 a_2 b_2$$

Vice versa, to every point P of the side AB we can make correspond a point Q of the square, carrying out the inverse procedure.

One inconvenient feature must be removed, though. Suppose, for example, that

$$t = 0 \cdot 330303 \ldots$$

Applying the preceding process we get:

$$x = 0 \cdot 300 \ldots$$
$$y = 0 \cdot 333 \ldots$$

(7) Waismann, 2, pp. 225–9.

But x would then be written in an inadmissible form. The difficulty is overcome by interpreting, as Koenig does, the digits or the sets of digits built up by a significant digit and from the possible zeros that precede it, as $a_1, a_2 \ldots b_1, b_2 \ldots$

In the example shown we have:

$$t = 0. \underset{a_1}{\lfloor 3 \rfloor} \; \underset{b_1}{\lfloor 3 \rfloor} \; \underset{a_2}{\lfloor 03 \rfloor} \; \underset{b_2}{\lfloor 03 \rfloor} \ldots$$
$$x = 0 \cdot 303 \ldots$$
$$y = 0 \cdot 303 \ldots$$

The correspondence we have studied is biunivocal but discontinuous. To work it out let us consider the value $t = 0 \cdot 5$, which, according to our convention, should be written in the form $t = 0 \cdot 4999 \ldots$

To this value t corresponds a point $Q(\frac{1}{2}, 1)$. Let us approach the value of t considered through a succession which has the limit $\frac{1}{2}$:

$$0 \cdot 5111 \ldots \qquad 0 \cdot 5011 \ldots \qquad 0 \cdot 5001 \ldots$$

The points of the square corresponding to the value of t of this succession do not lead to Q but to a point $R(\frac{1}{2}, 0)$. From this we see that the correspondence is not continuous.

Having established the biunivocal correspondence between points of the square and points of its side it is easy to extend this to the correspondence between points of a cube and those of its edge, and to the case of continuous sets of different dimensions.

Thus we establish that the character of a set of points in one or more dimensions does not involve the power of the set, but the order in which the points have been taken. Leibniz saw order in space, above all (8). According to Poincaré, Menger, and Urysohn the number of the dimensions of a figure has a topological character (9).

We must return to the set of real numbers. We have already seen, in the scholium to Euclid's Book X on the tragic fate of the man who revealed the secret of irrationals, that the ancients guessed at the existence of an inexpressible *quid* in the continuum. Descartes has a similar idea in a clearer form, as

(8) Sierpinski, p. 72.
(9) Poincaré, **2**, pp. 69–75; **4**, p. 65 ff., Waismann, **2**, pp. 234–5.

we have seen. Then there is Cantor's theorem, made clear by Richard and by Borel: 'There exist in the geometrical continuum (if it is not an abuse to use the verb 'exist') elements which cannot be defined'(10).

In fact let us consider a system of linguistic signs in a specific language, finite in number (letters, accents, punctuation marks, etc.) and the dispositions with repetitions of these signs, one by one, two by two...n by n.... The set of these dispositions is denumerable. Very many of these dispositions of signs do not make sense, but among those that do we may distinguish those which were definable as real numbers. Some of these definitions determine the real numbers themselves. But the infinite set of these numbers cannot have a greater power than the set of the definitions which is a part of a denumerable set and therefore denumerable. We conclude therefore that a set of real definable or expressible numbers is denumerable. But we know from a theorem of Cantor's that the set of real numbers has the power of the continuum which is greater than that of the denumerables. So real numbers exist which do not belong to the set of expressible numbers.

B. Levi observes (11): 'Starting from the aggregate of rational numbers, we can generate *some real numbers* but not the *aggregate of real numbers*. If for convenience we confine ourselves to speaking of Dedekind's method of cuts ... the notion of 'cut' is an *intuition* identifiable, in the simplest way, with that of a point of a Euclidean straight line. ...

'*In every mathematical theory,*' he continues, '*there are supposed to be certain aggregates, for each of which is postulated the possibility of choosing (fixing) elements arbitrarily, with the act of a single thought, which cannot be interchanged with, or reduced to, other and simpler elements.* I have called these aggregates the *first aggregates of the theory dealt with* and say that they define the *deductive dominion* in which it is developed. Thus ... the deductive dominion of elementary geometry is defined by *space*, a first aggregate of points; the greater part of the arithmetical theories developed in the definite *deductive dominion* from the first aggregate of *whole numbers*. ... Instead,

(10) Geymonat, **2**, pp. 292–4. The words between parentheses are Borel's, and refer to the existence of the sets we have hinted at.
(11) Levi, **3**, pp. 64–7; interesting remarks on Zermelo's axiom.

ordinary analysis has necessarily as its deductive dominion that in which the aggregate of *real numbers* comes first : all the same there exists a part of this analysis (for instance the calculus of variations or more generally the functional calculus) which cannot be contained in this dominion and which presupposes, as its first aggregate, the *aggregate of the functions* (possibly limited by some restrictive condition).'

We have met denumerable sets of gradually increasing size : whole numbers, rational numbers, algebraic numbers, expressible numbers.

Does an intermediate power exist among the power of the denumerables and the power of the continuum? Do sets of powers, that is, superior to that of the denumerables and inferior to that of the continuum, exist? This problem, called the problem of the continuum, which Sierpinski considers one of the most important and difficult problems of the theory of sets, has not yet been solved (12).

On the other hand it is easy to demonstrate the :

Theorem. Given a set it is always possible to construct one of a higher power.

The demonstration is a generalization of the one on the non-denumerability of the continuum.

Let I be the given set. One function of the set I is defined in the following way : to every element x of I corresponds an element y. Let us call f this function. I say that the set F of these functions is of a power superior to that of I. In fact, it can be seen at once that a biunivocal correspondence can be established between I and a part of F. To demonstrate that I and F have unequal powers, let us suppose that we can make a certain function $f_x(x)$ of F correspond to every element x of I. We will demonstrate that one can construct, contrary to the hypothesis, a function which is not one of the functions f_x considered. In fact we form a function $\varphi(x)$ in the following way : let φ for every x be different from $f_x(x)$. (It is possible, reasoning from any set whatever, to establish an exact rule for the functions $\varphi(x)$. Let us take two different objects : for example, u and v ; if $f_x(x) = u$, $\varphi(x) = v$, if $f_x(x) \neq u$ $\varphi(x) = u$.)

The function $\varphi(x)$ so constructed is certainly different from any f_x. So it is impossible to establish a biunivocal correspondence between the elements x of I and the functions f_x of F. Q.E.D.

Until now we have reasoned with infinite cardinal numbers. We will now deal with *transfinite ordinals*.

A set is called *ordered* when there exists among its elements a relationship expressed by the verbs to *precede* and to *follow* such that:

1. Given two elements a and b of the set it can always be established whether a precedes b or b precedes a; in the case of ordered sets, one of these cases always occurs and the one excludes the other; if a precedes b we say that b follows a.

2. If a precedes b and b precedes c then a precedes c.

We say that two sets are similar and belong to the same ordinal type if their elements can be placed in an ordered biunivocal correspondence which changes a generic element a that precedes b into an element a' that precedes b'.

An ordered set is called *well ordered* when every one of its parts has a first element.

'The concept of the ordinal type of a well-ordered set of an infinity of elements is identical with the notion of infinite ordinal number' (F. Severi).

Two sets of elements with equal power can have a different ordinal type (for instance, whole numbers and rational numbers which are ordered in increasing order have a different ordinal type). This cannot occur for finite sets.

Now let us examine the construction of Cantor's transfinites. Let us consider the succession:

$$1, 2 \ldots \omega$$

where ω represents the actual numerical infinite of the natural series. We can, besides, consider the succession:

$$\omega + 1, \omega + 2 \ldots 2\omega$$

and thus we can continue until we reach:

$$\omega^2, \omega^3 \ldots \omega^\omega$$

The transfinites considered are susceptible of an expressive geometric image (13), at least as far as the power of ω with a finite exponent. Let us consider the points of a straight line with whole numbers as their abscissae; we project this straight line in such a way that its point at infinity corresponds to a point at a finite distance on another straight line. Thus we have on a finite segment s a succession of points corresponding to $1, 2 \ldots \omega$. If, on the extension of s we make a succession of points congruent to the preceding, we will get the image of $\omega + 1, \omega + 2 \ldots 2\omega$, and so on.

ω^2 will now correspond to the point at infinity of the straight line. With a new projection we can take the image of ω^2 to be finite, and so on. . . .

To go back to the concept of well-ordered sets: real numbers taken in increasing order are not well ordered (for example, the real numbers greater than 1 have no first element).

Yet E. Zermelo, using a postulate he enunciated in 1904, which is known by his name, managed to demonstrate that the continuum can be well ordered. The postulate in question (14), which provoked interesting discussion at a high level, was enunciated in the following way: 'For every set M of which the elements are the non-empty sets P, without common elements, there exists at least one set N which contains one element, and one only, of each set P that belongs to M'.

What this postulate really means is that it is reasonable and possible to make an infinite choice without formulating a rule about the way in which the choice should be made.

Mathematicians with realist tendencies (cf. Chapter X, §5) agreed with Zermelo's postulate, which applies not just to the theory of sets but to analysis as well; but it has not been accepted by all mathematicians. Before Zermelo enunciated his postulate, Peano had already declared he was against the principle of infinite choice.

Cantor's theory of sets removes the paradoxes of the infinite (especially those relating to the axioms of inequality) but does not eliminate them completely (15). Disconcerting contradictions were still to appear, which we shall discuss later.

(13) Enriques, **9**, p. 270.
(14) Enriques, **9**, p. 271–2; Sierpinski, pp. 103 ff.; Cassina, 7; Geymonat, 2, pp. 215–16 and 301–5.
(15) Enriques, **12**, p. 244.

§2. *Functions and curves* (16)

The Greeks defined the curve, as Euclid did, as 'a length without width', and already knew a certain number of algebraic and transcendental curves, which we have met in studying the history of ancient mathematics.

The analytical geometry of Fermat and Descartes enormously widened the field of known curves, and there the concept of function was formed: Leibniz introduced the term in 1692.

At the time of Euler, who defined the function as 'an expression of calcul', it was thought that while a curve corresponded to every function (formed by means of algebraic or transcendental elementary operations), a function did not, vice versa, correspond to every curve (arbitrarily drawn); so it was thought that the concept of curve was more general than that of function.

Then under the influence of Fourier series (1807) the concept of function was generalized, and finally expressed by Dirichlet in the following way: 'A function of a variable x, in a certain field, is a quantity y which assumes a well-determined value for every value of x.' It was then realized that a curve in the intuitive sense of the word did not correspond as a diagram for all functions.

But how can this intuitive concept of curve be made more exact? Obviously there is always something arbitrary about any effort at precision. About 1880 C. Jordan suggested a definition that seemed to correspond fairly well to the intuitive idea of curve. This definition considered a curve as traced by a mobile point in a continuous movement. More precisely, the plane curve is the locus of the points defined by the parametric equation:

$$x = \varphi(t),$$
$$y = \psi(t),$$

where φ and ψ must be continuous one-valued functions.

Jordan's definition of a curve was at first accepted by all mathematicians, but in 1890, to everyone's astonishment, Peano showed that a curve could exist which conformed to Jordan's definition but which completely filled a square.

(16) Enriques, **12**, pp. 238–41; Waismann, pp. 209–36. On the topological meaning of the dimensions of a space, see Poincaré, **2**, pp. 69–75; **4**, p. 65 ff.

We will give the construction of Peano's curve with the simplifications suggested by Hilbert.

Divide a given square into four equal squares by joining the middle points of its sides, and do the same thing to the squares obtained and so on indefinitely. The given square will thus be divided into 4, 16, 64...$(2^n)^2 = 2^{2n}$ small squares.

Divide into the same number of parts the time during which we wish to make the point describe the whole surface. For a generic n we establish a correspondence between each small interval of time and each small square. The small squares must be taken in a certain order in such a way that they are all joined to one another, once only, and two successive squares always have a common side. For a generic n the line which

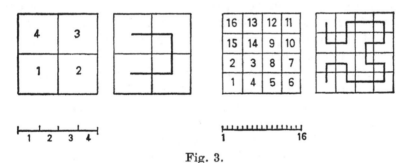

Fig. 3.

approximates to Peano's curve is given by the piece that joins the centres of the small squares taken in a determined order. Making n tend to infinity, the piece by definition tends to Peano's curve.

Let us indicate in fig. 3 the order of the small squares and the relative pieces for $n = 4$ and $n = 16$ from which comes the law of formation of the pieces in question for the successive values of n.

Peano's curve has the following properties:

1. To every value of t, or to every point of the segment which we make correspond to the interval of time during which the curve considered describes the square, a point in the square corresponds. In fact: if t always falls on the inside of the small intervals obtained in the successive divisions, to this succession of intervals in which the successor is always interior to the

preceding one there corresponds in the square a succession of small squares with similar properties which determine in the limit a point Q of the square corresponding to t. If t falls in one of the extremes common to two small intervals, we proceed as in the previous case, with the difference that instead of a simple small interval we consider pairs of small intervals separated by t. To these pairs will correspond pairs of small squares with one side in common, which in the limit will also define the point corresponding to t. In the sense from t to Q the correspondence is univocal.

2. The curve passes through all the points of the square; that is, to every point Q of the square corresponds at least one value t. If the point Q is always on the inside of the small squares of the subdivisions, the value t corresponding to Q is found through an inverse process to that followed to find Q, given t. If instead Q is found above a side or above a vertex of one of the small squares of the subdivision, the two and the four contiguous small squares to which Q belongs generally do not correspond to consecutive small intervals relating to t. Thus more values of t corresponding to Q will be found. In the sense from Q to t the correspondence is not univocal. The curve is therefore interwoven.

3. If we refer Peano's curve to a system of axes x and y, the parametric equations will be in the form $x = \varphi(t)$, $y = \psi(t)$, where from the construction of the curve, it results that $\varphi(t)$ and $\psi(t)$ are continuous functions.

We conclude that Peano's curve, although it fills the square, satisfies Jordan's definition.

It is interesting to compare the two correspondences we have examined between the points of a square and the points of a segment: Cantor's correspondence is biunivocal but is not continuous, Peano's is continuous, univocal in one sense, but not biunivocal.

Might we, in a third way, establish a correspondence, biunivocal and continuous at one and the same time, between points Q of a square and the values of a parameter t?

It has been demonstrated that this is impossible. We follow Jürgens' demonstration, reasoning *per absurdum*. Let us suppose (see fig. 4) that we have established a biunivocal and continuous correspondence between the points of a square and the points of a segment. Q_1 and Q_2, points of a square, correspond respectively to the values t_1 and t_2 of the parameter.

If we go from Q_1 to Q_2, following a continuous curve c, the parameter will pass with continuity from t_1 to t_2 through all the values between t_1 and t_2. The same thing will happen if we go from Q_1 to Q_2 by another continuous path c'. Meantime two points of the square, one on the curve c and the other on c', will correspond to a value of t contained between t_1 and t_2. The correspondence therefore is not biunivocal. Q.E.D.

There are other singular cases concerning the tangents to continuous curves, and corresponding to the derivative of a function.

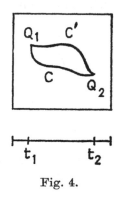

Fig. 4.

If a function is derivable it is also continuous (in fact if $\lim\limits_{\Delta x \to 0} \dfrac{\Delta y}{\Delta x}$ is a determined and finite quantity then

$$\lim_{\Delta x \to 0} \Delta y = \lim_{\Delta x \to 0} \frac{\Delta y}{\Delta x}\Delta x = 0).$$

But is the converse true? It was once thought that all continuous functions were derivable. But it has been established that this is not true. Bolzano made a discovery in this field (17).

Especially interesting is the curve which is everywhere continuous and everywhere without a tangent. Weierstrass gave an example of such curves; and H. von Koch a simpler one which we will reproduce here. Let us consider a segment

(17) Severi, 4.

AB (see fig. 5) and divide it into three equal parts at the points *C* and *D*.

Let us build on the base *CD* an equilateral triangle *CDE*, then suppressing *CD*. Let us do the same on the four segments *AC, CE, ED* and *BD*, and so on. If we continue this indefinitely the curve which we get in the limit will be the one we want. In fact, as the construction shows, the curve to which we thus approximate will be continuous, but it is easy to see that it has no tangents. To see this more clearly let us examine if the curve can have any tangent at *A*. Let us take a straight line that joins *A* to another point *P* on the curve, and then make *P* approach *A* along the curve; the straight line joining *A* with *P* will always oscillate between the extreme positions *AB* and *AE*, without tending to a limiting position.

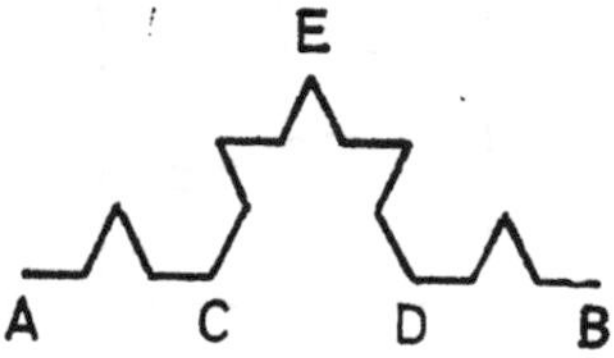

Fig. 5.

Similar results may be obtained for the other points of the curve which, although it is continuous, has no tangent at any point (18).

§3. *Postulates of continuity and non-Archimedean geometry* (19)

Their work on the infinitesimal calculus has led mathematicians to theories of real numbers, such as those of R. Dedekind (1831–1916), which are linked directly, as we have seen, with the Euclidean definition of equal ratios.

The postulate of continuity in Dedekind's form is in line with this theory of real numbers.

'If a segment of a straight line *AB* is divided into two parts in such a way that:

1. every point of the segment *AB* belongs to *one* of the two parts;

(18) A bibliography on this topic is in Pascal, pp. 98–103.
(19) Dedekind, **3**, Vitali, Enriques, **9**. The statements between quotation marks are from Vitali. For the 'monosenii' of Levi-Civita, and the interpretation of non-Archimedean numbers, see Enriques, **9**, pp. 373 and 377–83.

2. the end A belongs to the first part and B to the second;
3. any point on the first part precedes any point on the second, in the order AB of the segment:
   there exists a point C of the segment AB (which can belong to one part or the other), such that every point of AB which precedes C belongs to the first part and every point of AB which follows C belongs to the second part of the established division.'

This postulate can be put in a similar form by substituting the segments with a common origin for the points considered.

Many modern proofs of existence are based on Dedekind's postulate (as we have seen, in the ancient world the proofs of existence were based on an effective construction).

Cantor's theory of real numbers, however, is in line with the postulate of continuity, in the following form (which is, in fact, called Cantor's):

'If two classes of segments of straight lines are such that:
1. no segment of the first class is greater than any segment of the second;
2. having chosen a segment σ as small as we please, there exists a segment of the first and one of the second class for which the difference is less than σ; there exists *one* segment which is not less than any segment of the first class or greater than any of the second.'

The postulate can be enunciated in a similar way by referring suitably to the end points of the segments.

We might be tempted to think that the two postulates are substantially equivalent, but there is in fact a substantial difference between them: from Dedekind's postulate we can deduce the postulate of Archimedes, whereas we can construct a class of magnitudes to satisfy Cantor's postulate but not that of Archimedes.

Having admitted Dedekind's postulate we can prove the following, with Stolz:

Theorem. '*Given two segments there always exists a multiple of one greater than the other.*'

Let AB and AC be the two segments, on the same straight line with an end in common, and let $AB < AC$ (see fig. 6).

We must prove that there exists an integer n such that

$$n.AB < AC$$

Let us suppose that this is not true. Then there will exist, contained in the segment AC, some segments AH (with H different from A) such that for no n will we have $n.AH > AC$; on the other hand, there will exist some points K (at least the end C) such that $n.AK > AC$. This repartition of the points H and K satisfies the conditions of Dedekind's postulate, and so there will exist a point M such that the points of AM belong to the first class and the points of MC to the second. Let us take in MC (in the second class) a point Y such that $MY < AM$.

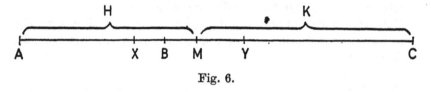

Fig. 6.

The mid-point X of AY will therefore fall at AM and will belong to the first class.

We can find an n such that

$$n.AY > AC$$

but $AY = 2AX$, so we conclude:

$$2n.AX > AC$$

which is absurd, because X belongs to the first class.

We now construct a set of magnitudes, continuous according to Cantor, but non-Archimedean.

Let us consider with Veronese (1854–1917) a system of infinite parallel straight lines succeeding one another at constant intervals (see fig. 7).

We order the points belonging to the two straight lines in the following way: if the point A is on a straight line above that on which B is found, A precedes B; if two points C and D belong to the same straight line and C is on the left of D, then C precedes D.

The set of points of the straight lines considered is thus ordered like the set of points of an open line, and in it one can define the congruence of finite or infinite segments, since it is possible to apply the system to itself by translating the plane, taking one point on to any other. This system satisfies the postulates of order and congruence, and, as an immediate result, Cantor's postulate too: but it does *not* satisfy the postulate of Archimedes nor that of Dedekind (since it would then satisfy Archimedes' also).

It can be shown too that Cantor's and Archimedes' postulates are logically equivalent to Dedekind's, but for reasons of space we will not give this demonstration.

This is one of the simplest examples of a non-Archimedean continuum; non-Archimedean magnitudes developed for a wide variety of reasons which can be found in the ancient world—among them Euclid's discoveries on the angle of contingency and Archimedes' own reflection on the postulate called after him. The work done on the angle of contingency

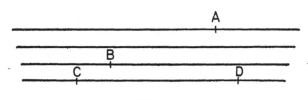

Fig. 7.

(also by Giordano Nemorario and by Campano) in the direction of the infinitesimal calculus induced Newton to compare these angles by means of the derivatives of higher order of the functions of which the curves in question are graphs. But from another point of view the angles of contingency, considered beside rectilinear angles, give us elements of a class of non-Archimedean magnitudes.

Many non-Archimedean theories have been developed, such as Veronese's and Levi-Civita's.

Interpretations of non-Archimedean numbers have been made apart from those of the angles, between two curves, as we have already mentioned, from Du Bois Reymond's orders of infinity, and from certain cases of probability in the continuum.

In so far as the order of infinity is concerned we recall that

313

being given two functions $f(t)$ and $\varphi(t)$, increasing beyond all limits, for example as $t \to \infty$, we can have three cases :

$$\lim_{t \to \infty} \frac{f(t)}{\varphi(t)} = \infty, \ \lim_{t \to \infty} \frac{f(t)}{\varphi(t)} = \text{const.} \ (\neq 0), \ \lim_{t \to \infty} \frac{f(t)}{\varphi(t)} = 0$$

In the first case the order of infinity of f is greater than the order of infinity of φ, in the second it is equal, and in the third it is less.

The infinities of the functions $a_1 t + a_0, a_2 t^2 + a_1 t + a_0 \ldots$ are measured respectively by the numbers $1, 2 \ldots$

The constant a_0 can be considered of order 0, the functions $\frac{1}{t}, \frac{1}{t^2} \ldots$ of the negative orders $-1, -2 \ldots$ But

$$\lim_{t \to \infty} \frac{e^t}{t^n} = \infty$$

whatever n is, therefore the order of infinity of e^t can be considered a non-Archimedean number.

In so far as the example taken from the calculus of probability is concerned, let us suppose that we let fall a point P on to the inside of a square, and that there, given two equivalent areas, or two lines of equal length, P is equally likely to fall on either. Having established this, we consider the probability that P falls in a space made up of a surface S with a boundary included, and a curve, without points in common. This probability can be expressed by a number of the form

$$a + b\eta$$

where $0 < a < 1$ and η is actually infinitely small.

CHAPTER XVIII

Symbolic Logic and the Fundamentals of Arithmetic

§1. *The 'characteristica' of Leibniz and the followers of Leibniz's logical tradition*

Symbolic logic, although some, Raymond Lully among them, foreshadowed it (1), substantially began with Leibniz.

The use of an adequate symbolism, *characteristica universalis*, suitable for expressing logical relations, should, according to Leibniz, constitute the basis of a logical algebra, *calculus ratiocinator*, which could be applied to all rational knowledge.

In a project for an encyclopaedia Leibniz wrote: 'De iudice controversarium humanarum, seu Methodo infallibilitatis et quomodo effici possit, ut omnes nostri errores sint tantum errores calculi, et per examina quaedam facile possint justificari'(2).

That is, the *characteristica* should become the judge of all human disagreement, and all our mistakes should be reduced to errors of calculation, which careful examination would easily correct.

When two philosophers had an argument to settle, a calculation should be able to decide between them:

'Quo facto, quando orientur controversiae, non magis disputatione opus erit inter duos philosophos, quam inter duos computistas. Sufficiet enim calamos in manus sumere sedereque ad abbacos et sibi mutuo (accito si placet amico) dicere, calculemus!'(3).

Leibniz's wish to construct the *characteristica* was only

(1) On the ancient history of symbolic logic, see Enriques, **2**, p. 91: following Couturat, Philodemus is named, for the classical times, and also Raymond Lully, Dalgarnus (1661), Wilkins (1668), pp. 96–8 refer to the 'characteristica' of Leibniz. See also Bocheński, **4**.
(2) Couturat, p. 98.
(3) Couturat, p. 98.

partly fulfilled by himself; we have already met the best realization of his idea in the symbolism of the differential and integral calculus. His other work concerns logical operations proper (4), such as the property of the sign of negation, the identity of the sign of deduction among classes and propositions, some similarities between logical relations and propositions on the divisibility of integers, use of the so-called circles of Euler, etc.

Leibniz's researches on formal logic have an important place in the framework of his philosophy (5): his wish (criticized by Locke) to construct science on the basis of identical propositions, his proposal to resolve every concept into its simplest elements, his search for criteria for the coherence of a system, his theory of monads.

Unfortunately, interesting though they are, we have not the space to go into them here.

Leibniz's idea of finding the simple concepts that constitute the elements of knowledge was taken up by Lambert, whom we have already met with among the forerunners of non-Euclidian geometry; the work of J. A. Segner (1740) and of other thinkers of the time was also inspired by Leibniz (6).

The idea of using a better logical symbolism instead of ordinary language reappears independently among the English-speaking logicians of the nineteenth century: A. de Morgan (1806–76), G. Boole (1815–64), W. S. Jevons (1835–82), and C. S. Peirce (1839–1914), while with the developments of E. Schröder we make contact with Leibniz again (7).

§2. *The mathematical logic of Peano* (8)

Although it would be interesting to trace in detail the development of the ideas and logical symbolism in these authors, it seems better, from the point of view of this book, to go straight on to one of the most perfect formulations of symbolic logic, that of G. Peano (1858–1932), who begins his study of

(4) Vacca, **2**.
(5) Enriques, **2**, pp. 85–105, 192–3.
(6) Enriques, **2**, pp. 110–12.
(7) On the history of symbolic logic, see Enriques, **2**, pp. 174–6; Levi, **4**; Bocheński, **1**, **2**, **3**, **4**; Church, **1**, **2**.
(8) Peano, **3**, and **5**, **4**, pp. III and 1–7. On Peano's work, see Cassina, **2**, **3**, **9**.

symbolic logic with his *Geometrical calculus according to the Ausdehnungslehre of H. Grassmann, preceded by the operations of Deductive Logic* (Turin, 1888).

Peano declares himself inspired by Schröder's *Der Operationskreis der Logikkalkuls* (Leipzig 1877), but says that he has changed those of *Schröder's* symbols, which might cause confusion between the signs of logic and those of arithmetic. Peano's signs are ∩, ∪, −A, ○, ◉, which he substituted respectively for Schröder's signs ✕, +, A_1, 0, 1.

We will now deal with Peano's calculus of classes.

Consider a system of entities (for instance real numbers), and indicate by A, B,... the classes of this system.

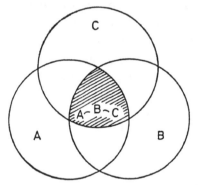 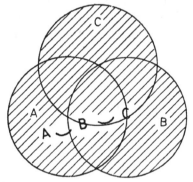

Fig. 1.

When we write $A = B$ we indicate the identity of two classes. For instance 'rational number' = 'number that can be developed in a finite continuous fraction'.

When we write $A \cap B \cap C$... or else ABC, we indicate the largest class contained within the classes A, B, and C, ... that is the class formed of all the members that are at once A and B and C. The sign ∩ means 'and'; the operation in question is called *logical conjunction* or *logical multiplication*. For instance, 'multiple of 6' = 'multiple of 2'∩'multiple of 3'.

$A \cup B \cup C$, *disjunction or logical sum*, indicates the smallest class that contains the classes A B C, that is the class formed by the members which are either A, or B, or C... The sign ∪ means 'or'; for instance 'rational number' = 'integer' ∪ 'fraction'.

Schematic figures of the type drawn here to illustrate the

sum and the logical product are not necessary, but may be useful to clarify concepts and formulae of the calculus of classes.

The sign $-A$ or $\overline{A}$ indicates the class formed by all the members which are not A. For instance, ' $-$ rational' = 'irrational'.

The sign ● indicates all the system of members considered, while ◯ indicates nothing, the empty class, successively indicated by $\wedge$.

$A = \bigcirc$ means: there is no A

$AB = \bigcirc$ means: no A is B

$A\overline{B} = \bigcirc$ means: there do not exist members which are A and contemporaneously are not B: that means, every A is B.

The following fundamental identities which are obvious in themselves are used for the calculus of classes; some of them can be deduced from the rest, which are taken as axioms:

1. $AB = BA$

1'. $A \cup B = B \cup A$

2. $A(BC) = ABC$

2'. $A \cup (B \cup C) = A \cup B \cup C$

3. $AA = A$ (9)

3'. $A \cup A = A$

4. $A(B \cup C) = AB \cup AC$

4'. $A \cup (BC) = (A \cup B)(A \cup C)$

5. $A● = A$

5'. $A \cup \bigcirc = A$

6. $A\bigcirc = \bigcirc$

6'. $A \cup ● = ●$

7. $-(-A) = A$

8. $-(AB) = (-A) \cup (-B)$

8'. $-(A \cup B) = (-A) \cap (-B)$

9. $A \cap -A = \bigcirc$

9'. $A \cup -A = ●$

10. $-\bigcirc = ●$

10'. $-● = \bigcirc$

$A < B$ means that every A is B, in which case we also write that $B > A$.

If we operate on a variable class X and on other classes considered as fixed, through the signs $\cap \cup -$ we obtain a new variable class which is indicated by $f(X)$.

$f(X)$ is written in a separate form when it is expressed in the form

$$f(X) = PX \cup Q\overline{X}$$

where P and Q are independent of X.

(9) Formula (3) does not hold in V. A. Pastore's logic (1868–1956): it is substituted by: $a.a = a^2$, as in algebra, and generally by $\dfrac{a}{1} \dfrac{a}{2} \cdots \dfrac{a}{n} = a^n$, where a^n means that we are dealing with n entities, of which one is a, and we consider only the one which is different from the $n-1$ others, though it depends on them. See Pastore.

Every logical function f(X) can always be put in the separate form indicated.

In fact if A is independent of X we can always write:

$$A = AX \cup A\bar{X}$$

Besides:

$$X = \bullet X \cup \circ \bar{X}$$

It is also stated that the sum of two separate expressions and the product of two separate expressions are still separate expressions. Moreover, the negation of a separate expression is still a separate expression: it is concluded that the thesis is valid.

We now go on to the calculus of propositions. A relation between determinate entities constitutes, according to Peano, a *categorical proposition* which is true or false, whereas a relation which contains variables and is verified by certain entities but not by certain others is called a *conditional proposition*. This, for instance, is a *conditional proposition*:

$$x^2 - 3x + 2 = 0$$

whereas this is a categorical proposition: 'the equation $x^2 - 3x + 2 = 0$ has 1 and 2 for roots'.

If α is a conditional relation or proposition 'x:α' indicates the class formed by all the elements for which the proposition α is true.

If the proposition α contains several variables x and y then $(x, y \ldots)$:α indicates the class formed by all the elements $(x, y \ldots)$ for which α is true.

Sometimes to simplify the writing we omit the sign $(x, y \ldots)$ when we consider as variables all the letters which appear in the formula.

There are ways of treating propositions, too, similar to those for the classes.

For propositions of the type $A = B$, and $A < B$, $A > B$ there are the following identities:

(I) $(A = B) = (B = A)$
(II) $(A < B) = (A\bar{B} = \circ)$
(III) $(A > B) = (B < A)$
(IV) $(A = B) = (A > B) \cap (A < B)$

319

The identities (I) and (III) mean that 'Every logical equation is transformed into another equal one where the two members are exchanged, and the signs $=$ $<$ $>$ that join them are changed into $=$ $>$ $<$.'

It is equally easily demonstrated and is, besides, obvious that 'From a system of logical equations, all true, and all containing the same signs '$=$' or '$<$' or '$>$' are deduced new equations which are true by multiplying both the members by the same factor and adding the same term, or adding them up member by member, or multiplying them member by member.'

The same is said for the identity

$$(A \cup B = \bigcirc) = (A = \bigcirc) \cup (B = \bigcirc)$$

The four fundamental propositions of classical logic, every A is B, no A is B, some A is B, some A is not B, can be expressed respectively with the formulae of Peano's logic:

$$A\bar{B} = \bigcirc, \quad AB = \bigcirc, \quad -(AB = \bigcirc), \quad -(A\bar{B} = \bigcirc)$$

On this basis we can reconstruct the theory of conversion and of the syllogism, with the variants in respect of the classical theory which we have already mentioned in dealing with Aristotle's logic (Chapter IV, §6).

We should note that it is always possible to transform a logical equation of the type $A < B$ or $A = B$ into another in which the second member is equal to $\bigcirc$.

In fact

$$(A < B) = (B > A) = (A\bar{B} = \bigcirc)$$

and

$$(A = B) = (A < B) \cap (A > B)$$
$$= (A\bar{B} = \bigcirc) \cap (\bar{A}B = \bigcirc)$$
$$= (A\bar{B} \cup \bar{A}B = \bigcirc)$$

If in a given equation an unknown class x appears, this can be written in the form $f(X) = \bigcirc$ which can be written in the separate form

$$(AX \cup B\bar{X}) = \bigcirc$$

that is

$$(AX = \bigcirc) \cap (B\bar{X} = \bigcirc)$$

which can also be written

$$(X < \bar{A}) \cap (B < X)$$

320

that is

$$B < X < \bar{A}$$

These latter relations give us the solution of the logical equation $f(X) = 0$.

We also get at once the condition of the solubility of the logical equation we have mentioned:

$$B < \bar{A} \text{ that is } AB = \bigcirc$$

'To eliminate from an equation (or system of equations) an unknown means to write, if it exists, an equation which no longer contains the unknown, but the other variables, expressing the condition that the equation proposed can be satisfied by some value of the unknown.'

The problem has been solved in the case of the equation $AX \cup B\bar{X} = \bigcirc$ to which any logical equation in a single unknown can be reduced.

In the case of several unknowns, one unknown is eliminated at a time, as in algebra.

Peano introduced other important logical symbols, among which we should remember $\in$, ɜ and $\supset$. The symbol $\in$ indicates the belonging of a member to a class, which in general is indicated by 'Cls'. For example 'I $\in$ Italian'. The symbol $\in$ is not transitive. For instance, from the proposition: Peter and Paul were Apostles, the Apostles were twelve, we cannot conclude: Peter and Paul were twelve. Expressing this by using the symbol $\in$ we shall have: (Peter and Paul) $\in$ Apostle. Apostle $\in$ (Class of 12 persons). We cannot deduce (Peter and Paul) $\in$ (Class of 12 persons).

According to Enriques (10) the fallacy of the syllogism depends instead on the ambiguity of the middle term (apostle) taken once as an abstract concept and the other time as a class.

The symbol $\in$ should not be confused with the sign $\supset$ which indicates that a class is included in another class. That is, if A and B are classes, $A \supset B$ means that every element of A is an element of B; e.g. triangle $\supset$ polygon. The sign $\supset$ contrary to $\in$, is transitive. E.g. Piedmontese $\supset$ Italian. Italian $\supset$ European. $\supset$ Piedmontese $\supset$ European. (Note that

(10) Enriques, **2**, pp. 187–8.

21   321

the third sign $\supset$ does not indicate inclusion among the classes, but implication through propositions.)

The symbol $\ni$ is defined thus: 'Let p be a proposition containing a letter x. Then '$x \ni p$', which is read 'the x's such that p' indicates the classes of the individuals which satisfy the condition p. $\ni$ is inverse of $\in$:

$$a \in \mathrm{Cls}. \supset .x \ni (x \in a) = a$$

The use of the two symbols $\in$ and $\ni$ is of the highest importance, since it allows us to pass from the calculus of classes to the calculus of propositions, and vice versa (11).

The signs which we have met in the calculus of classes are also used by Peano in the calculus of propositions in the following way.

If p and q are propositions containing the variables $x \ldots z$ the formula

$$p \supset_{x \cdots z} q$$

means: from p we deduce q, for any $x \ldots z$ (implication). The simultaneous affirmation of p and q is indicated by $p \cap q$, or by pq.

The affirmation of at least one of the two propositions p and q, that is p or q, is indicated by $p \cup q$.

If p is a proposition, its negation is indicated by $-p$.

We also owe to Peano a rational system of logic based on the following primitive propositions (Pp) expressed in symbolical form (12), where 'simpl' means simplifying; 'syll', syllogism; 'cmp', composing; 'export', exporting.

$a \in \mathrm{Cls}. \supset_a . a \supset a \;\; Pp$

$a, b, \in \mathrm{Cls}. \supset_{a,b} . ab \in \mathrm{Cls} \;\; Pp$

$\qquad\qquad\qquad . ab \supset a \;\; Pp$ ⎫
$\qquad\qquad\qquad . ab \supset b \;\; Pp$ ⎬ Simpl

$a, b \in \mathrm{Cls}. a \supset b . x \in a . \supset_{a,b,x} . x \in b \;\; Pp$ ⎫ Syll
$a, b, c \in \mathrm{Cls}. a \supset b . b \supset c . \supset_{a,b,c} . a \supset c \;\; Pp$ ⎭

(11) For the sake of completeness, we recall three other symbols of Peano: Let x be an element of a class; then, ιx is the class containing x, and only x. Let a be a class, containing one element: ιa is the element pertaining to the class a. Let x, y, z be any elements; then $x;y$ is the ordered pair, where x is the first element. So also with $x;y;z$.

(12) **Peano, 3**, t. II (1897), formulas 12–16, 21–7; 72; 105–7; pp. 26–35.

$\underline{\hspace{3cm}}.a \supset b.a \supset c.\supset_{a,b,c}.a\supset bc$ Pp Cmp

$a, b, c \in \text{Cls}: x \in a.x\,; y \in b \supset_{x,y}.x\,; y \in c\,\therefore\, \supset_{a,b,c}\,\therefore$

$\quad x \in a \supset_x: x\,; y \in b.\supset_y.x\,; y \in c$ Pp Export

$a \in \text{Cls}.\supset .(-a) \in \text{Cls}$ Pp

$\underline{\hspace{3cm}}. -(-a) = a$ Pp

$a, b, c \in \text{Cls}.ab \supset c.x \in a.x- \in c: \supset_{a,b,c,x}.x- \in b$ $Pp.$

Let us substitute for the signs $\supset$, $\cap$, $=$ among the classes their definitions (Df)

$$a, b \in \text{Cls}.\supset_{a,b}\therefore a \supset b. = .x \in a \supset_x.x \in b \qquad \text{Df}$$

$$\underline{\hspace{2cm}}.a \cap b = .x\,3\,(x \in a.x \in b) \qquad \text{Df}$$

$$\underline{\hspace{2cm}}\therefore a = b = a \supset b.b \supset a \qquad \text{Df}$$

The original propositions shown above should now be read as :

If p is a condition *then* from p we deduce p.

If p and q are conditions *then* their simultaneous affirmation is a condition, from which we can deduce p, and from which we can deduce q.

If p and q are conditions and from p we deduce q, and x is a solution of p, *then* x is also a solution of q.

If p, q and r are conditions and from p we deduce q and from p we deduce r, *then* from p we deduce the simultaneous affirmation of q and of r.

If p is a condition in x and q and r are conditions in x and y, and from the simultaneous affirmation of p and q we deduce r, whatever x and y are, *then* from p we get, whatever x is, that q implies r in respect of y.

If p is a condition, *then* its negation is also a condition.

If p is a condition, *then* the negation of the negation of p is equal to p.

If p, q and r are conditions and the simultaneous affirmation of p and q implies r, *then* the simultaneous affirmation of p and the negation of r implies the negation of q.

To these eleven basic propositions expressed above in a symbolic form, Peano adds another two expressed in ordinary language.

The twelfth proposition is *the principle of substitution of apparent variables*: one calls apparent variables the indices a (a, b), etc. which follow the symbol $\supset$; the formulae in which these appear remain still true if for the said variables any variable or constant symbols are substituted.

The thirteenth basic proposition is *the principle of the invertibility of the order of primitive propositions*.

U. Cassina remarks (13): 'Peano's eleven basic propositions, and the propositions demonstrated through them, take the form of deductions, and cannot be reduced to affirmations of *simple* propositions (that is, which are not the simultaneous affirmations of other propositions). The result is that to him *logic* appeared (in accordance with its etymological meaning) essentially the science that studies the forms of reasoning, that is as the *theory of deductions*: or else of propositions of the type "If *a* is true, then *b* is also true".'

In Peano's ideographic creation and in his translation of the propositions of logic into symbols, Leibniz's dream came true, in a way that Peano himself made clear (Introduction to volume two of the *Formulaire de Mathématiques, Révue des Mathématiques*, volume six, Turin, 1896–9):

'Après deux siècles, ce "songe" de l'inventeur du Calcul infinitésimal est devenu une réalité. . . . Enfin nous sommes arrivés à terminer l'analyse des idées de logique, exprimées dans le langage ordinaire par une foule de termes . . . en les exprimant toutes au moyen des idées représentées par les signes $\in, \supset, =, \cup, \cap, -, \wedge$, lesquelles sont encore réductibles.

'Dans le petit livre "Arithmetices principia, nova methodo exposita, a. 1889" nous avons pour la première fois exposé toute une théorie, théorèmes, définitions et démonstrations, en symboles qui remplacent tout à fait le langage ordinaire.

'Nous avons donc la solution du problème proposé par Leibniz. Je dis "la solution" et non "une solution" car elle est unique. La logique mathématique, la nouvelle science composée de ces recherches, a pour objet les propriétés des opérations et des relations de logique. Son objet est donc un ensemble de vérités et non de conventions.'

§3. *The foundations of arithmetic according to Peano* (14)

Peano's arithmetic is based on three primitive concepts, zero: 0, number: N_0, successive to a: $a+$, and on postulates of which follows the expression in symbols and the translation into ordinary language:

(13) Cassina, **4**.
(14) Peano, **4**.

·0 $N_0 \in \mathrm{Cls}$
·1 $0 \in N_0$
·2 $a \in N_0. \supset .a^+ \in N_0$
·3 $s \in \mathrm{Cls}.0 \in s : x \in N_0 \cap s. \supset_x. x^+ \in s : \supset N_0 \supset s$
·4 $a, b \in N_0.a^+ = b^+. \supset .a = b$
·5 $a \in N_0. \supset a^+ - = 0.$

·0 Number is a class . . .
·1 Zero is a number.
·2 If a is a number the successive of a is a number.
·3 Let s be a class; and let 0 be a member of this class; and
    let us suppose also that if x is a number belonging to this
    class s we deduce, whatever x is, that its successive x^+ also
    belongs to this class; then every number is an s.

This proposition is called the 'principle of induction' . . .

We may also read: 'If s is a property; if 0 has this property;
if every time a number has this property its successive also has
the same property; then every number has this property.'

·4 Two numbers followed by equal numbers are equal.
·5 The successive of a number is never equal to zero.

Starting from these logical principles and from the system of
postulates now expressed, Peano built up his arithmetic and
analysis with wonderful clarity and rigour. With the help of
his collaborators he translated an enormous amount of mathematical theory with his symbolism in several editions of his
valuable *Formulario*.

§4. *Critical considerations on the concept of number* (15)

In Peano's arithmetical system, the natural number appears
as a primitive concept. Can we say that this concept is
defined implicitly from the postulates in question? It has
been said that the axioms we have mentioned do not characterize completely the series of whole numbers; but every succession,
interpreting suitably the concepts of zero, number, and successor, satisfies the postulates.

It might appear that the difficulty could be overcome by
adding some new postulates, but (16): 'Skolem has demon-

(15) Frege, **2**, Waismann, **2**, pp. 145–68; Geymonat, **2**, pp. 188–91 and
202–9.
(16) Waismann, **2**, pp. 147–8; Ladrière, pp. 354–7.

strated that every hope of this kind is vain. He has in fact proved ... that no one will ever succeed in characterizing the series of integers with a finite group of axioms. Every valid enunciation in the arithmetic of natural numbers is also valid in systems which are quite different from it, and so it is absurd to try to discover some property internal to the series of integers which is in a position to distinguish it from any other series.'

So mathematicians tried to define number not in the field of arithmetic but in that of a more general doctrine : logic. This was first undertaken by Frege. 'While Peano reduced mathematics to arithmetic,' writes Waismann, 'Frege starts reducing it to logic.'

'After such a long separation from Euclid's rigour,' Frege observes, 'mathematics has now returned to it and is tending to supersede it. ... The concepts of function, of continuity, of limits, of the infinite, showed the need for the most precise determination, the negative number and the irrational, so long a part of mathematics, had to be more exactly examined. ...' But in regard to the definition of the number one, even today 'the majority of mathematicians have no satisfactory answer ... and yet is it not shameful for science to be so ignorant of an object which is so near and seems so simple? We must have no illusions : since we cannot define the number one, we cannot expect to say what an integer in general is. ...' According to Frege, all the fundamental concepts of arithmetic and analysis are obscure, lacking a definition of number. According to the logic mentioned so far, number expresses a property relating not to the objects numbered, but to the *concept* of these objects. (When we say that Venus has zero satellites, we do not refer to the satellites of Venus which do not exist but to the relative concepts.)

Frege defines right away what is meant by 'equally numerous', then he defines number. 'We will establish that the concept F is equally numerous to the concept G, since there exists the possibility of placing in a one-to-one correspondence the objects which fall under one with those which fall under the other concept. ... The natural number which belongs to the concept F is none other than the extension of the concept "equally numerous" to F.'

Russell takes up Frege's idea and speaks of classes instead of

concepts of extension, and defines number in the following way :
'the number belonging to a class C is the class of all the classes
in a one-to-one correspondence with C' (17).

This definition also refers to infinite cardinal numbers.

But does this definition correspond to our intuition of number? Is it not dangerous, perhaps, to have recourse to an
object of thought as hard to grasp as the class of all the classes
having a given property?

All the same, on this basis Whitehead and Russell in the
Principia Mathematica gave a wonderful reconstruction of
mathematics, of which we shall speak later, expressed in a
symbolism rather close to that of Peano.

Burali Forti, having seen the disadvantages of the definition
of the number of objects of a class C, as the class of all the
classes equivalent to C, that is which can be placed in a one-to-
one correspondence with C, introduced a function $f(C)$ to define
the number considered but did not manage to determine the
function itself. In fact if C and C' are equivalent

$$f(C) = f(C')$$

but also, Enriques notes,

$$\varphi f(C) = \varphi f(C')$$

To define the number of members of a class, Enriques
therefore made use of a definition by abstraction, with a process
similar to that Euclid followed in defining ratios (compare
Chapter V, §7) and defined the number of members of a class,
establishing when it is that the numbers corresponding to two
classes C and C' are equal (when, that is, the two classes are
equivalent) (18).

§5. *The theoretical logic of Hilbert*

A systemization of symbolic logic, an agile and powerful
instrument for the most subtle researches on the fundamentals
of mathematics, was given to us in the valuable work of
D. Hilbert and W. Ackermann, *Grundzüge der theoretischen
Logik* (Berlin, editions of 1928, 1937, 1949, 1959), from which

(17) Russell, **1**, p. 114.
(18) Enriques, **9**, pp. 242–8; Geymonat, **4**, pp. 51–63.

the contents of this section, except where otherwise stated, are taken (19).

Hilbert's logic is based on the primitive concept of proposition. At least at the beginning the proposition is introduced without an analysis of its structure, but it is said only that by proposition (*Aussage*) we mean any affirmation (*Satz*), of which it is meaningful to say that it is true or false. (Compare the definition 'judgment' according to Aristotle, Chapter IV, §1.) The subtle idea of what is 'meaningful' is not analysed; we will return to this subject later.

Let us define the fundamental operations on the propositions X, Y, Z... and establish the most important properties of these operations. (Remember the scholastics' calculus of propositions in Chapter X, §4.)

$\bar{X}$ (read *non X*) indicates the contradictory of X, the proposition, that is, that is true if X is false and false if X is true.

X & Y (read X and Y) indicates the proposition which is true when, and only when, X and Y are both true, and is called conjunction.

$X \vee Y$ (read X or Y) indicates the proposition which is true when at least one of the two propositions X and Y is true, and is called disjunction. This operation should not be confused with *or* in the Latin sense *aut ... aut*, in which the contemporaneous existence of X and Y is excluded, but corresponds rather to *vel*.

$X \rightarrow Y$ (read *if X then Y*, or else X *implies Y*) indicates the proposition which is false when, and only when, X is true and Y is false, and is called implication.

$X \leftrightarrow Y$ (read X is equivalent, *gleichwertig*, to Y) indicates the proposition which is true when, and only when, X and Y are both true or both false. For example (2 plus 2 equals 4)$\leftrightarrow$(the snow is white) is a true proposition, (2 plus 2 equals 4)$\leftrightarrow$(the snow is black) is a false proposition, $(2 > 3)\leftrightarrow$(the snow is black) is a true proposition.

The sign $\leftrightarrow$ should not be confused with äq: X äq Y (read: X signifies Y), that is X and Y have the same meaning.

The calculus of propositions can be developed in an axiomatic way on the basis of some fundamental formulae taken as

(19) For a comparison between Peano's and Hilbert's logic and arithmetic, see Cassina, **8**, pp. 129–38.

postulates, working through some fundamental rules for deducing them.

The fundamental logical formulae are the following:

$$X \lor X \to X$$
$$X \to X \lor Y$$
$$X \lor Y \to Y \lor X$$
$$(X \to Y) \to [Z \lor X \to Z \lor Y]$$

The rules for deduction are:

(a) Rule of substitution. In place of a variable proposition the same combination of propositions can be substituted whereever it appears.

(b) A scheme of deduction, which we owe to Crisippus (see Chapter V, §1, note (6)). From two formulae A and $A \to B$ we deduce the new formula B.

There exist various combinations of symbols with the same meaning. Here are some examples:

(I) $\qquad\qquad\qquad \overline{\overline{X}} \text{ äq } X$

(The negation of the negation of the proposition X means the proposition X.)

(II) $\qquad\qquad\qquad X \,\&\, Y \text{ äq } Y \,\&\, X$

(The logical conjunction enjoys the commutative property.)

(III) $\qquad\qquad X \,\&\, (Y \,\&\, Z) \text{ äq } (X \,\&\, Y) \,\&\, Z$

(The logical conjunction enjoys the associative property.)

(IV) $\qquad\qquad\qquad X \lor Y \text{ äq } Y \lor X$

(The logical disjunction enjoys the commutative property.)

(V) $\qquad\qquad X \lor (Y \lor Z) \text{ äq } (X \lor Y) \lor Z$

(The logical disjunction enjoys the associative property.)

(VI) $\qquad\qquad X \lor (Y \,\&\, Z) \text{ äq } (X \lor Y) \,\&\, (X \lor Z)$

(The logical disjunction enjoys the distributive property in respect of the logical conjunction.)

The first five of these equalities of meaning are obvious, yet they can be demonstrated, like (IV), on the basis of the following principles given as obvious by Hilbert:

Two combinations of fundamental propositions relating to $XYZ\ldots$ are of equal meaning when, and only when, for every equal substitution of true or false propositions in place of $XYZ\ldots$, both the propositions are both true or both false.

We will consider only the demonstration of (IV). Let us indicate a true proposition with R and a false proposition with F and substitute, for example, for XYZ, respectively, RFF. If (VI) is right, it should be

$$R \vee (F \mathbin{\&} F) \leftrightarrow (R \vee F) \mathbin{\&} (R \vee F)$$

Now in the first member $F\mathbin{\&}F \leftrightarrow F$, $R\vee F \leftrightarrow R$, that is $R\vee(F\mathbin{\&}F) \leftrightarrow R$; in the second $R\vee F \leftrightarrow R$, $R \vee R \leftrightarrow R$, therefore $(R \vee F)\mathbin{\&}(R \vee F) \leftrightarrow R$. The two members are both true. Similarly we must check all the possible attributions of the values of truth or falsity to $X\,Y\,Z$.

For conjunction and disjunction the terms *logical sum* and *logical product* are used respectively, but it would also be allowable to call $X\mathbin{\&}Y$ the logical product and $X\vee Y$ the logical sum, since the second distributive property is also valid.

(VII) $\qquad X \mathbin{\&} (Y \vee Z)\, \text{äq}\, (X \mathbin{\&} Y) \vee (X \mathbin{\&} Z)$

which is obvious from the second illustrative example given by Hilbert: he uses a weather forecast: today it rains and tomorrow or the day after tomorrow the sun will shine; this affirmation can also be expressed thus: today it rains and tomorrow the sun will shine, or else today it rains and the day after tomorrow the sun will shine.

As far as the notation is concerned we need not use parentheses, since it is a convention that $\vee$ binds more strongly than $\&$, and $\&$ in its turn binds more strongly than $\rightarrow$ and $\leftrightarrow$; the sign $\vee$ can be omitted, too, like the multiplication sign . in algebra.

It is possible to express the signs $\rightarrow$ and $\leftrightarrow$ by $-\mathbin{\&}$ on the basis of the following considerations: the proposition $X \rightarrow Y$ means, by definition, that X cannot be true and Y false together; this is expressed by writing

(α) $\qquad\qquad X \rightarrow Y\, \text{äq}\, \overline{X \mathbin{\&} \overline{\overline{Y}}}$

This relation allows us to express the sign $\rightarrow$ by $\&-$.

Besides, the co-existence of the two propositions $X \rightarrow Y$ and

330

$Y \rightarrow X$ means that we cannot have X true and Y false together, and also that we cannot have Y true and X false together. In other terms this can be expressed by:

(β) $\qquad\qquad X \leftrightarrow Y$ äq $(X \rightarrow Y) \,\&\, (Y \rightarrow X)$

In this way we can eliminate the sign $\leftrightarrow$ from the formula, and with this process we can reduce the signs used to $-$, $\&$, and $\vee$. Or they could be reduced to just $\&$ and $-$, as in Brentano's theory of judgment, or else $\rightarrow$ and $-$ can be used with the addition of other symbols, such as those Frege uses, or $\vee$ and $-$, which Bertrand Russell uses.

For the later developments the following relations are important:

(VIII) $\qquad\qquad \overline{X \,\&\, Y}$ äq $\overline{X} \vee \overline{Y}$

(denying the simultaneous validity of X and Y means denying X or denying Y.) Hilbert illustrates this with the following example: X indicates the affirmation: the triangle $\triangle$ is right-angled; Y means: $\triangle$ is equiangular. X and Y then mean the triangle $\triangle$ is right-angled and equiangular; the contradictoriness of this proposition, that is $\overline{X \,\&\, Y}$, means that the triangle $\triangle$ is not right-angled or else is not equiangular: $\overline{X} \vee \overline{Y}$.

Similarly we have:

(IX) $\qquad\qquad \overline{X \vee Y}$ äq $\overline{X} \,\&\, \overline{Y}$

(Denying that at least one of the two propositions X and Y is valid means denying X and denying Y.)

This seems obvious from the following example which Hilbert gives: In an examination in mathematics the candidate is asked to be expert in at least one of two subjects, arithmetic and geometry. X indicates the proposition 'The candidate knows arithmetic', Y means 'The candidate knows geometry'. The requirements of the examination are satisfied by the candidate if $X \vee Y$ is true. The candidate fails the examination; then $\overline{X \vee Y}$ is verified, which means: 'The candidate does not know arithmetic and does not know geometry', which is represented by the formula $\overline{X} \,\&\, \overline{Y}$.

This allows us to use any expression formed from the fundamental propositions X, Y, Z ... through as many applications as we like of the operations $- \,\&\, \vee \rightarrow \leftrightarrow$, under a determined

331

normal form, in which appear only conjunctions or disjunctions in which every factor is a fundamental proposition or the negation of one of these.

To reduce any expression formed from the fundamental proposition X, Y, Z ... to the normal form through the operations $-$ & $\vee$ $\rightarrow$ $\leftrightarrow$, we proceed in the following way:

First of all the signs $\rightarrow$ and $\leftrightarrow$ are eliminated by applying formulae (α) and (β). Then by applying formulae (VIII) and (IX), it is always possible to make the larger sign of negation move towards the inside and finally remain above the fundamental propositions. The disjunctions are found in the same way as is done in algebra, by applying the distributive law.

Remember where (I) is needed.

For instance, from

$$\overline{(XY\ \&\ \bar{Y})\ \vee\ (Z\ \&\ Y)}$$

we obtain, in the first place, through (IX) by applying (VIII)

$$\overline{XY}\ \vee\ \bar{\bar{Y}}\ \&\ \bar{Z}\ \vee\ \bar{Y}$$

and by applying (IX) again

$$(\bar{X}\ \&\ \bar{Y})\ \vee\ \bar{\bar{Y}}\ \&\ \bar{Z}\bar{Y}$$

Forming the product as in algebra we obtain

$$\bar{X}\bar{\bar{Y}}\ \&\ \bar{Y}\bar{\bar{Y}}\ \&\ \bar{Z}\bar{Y}$$

Recalling (I) we have finally

$$\bar{X}Y\ \&\ \bar{Y}Y\ \&\ \bar{Z}\bar{Y}$$

The truth or falsity of an expression built from the fundamental propositions through the signs $-$ & $\vee$ $\rightarrow$ $\leftrightarrow$ depends only on the way in which the truth or falsity of the fundamental proposition is distributed. The value of the truth of an expression of the type we have considered does not vary if for any proposition an equivalent one is substituted. This means that the sign $\leftrightarrow$ has a function similar to that of the sign $=$ in algebra.

We must now solve a problem whose importance will be shown by the later developments of the theory: to find the combinations of propositions that are always valid, independently of the truth or falsehood of the fundamental propositions. For instance, the following obviously has this property:

$$X \leftrightarrow X$$

Given that every logical expression can always be put in a normal form, all we need do is establish when it is that an expression in a normal form represents a combination of propositions that is invariably true.

The problem is solved by bearing in mind the following obvious rules:

(b1) $X\bar{X}$ is always true.

(b2) If X is true and Y indicates any proposition, XY is also true.

(b3) If X is true and Y is true then $X\&Y$ are true.

These rules should be taken to mean that any proposition or combination of propositions can be substituted in place of X and Y.

According to these rules, bearing in mind the formal properties of logical calculus we have already explained, it is obvious that *all expressions are true which, in a normal form, are characterized by the fact that, in every disjunction, there appears as a factor at least one of the fundamental propositions, together with its negation.*

But it can be shown that the expressions characterized by this property are the only ones which are always true.

In fact, let us suppose that in a logical disjunction, which is a term of an expression written in a normal form, no proposition X appears together with its negation $\bar{X}$.

To make this product false all one needs is to substitute a false proposition for every undenied proposition and a true proposition for every denied proposition. Since one term of the logical sum is false, the whole expression will be false, independently of the value of the other term's truth. The condition we have mentioned is therefore necessary and sufficient.

The *problem* solved by expressions in which the individual signs do not appear is called *of general validity*. Besides this problem Hilbert poses another: to determine whether a given expression E, without signs designating elements, constructed with the means furnished by the calculus of propositions, is always false, or if it can become true through particular values of the propositions which appear in it; let us say in such a case that E is verifiable.

This *problem of verifiability* can be reduced to the preceding one.

Let us consider $\bar{E}$. If this expression is always true, then E is always false, but if $\bar{E}$ is not always true, that means that E is satisfiable.

The two problems of general validity and verifiability form the *Entscheidungsproblem*, or problem of decision, considered by Hilbert to be the main problem of mathematical logic.

In the field of the calculus of propositions the problem of decision is fully solved, as we have demonstrated.

The importance of this problem arises from the two following applications. (1) Suppose that we are given the postulates $P_1 P_2 \ldots P_n$, constructed, of course, with the means offered by the calculus of propositions.

We ask if the given system is non-contradictory or coherent. The given system is coherent when and only when the expression

$$P_1 \,\&\, P_2 \ldots \,\&\, P_n$$

is satisfiable.

Given the compatible postulates $P_1 \ldots P_n$, and a proposition X to be proved, we ask which of the three possibilities is verified: X is a consequence of $P_1 \ldots P_n$; X is incompatible with $P_1 \ldots P_n$; X is independent of $P_1 \ldots P_n$.

To solve this question we form, with Hilbert, the expression

$$(P_1 \,\&\, P_2 \,\&\, \ldots \,\&\, P_n) \to X$$

If this expression is always true X is a consequence of $P_1 \ldots P_n$.

If instead the following expression is always true:

$$(P_1 \,\&\, P_2 \,\&\, \ldots \,\&\, P_n) \to \bar{X}$$

X is incompatible with $P_1 \ldots P_n$.

If neither of the last two expressions considered is always true, then X is independent of $P_1 \ldots P_n$.

The last two problems (remembering what we said in §1 of this chapter) might together be called the *problem of Leibniz*.

These results, although remarkable, should not be over-valued, since, as Hilbert observes, not all logical relations can be expressed in the field of the calculus of propositions. For instance, in that field we should not be able to demonstrate affirmations of the type: 'If *B* lies between *A* and *C* then *B* lies between *C* and *A*', or else: 'If there is a son there is also a father'.

So we will give up the idea of confining ourselves to the calculus of propositions, and take up a new symbolism in which we can distinguish subject and predicate. This new method is called the *calculus of logical functions*, or the *calculus of predicates*, and is based on the representation of the predicate by the functional sign, where, in the empty place destined for independent variables, we put the subject. For instance if *P*( ) indicates the predicate 'is a prime number' *P*(5) is the expression of the proposition '5 is a prime number'.

In the same way relations between two or more objects with the sign of the functions of two or more variables are expressed. For instance $< (2,3)$ means '2 is less than 3'; $Z(ABC)$ can mean '*B* is contained between *A* and *C*; to the formula $x+y=z$ corresponds a predicate of 3 terms $S(x,y,z)$ etc.

Instead of given values for the subjects, we can also consider variables, for which values taken in determined fields can be substituted. For instance, if $L(x,y)$ means: 'the point *x* belongs to the straight line *y*' *x* varies among the points, and *y* among the straight lines.

If for the variables we substitute determinate values, then we have determinate propositions which are either true or false.

In using $\overline{P(x)}$ we must distinguish between: '$\overline{P(x)}$ is valid for every *x*' or 'it is not true that $P(x)$ is valid for every *x*'. To make this distinction exact we introduce the so-called *sign of totality*: $(x)A(x)$ reads: 'for every *x*, $A(x)$ is valid'. Thus the ambiguous expression is made clear in the following way:

$(x)\overline{P(x)}$ $\overline{P(x)}$ is valid for every *x*

$\overline{(x)P(x)}$ it is not true that $P(x)$ is valid for every *x*

For reasons of symmetry Hilbert introduces another sign, called the sign of existence.

$(Ex) A(x)$ means: There exists an x for which $A(x)$ is valid.

The signs (x) and (Ex) are called signs in parentheses.

The sign of totality and of existence can be combined in various ways: for instance:

$(x)(y)A(x,y)$, which can also be written for greater clarity as $(x)[(y)A(x,y)]$, means that for every x and every y $A(x,y)$ is valid.

$(x)(Ey)A(x,y)$, which can also be written $(x)[(Ey)A(x,y)]$ means: for every x there exists a y for which $A(x,y)$ is valid.

But the parentheses can be omitted when there is no danger of confusion.

Two signs of totality or two signs of existence can change their order, but not one sign of totality and one sign of existence. For instance, if x and y are real numbers

$(x)(Ey) < (x,y)$ true (for every number there is a greater number)
$(Ey)(x) < (x,y)$ false (there is a number greater than all others)

The calculus of predicates can also be arranged in an axiomatic way. We can still choose four axioms of the calculus of propositions, adding to them two others which characterize the signs (x) and (Ex):

(e) $\qquad\qquad (x)F(x) \rightarrow F(y)$
(f) $\qquad\qquad F(y) \rightarrow (Ex)F(x)$

Having admitted that these axioms, and other characteristic formulae of the theory to be developed, are true, these fundamental relations are worked on by means of the following rules:

(α) rule of substitution, (β) scheme of deduction, like, 'mutatis mutandis', the corresponding rules of the calculus of propositions.

To the preceding rules we add the scheme (γ) for the symbols (x) and (Ex):

Let $B(x)$ be any logical expression depending on x, while A is an expression independent of x. Then if $A \rightarrow B(x)$ is a true formula, then $A \rightarrow (x)B(x)$ is a true formula too. Thus from a true formula $B(x) \rightarrow A$ we get the new formula $(Ex)B(x) \rightarrow A$.

The axioms (e) and (f) and the rule (γ) which characterize the signs (x) and (Ex) were introduced by P. Bernays.

For the calculus of predicates Hilbert proposed the *Entscheidungsproblem*, and various authors have obtained interesting results on the subject, but the problem in general has remained unsolved, in fact recent researches have led us to believe that it is insoluble.

In the calculus of predicates propositions and predicates are held to be distinct from the objects which are being considered. But this restriction in the choice of subjects which are considered essentially different from predicates does not allow us to develop some important mathematical theories adequately, for instance the theory of sets. For this reason we shall pass from the calculus of predicates, considered narrowly, to the general calculus of predicates in which propositions and predicates also figure among the variables.

Some logical laws need an extension of this kind for their expression. For example,

$$(X)(EY)\,(X \lor Y \,\&\, \overline{X \,\&\, Y}),$$

that is: 'For every proposition X there exists a proposition Y (the contradiction of X) such that of the two propositions at least one, and only one, is true.'

But the general calculus of predicates raises very serious problems, which we shall deal with in the next chapter.

§6. *Other developments of symbolic logic* (20)

In contemporary mathematical logic, apart from the symbolism of Peano–Russell and of Hilbert, we should remember the notation of J. Łukasiewicz, adopted by the Polish school, where, by using the symbols indicating the operations immediately before the corresponding arguments, we have the advantage of making the points and the parentheses superfluous.

The following table sums up the symbolism we have considered, of which Hilbert's has been illustrated in the last section.

Another way of representing these operations was shown

---

(20) See Bocheński, **1**, **2**, **4**; Casari, Vaccarino, **1**. The review *Fundamenta Mathematicae* is published in Warsaw.

Logical operations	Notations of			
	Peano–Russell	Łukasiewicz	Hilbert–Ackermann 1928–49	Hilbert–Ackermann 1959
Negation	$\backsim p$	Np	$\bar{P}$	$\daleth p$
Disjunction	$p \lor q$	Apq	$P \lor Q$	$P \lor Q$
Conjunction	$p \cdot q$	Kpq	$P \,\&\, Q$	$P \land Q$
Material implication	$p \supset q$	Cpq	$P \to Q$	$P \to Q$

by the logical matrices introduced by C. S. Peirce (21), the values of truth 1 and 0 indicating respectively true and false.

Negation:

$$
\begin{array}{c|c}
p & Np \\
\hline
1 & 0 \\
0 & 1
\end{array}
$$

Disjunction:

$$
\begin{array}{c|c}
pq & Apq \\
\hline
1\,1 & 1 \\
1\,0 & 1 \\
0\,1 & 1 \\
0\,0 & 0
\end{array}
\qquad \text{or} \qquad
\begin{array}{c|c}
A & 1\,0 \\
\hline
1 & 1\,1 \\
0 & 1\,0
\end{array}
$$

(21) Vaccarino, **2**, p. 243.

Conjunction:

pq	kpq		k	10
1 1	1	or	1	1 0
1 0	0		0	0 0
0 1	0			
0 0	0			

Material implication:

pq	Cpq		C	10
1 1	1	or	1	1 0
1 0	0		0	1 1
0 1	1			
0 0	1			

Among the most important pointers in contemporary mathematical logic is the American School of A. Church, editor of the *Journal of Symbolic Logic* (Fine Hall, Princetown, New Jersey). We should recall too the school of Münster.

Algebra, infinitesimal analysis and chemistry have been given the unity of a language universally accepted throughout the world, but symbolic logic has no universal language as yet. B. Levi made an effort to unify the symbolism of logic by using contemporaneously (in the same formulae) Peano's notation for operations on classes, and Hilbert's for operations on propositions (22). And for the development of polyvalent logic (which we will deal with in Chapter XX, §2) Łukasiewicz's work on matrices and symbolism may be advantageous.

§7. *Mathematical logic and its applications*

Consideration of the different trends of logical research followed by mathematical thinkers leads us to examine the exact meaning and contents of the body of doctrines, drawn up

(22) Levi, **5**; there is also a bibliography of Levi's work on logic.

towards the end of the nineteenth and in the twentieth century, known as 'mathematical logic'.

This term has not always had the same meaning. According to Peano (23): 'mathematical logic ... studies the properties of the operations and the relations of logic'; its object (24) is to formulate the simplest system of logical notions, necessary and sufficient to represent symbolically mathematical truths and their proofs.

Hilbert and Ackermann (25) express themselves on the subject rather differently: 'Theoretical logic, also called mathematical and symbolic logic, is an extension of the formal methods of mathematics to the object of logic'.

The term 'logistics', proposed by Couturat in 1904 (26), has until now been used to indicate symbolic logic. But it seems inadvisable to use this term, because it has been employed at other times with very different meanings: Leibniz (27), for instance, uses it to mean the science of magnitudes or of proportion in general. Today logistics also means a branch of military science.

I propose to use the term mathematical logic in a rather wider sense than that given it by Peano and Hilbert: that is, I will use it to mean the body of doctrines concerned with the rational structure of mathematical theories.

Understood in this sense, mathematical logic (28), although it has its own particular character, cannot be completely distinguished from the traditional logic to which it is historically linked. It has had fruitful relations with other branches of knowledge, among them the philosophy of mathematics (reflections on the relations between mathematics and the more general problems of being, knowledge, and action), research on the fundamentals of mathematics in history and in present-day problems, study of the rational systemization of various sciences, physics, chemistry, biology, sociology. Technology has also used it, for instance, in applying the principles of programming to electronic computers (29).

(23) Peano, **2**, p. 1.
(24) Vacca, **12**, p. 32.
(25) Hilbert and Ackermann, 1949, p. 1.
(26) Bocheński, **1**, p. 167.
(27) Leibniz, **1**, II Abth., I Band, p. 178.
(28) Carruccio, **22**.
(29) De Finetti, **5**; Whitesitt.

CHAPTER XIX

Hypothetical-Deductive Systems, Non-Contradictoriness, Antinomies

§1. *Concept of the hypothetical-deductive system* (1)

Various themes in the development of modern mathematical thought with which we have already dealt have influenced mathematicians profoundly with regard to the concept of the structure and meaning of their theories: abstract geometry, for instance (with the elements that contributed to its formation: laws of duality, Plücker's co-ordinates, hyperspace), non-Euclidean geometry, the theory of infinite sets and non-Archimedean geometry, symbolic logic; as well as criticism of the fundamental principles of geometry in reference to physical facts (Mach, Maxwell, Helmholtz, Enriques), and expressed in the form of purely logical relations (M. Pasch).

All these together brought about what Enriques called *the reform of contemporary logic*, by which we have come to consider every mathematical theory as a *hypothetical-deductive system*.

This term, introduced by M. Pieri (1860–1913) (2), indicates a system of propositions consisting of postulates and theorems: the postulates (which from the purely logical point of view are arbitrary) define implicitly the primitive concepts introduced in the theory without explicit definitions; but these concepts are defined up to a point, because they can still be interpreted concretely in several ways, for instance, physico-mathematically, geometrically, purely analytically; the theorems are obtained from the postulates by applying the laws of formal logic without an appeal to intuition. Bertrand Russell has

(1) Enriques, **2**, ch. III, 123–215, and chiefly pp. 196–204; Russell, **2**.
(2) Pieri.

written paradoxically : 'Mathematics is the science in which we do not know what we are talking about nor whether what we are saying is true or false.'

This sentence, which might appear to be dictated by extreme scepticism, means, however, that the primitive concepts introduced in mathematical theory cannot be defined explicitly and can be interpreted in various ways (we do not know what we are talking about); and that there is no point in wondering if theorems are true or false in respect of a reality external to the hypothetical-deductive system : the mathematician does not worry, for instance, about establishing whether a certain theorem is or is not verified by experience (we do not know if what we say is true or false).

This point of view abandons Aristotle's conception (expressed in Chapter IV, §9) of a deductive science, and recognizes that it is logically arbitrary to choose the group of primitive propositions which are the basis of the theory to be constructed. This thought is lucidly expressed by Vailati, who takes it from Enriques (3):

'Instead of conceiving the difference between the postulates and the other propositions as consisting in the possession, on the part of the first, of some special character which makes them "in themselves" more acceptable, more evident, less disputable, mathematical logicians consider postulates as propositions like all the others, the choice of which can be varied according to the objective.

'. . . If the relationship between the postulates and the propositions depending on them can be compared to those which, in an autocracy, exist between the monarchy and the privileged classes on the one hand and the rest of society on the other, the work of mathematical logicians is in a way similar to that of those who introduce a constitutional or democratic regime, in which the choice and election of the leaders depends, at least ideally, on their known capacity to exercise temporarily determinate functions in the interests of the public.

'The postulates have had to renounce that kind of "divine right" which their "self-evidence" seemed to invest them with, and resign themselves to becoming no longer the "arbiters", but the "servi servorum", simple "employees", of the great "unions"

(3) Vailati, 1, p. 699; Enriques, 3, p. 202.

of propositions which constitute the various branches of mathematics.'

We are still keeping close to Vailati's thought, as Enriques notes, if in reading the passage for the words 'postulates' and 'propositions which depend on them' we substitute the words 'primitive concepts', and 'concepts defined by means of them'.

§2. *The compatibility of a system of postulates* (4)

This new conception of mathematics raises new problems, above all that of the compatibility of postulates. When the basis of mathematical theory was axioms, postulates and definitions, thought to reflect a transcendent reality superior to the human mind as in the conceptions of Plato, Aristotle, and St. Augustine, there was, *a priori*, no danger of finding contradictions in their logical consequences, which, since they were true, could not lead to absurdities.

But when, with the evolution of modern mathematical thought, the basis of every hypothetical-deductive system was changed to propositions that from the logical point of view were arbitrary, the problem of guaranteeing the non-contradictoriness or compatibility of a system of postulates became of fundamental interest.

Some remarkable results have been obtained in connection with this problem: for instance the non-contradictoriness of postulates at the basis of non-Euclidean geometry has been reduced to the non-contradictoriness of Euclidean geometry (see Chapter XV), and the latter's non-contradictoriness to that of analysis (see Chapter XII, §3).

But the general problem is not solved; in fact the famous paradoxes of logic have made things worse.

§3. *Antinomies, and the efforts to overcome them* (5)

First of all there is the paradox of the liar, which goes back to classical antiquity (6): is it possible for a man who is telling the truth to lie, and if he lies to tell the truth? It seems impossible; yet it is just what happens when someone says 'I am lying'.

(4) Carruccio, **9**.
(5) Whitehead and Russell; Poincaré, **3**, pp. 192–214 (highly controversial towards 'Logisticiens' and 'Cantoriens'); Hilbert and Ackermann, 1928; Geymonat, **3**, pp. 111–35.
(6) For ample historical notions, see P. Rustow.

A second paradox is known by the name of Richard. Here is the exposition given by L. Geymonat (7): 'Consider the class K of the definable natural numbers in the English language, with no less than a thousand linguistic signs. Since K is an aggregate of natural numbers it certainly has a minimum: let that be m. Let us consider the following definition of m: "the smallest natural number definable in the English language, with not less than a thousand linguistic signs". This proposition has an obviously antinomical character because the number m, defined by it, on the one hand should belong to K, being the smallest element of this class, and on the other cannot do so because our own proposition shows that m is definable with less than a thousand linguistic signs.'

We should also remember Bertrand Russell's famous paradox: Does the set of all the sets that do not contain themselves as an element at the same time contain or not contain itself as an element? However this is answered we always fall into a contradiction.

From these paradoxes, in which a disconcerting enigma is hidden behind a simple play of words, it turns out that the usual rules of logic do not apply to all sets and affirmations. The construction of logical expressions must be suitably limited if we do not want to fall into absurdity.

With this object in mind Whitehead and Russell, who belonged to the logisticians, built up the theory of logical types and degrees, which Hilbert took up in his writings. L. Geymonat expounded the theory and it is chiefly his writings on the subject which we will rely on here.

The antinomies, of which we have given three significant examples, were divided by the logicians of the first ten years of the twentieth century into two classes:

(1) logical antinomies in the strict sense;
(2) syntactical antinomies.

The first are concerned with the concepts (for instance Russell's paradox); the second with the linguistic expression of the concepts themselves (for instance, the paradox of the liar and that of Richard).

Whitehead and Russell explain the logical antinomies in the

(7) Geymonat, 2, p. 194.

strict sense by attributing their origin to a vicious circle arising from the supposition that a collection of objects can contain elements which can be defined only by means of the same collection considered as a whole. For instance, they consider illegitimate a proposition *p* in which we take all the propositions, *p* included: it is not allowed to speak of *all the propositions*, if new propositions take their origin from affirmations relating to all the propositions.

To avoid vicious circles and contradictions of the kind, the concepts are divided into various types, and we must work on them particularly carefully.

We call *type zero* the individual concepts, that is the proper names, which in the propositions can occupy only the place of subject.

We call *type one* the concepts that express the property of individuals. In a proposition they can occupy the place of subject or predicate, but their subject should be a concept of type zero.

We call *type two* the concepts that express a property of a property; in a proposition there can be subjects and predicates, but in the second case their subject should be a concept of zero or one type.

Higher types are similarly defined. Bertrand Russell's fundamental rule is this: 'A concept can never function as a predicate in a proposition in which the subject is of a type equal to or greater than that of the concept itself.'

In Russell's paradox this rule is not observed: when we consider a class *I* that contains itself as an element, that is *I* (subject) and *I* (predicate), we do not respect the theory of types, because the subject should have a type less than that of the predicate, which does not happen here (if this proposition is nonsense it is so also for the contradictory proposition).

But this theory cannot overcome the syntactic antinomies. In fact in the proposition 'I am a liar', 'I' and 'liar' are concepts respectively of degree 0 and 1.

To solve these antinomies as well logicians use a theory of degrees regarding the propositional function. We call *propositional function* an expression that contains variables which, replaced by suitable terms, give rise to propositions (a propositional function is not in itself either true or false: for instance,

'x is greater than 9'). If before a propositional function which contains the variable x we put the sign (x) and the sign (Ex), x takes the name of apparent variable.

We call of *degree zero* a propositional function which does not contain apparent variables.

We call of *degree one* a propositional function which contains apparent variables in which the field of variability is the set of individual objects.

We call of *degree two* a propositional function which contains apparent variables that can be substituted for the propositional functions of degree one but not of a higher degree; and so on.

Russell's rule on degrees is the following: It is not allowed to consider as belonging to the same set propositions which are obtainable from propositional functions of various degrees.

This rule is not observed in the so-called impredicative processes, that is in those in which a property p is defined through a totality in which p belongs.

As we can see easily in the antinomies of the liar and of Richard we do in fact use impredicative processes.

When the theory of degree is grafted on to the theory of types it gives place to the perfected and complex theory of branched types.

This avoids the antinomies we have mentioned, but there are serious disadvantages in applying the theory to classical mathematics, in which many fundamental theorems need highly complex proofs; in fact some of these would cease to hold. If, with Dedekind, we define a real number as a section of the rationals, because of the rule of degrees it would no longer be possible to consider the rational and irrational numbers as belonging to the same totality; in general the definition of real numbers would result in a concept of different degree. It is also easy to see that the definition of the upper bound U of a group of real numbers is impredicative; the upper bound of a set of real numbers is a real number which is not exceeded by any number of the set, while every real number less than U is exceeded by some number of the set.

The proof of Cantor's result that the continuum is not denumerable also becomes illegitimate.

To repair this damage Russell thought up a new *axiom* called *the axiom of reducibility*. To understand what he says we give the following definition: two logical functions $\Phi(x)$ and $\Psi(x)$ are called formally equivalent, when for every possible argument x, $\Phi(x)$ and $\Psi(x)$ are both true or both false: 'a predicative function of a variable argument is one which involves no totality, except that of the possible values of the argument, and those that are presupposed by any one of the possible arguments' (8).

The axiom of reducibility says that, given any logical function $\Phi(x)$, there exists a predicative logical function equivalent to $\Phi(x)$.

In spite of Whitehead and Russell's interesting defence (9) of this axiom, in which the principle of Leibniz's indiscernibles is also examined, and the excellent results obtained from them, without falling into the known antinomies, the axiom makes many mathematicians perplexed or hostile, since it is not obvious, and even leaves a doubt that it might lead to new antinomies. In fact, its character is so artificial that Weyl writes: 'Russell, with his axiom of reducibility, has made reason commit suicide', and Russell himself finally abandoned the axiom of reducibility as inessential to his system (10).

Some logicians, taking up an idea of Peano, say that for the rigorous construction of mathematics a theory of types without degrees and without the axiom of reducibility is enough.

Peano had observed that the syntactical antinomies which could not be avoided with the theory of types are not concerned with concepts but with their expression, and so have nothing directly to do with mathematics, and mathematicians must not take them into account.

L. Geymonat writes on this subtle and serious question: 'I must say plainly that today this outlook is absolutely unacceptable. Peano could be content with this explanation, because he thought the symbolic language, which could express mathematics, was incapable of expressing the comments and subtleties inherent in the word "syntax". He may have thought the symbolic language of mathematics was incapable of expressing syntactical antinomies, and so was not worried by them. But in 1931 the situation changed radically. In

(8) Whitehead and Russell, p. 57.
(9) Whitehead and Russell, pp. 58–62.
(10) Barone, p. 16.

that year the mathematical logician Kurt Gödel, followed later by Rudolf Carnap, showed, in an important note published in the *Monatshefte für Mathematik und Physik*, that under certain very general conditions the synthesis of a formalized language could be translated into arithmetical propositions, so that if a language of the kind contained an arithmetic, the arithmetic could also express its own syntax. Thus a formalized language can, though Peano did not realize it, also express the syntactical antinomies, and so should not be considered as immune from them.'

Yet the idea that syntactical antinomies which bear only on the expression can be neglected appears again in the third edition (1945), edited by Ackermann, of Hilbert and Ackermann's *Grundzüge der theoretischen Logik* (11).

Not all the mathematical logicians of our time believe that antinomies can be overcome by the theory of types: according to B. Levi, antinomies, properly so-called, do not exist in logic, and what look like antinomies in every case conceal some infraction of traditional logic (which itself needs no correction (12)). From this point of view he examines the most famous antinomies, showing the origins of the contradictions and how cautiously one must move in order to avoid others of the kind.

A class, he says, cannot be identified with an element, but a class can be subjected to a *process of elementation* (13). This term means every operation that makes an individual (which in some way represents the class itself, considered as a whole) correspond to a class. This process, according to Levi, is not always the same, nor can it be applied to every class; sometimes it is expressed by 'the idea of'. 'Can we say that the properties which result when "the idea of" is *applied to any aggregate* can be determined? If, for example, *a* is an aggregate made up of a single element, is "the idea of *a*" the same thing

(11) 'Die Paradoxien dieser zweiter Art, für die die Bezeichnung "semantische Paradoxien" gebräuchlich ist, treffen also gar nicht unseren Kalkül, da dieser nicht imstande ist, ihren rein logischen Charakter zum Ausdruck zu bringen. Vielmehr mussten wir zu der teilweisen Formalisierung inhaltiliche Gedankengänge zu Hilfe nehmen. Wir brauchen daher für unseren prädikatenkalkül keine Konsequenzen aus den Widersprüchen der letzten Art zu ziehen und wollen daher auch nicht näher auf sie eingehen.' (H.A., p. 128.)
(12) Levi, **6**.
(13) For the procedure of elementation, and the elimination of the antinomies of Russell and of Richard, see Levi, **1**, pp. 187–216.

as the individual *a* or something different? This is the principle of a series of indeterminations which must be solved one by one: we can try a general solution: Russell's antinomy shows that the effort must necessarily be in vain. We must first of all remember what it consists in: let us consider the aggregate *E* of the individuals which we obtain by elementing aggregates which do not contain themselves as element: we ask: when we element *E* do we obtain an individual of *E* or an individual not belonging to *E*? However we answer this, we fall into a contradiction. The origin of the contradiction lies in this, that we think a process of elementation is rigidly fixed and admit that this process applies to every aggregate, in particular to *E*: but there is no contradiction if *E* cannot be elemented with this process.' In fact, however we arrange the process of elementation, it cannot apply to *E*.

B. Levi gives a different version of Richard's antinomy from the one we have considered in this section, but from our point of view it is substantially equivalent. According to B. Levi, we overcome the antinomy by distinguishing suitably between the *primitive idea* and the *definite idea*, and observing that a totality cannot be defined which contains numbers that are defined through the totality itself. We deduce from this that the set considered in Richard's antinomy is a primitive idea, but this primitive idea and the others that appear in the same antinomy are linked by contradictory postulates. 'Now it is not in the least surprising that, when we try to satisfy postulates chosen without due caution, we get contradictory primitive ideas. What made the contradiction repugnant was the fact that it seemed as if the only ideas used to construct it were those of common logic, together with the primitive ideas of arithmetic. . . .'

When he deals with the antinomy of the liar (14), Levi says that *logic* does not deal with the truth of the propositions in themselves; '*it has to do with the form, not the substance, of the judgments*'. More precisely: the propositions are divided into classes: true and false. Logic cannot demand this classification in any absolute sense, but can only say: 'The propositions are placed in systems, such that, when the classes have been assigned ("true" or else "false") to which the

(14) Levi, **2**, pp. 239–52.

proposition of a certain system belongs [or certain propositions belong], through the rules of logic, the classes for all the other propositions of the same system are determined.' Now let us consider the proposition 'I lie', the logical rules (identity, non-contradiction) and the propositional functions with A as the variable proposition:

(1) A is true (true if A is true, false if A is false).

(2) A is false (false if A is true, true if A is false).

Let us say $A =$ 'I lie' in (1) and (2), classifying A freely as either true or false. According to Levi, the propositions thus obtained are not antinomous.

We are shown other ways of overcoming antinomies by intuitionists, such as Brouwer and Heyting (15). These believe that in the world of mathematics all that belongs by right is the set of intuitive elements which the mathematician has at the beginning of his studies, and others which are clearly and precisely derived from them. This rejects the principle of the excluded middle with its consequences, among them all indirect proofs.

Thus the antinomies are eliminated, since they originate in non-constructive concepts; and at a high price, too, for many fundamental theorems of classical mathematics are no longer valid for intuitionist mathematicians.

We shall deal with the neo-empiricists' position in the next chapter.

All this is very far from exhausting the arduous, complex and delicate subject of the foundations of analysis, bearing antinomies and the recent developments of logic in mind, but it shows something of its interest and the fundamental character of its problems, on which, of course, much study and research is still needed (16).

(15) On intuitionistic mathematics, see the Swiss review *L'Enseignement Mathématique*, 1935, chiefly p. 103. See also ch. XX, §2 of this book.

(16) Ackermann.

CHAPTER XX

Problems of Contemporary Logic

§1. *Croce's idealism in the face of logic and mathematics*

From Kant's philosophical position, there evolved eventually the idealism of J. G. Fichte (1762–1814), G. W. F. Hegel (1770–1831), and F. W. J. Schelling (1775–1854); which, after a positivistic interval, inspired contemporary neo-idealism, with Croce as one of its most eminent upholders.

Benedetto Croce (1866–1952) (1) distinguishes two pure theoretical forms in the cognitive spirit: intuition and concept (concrete universal); while the first gives rise to art, the second, when expressed in the form of a definition, constitutes philosophy properly so-called, and when expressed in the form of individual judgments constitutes history. Besides these Croce distinguishes two forms of practical elaboration of knowledge, that is: the formation of empirical classificatory pseudo-concepts, and of abstract numerative and measurative pseudo-concepts; the first are concrete but not universal and make up the natural sciences; the second are universal but not concrete and constitute the mathematical sciences. Thus, according to Croce, the mathematical concepts (pseudo-concepts to him) have no reality (2): 'there is never a geometrical triangle in reality, because in reality there are no straight lines, right angles, and sums of angles equal to two right angles. . . . A thought which has nothing real for its object is not a thought: and these concepts are not concepts but conceptual fictions.' 'The geometrical triangle is no use either to fantasy or to thought . . . but is indispensable to a man measuring a field.' Croce mentions Bertrand Russell's definition of mathematics which we have already examined and comments (3): 'A science that does not affirm anything does not

(1) Croce, **1**, p. 173.
(2) Croce, **1**, pp. 19–20 and 26.
(3) Croce, **1**, pp. 251–4.

belong to the theoretical spirit, it is not even poetry: and a science that does not refer to anything is not even an empirical science, which always refers to a determinate group of representations ... and ... it is impossible to believe that the principles of mathematics are real. In fact, strictly speaking, they are all entirely false. The numerical series is obtained by starting from unity and always adding another unit; but in reality there is nothing that can function as first term of a series and no way of generating a discontinuous series. ...'

Croce's condemnation of mathematical logic (4) also includes the formal logic called 'formalistic', which he accuses of being empirical. It is not at all surprising to find an idealistic logician condemning formal logic, since in idealistic logic it is admitted that (5) '*A* is at the same time not *A*' or else 'everything contradicts itself'; this latter principle, which has no place in formal traditional or in mathematical logic, being derived from Heraclitus.

But Croce's work shows his attitude towards mathematics is not entirely negative, although he thinks it has no place in the world of art or of philosophy, but only in the practical field. On page 26, and more explicitly on page 251 of *Logic as the science of pure concept*, he denies that mathematics is an art, but on page 198 he writes: 'Every thought, even the most abstruse philosophical and mathematical thought, is made concrete in an artistic form', thus admitting that there exists a concrete mathematical thought expressed in an artistic form.

As we see from the following passage, Croce, who was much influenced by G. B. Vico (see Chapter XII, §7 of this book), finally allows mathematics a place in theoretical thought since it exists in history (6): 'Some ... would wish the philosopher to be a physiologist, a physicist, a mathematician, that is, that his brain should be full of abstractions, which are certainly not useless ... but which are in no direct relation to that form of knowledge which must be the condition of philosophy. This form of knowledge is, on the contrary, history; or, as it is said (with an *a potiori* intention), the history of philosophy, which of necessity, as the history of a moment of the spirit,

(4) Croce, **1**, pp. 95, 99, 420.
(5) Croce, **1**, pp. 70–1.
(6) Croce, **1**, p. 217.

includes all history in itself. . . . And to the extent that they can be of use according to the requirements of the problem, we must know also the natural, physical and mathematical sciences. But we must not know them *as such* and develop them as such, but rather as *historical knowledge* concerning the state of the natural sciences, of physics, and of mathematics, in order to understand the problems that they help to raise for philosophy.'

Enriques has dealt with the position of mathematics and science in the history of thought, and his view of idealism is shown in the following passage (7): 'Romantic idealism has arisen again in our generation in contrast to positivistic particularism, and has appealed to many by seeming to listen to all human voices, and to find in them the harmony of contrasts we find in history. But what should we now say if prejudice or any other reason should cause this humanistic universalism to degenerate into a philosophy of watertight compartments, if we hearken to the word of the Spirit only if it blows from one particular side of the mountain?'

§2. *The new logics*

We have examined (see Chapter XIX, §1) the concept of the *hypothetical-deductive system* in the world of mathematics, a system based on arbitrary postulates, but in which the processes for deducing the theorems from the postulates are still those of a single universal logic. But this has given rise to curious and attractive developments in contemporary logical-mathematical thought : the new logics (8) (which differ from traditional logic) the beginnings of which we have already met in Aristotle, in the Epicureans, in the scholastics, particularly in Occam (see especially Chapter IV, §1; Chapter V, §1; Chapter X, §5). The ethnologist Levy-Bruhl announced, too, the discovery of non-Aristotelean logics. One of the new logics is that of the intuitionists Brouwer and Heyting, which we have already mentioned (see Chapter XIX, §3), in relation to the efforts directed towards overcoming antinomies.

Mathematics, according to the intuitionists, is the exact part

(7) Enriques, **12**, p. 138.
(8) On many-valued logics, see: Bocheński, **1**, pp. 40–1 ; **2**, p. 75 ; **4**, pp. 469–472 ; Vaccarino, **1**.

of our thoughts; it is not a collection of symbols but an activity of the spirit, which operates only in the following cases (9).

(a) To undertake a mental act, taking away all its particular aspects, to become conscious of its simple unity. . . . A mental act of this type means the realization of a *bi-unity* (*unity* is the individuation of an act of the mind, *bi-unity* since this individuation is made possible only by a *process of contraposition* in the face of a 'remaining activity of the mind').

(b) To realize a finite number of bi-unities, external to one another in time, with respect to which they are well ordered.

These acts are based on the *a priori* notion of time, substantially Kant's. 'Mathematics admits only one source: intuition' does not consist in the classical processes of proof but in the construction of a finite succession (illimitably proceedable) of single mental acts. 'For the intuitionists every mathematical assertion represents the *intention* of a construction which satisfies determinate conditions; to prove a mathematical assertion means to effect a required construction. So the principle of the excluded third is particularly inacceptable as an existential demonstrative criterion: the absurdity of the absurdity of an exact construction does not guarantee its existence'.

A typical case of the non-constructive demonstration rejected by the intuitionists is the Cantor–Bernstein theorem stated in Chapter XVII, §1. According to the intuitionists, it is impossible to exhaust the possibilities of exact thought with a determinate system of logical rules. All the same, a formalization of the most important principles of intuitionist logic and mathematics has been made by Heyting, who for the calculus of propositions introduced the following terms:

$\neg$ negation
$\wedge$ conjunction
$\vee$ disjunction
$\supset$ implication

We cannot deduce the principle of the excluded third, or even its negation, from the principles Heyting formalized.

(9) We quote from Lerda's paper on intuitionism.

In Heyting's logic of predicates the following proposition (we should remember, too, Hilbert's symbolism) is valid:

$$(Ex)\{\neg A(x)\} \supset \neg\{(x)A(x)\};$$

But the inverse implication is not valid:

$$\neg\{(x)A(x)\} \supset (Ex)\{\neg A(x)\}.$$

This latter result shows differences between classical and intuitional logic.

According to Kolmogoroff the intuitionists' calculus of propositions can be interpreted as a calculus of problems.

In the original intuitionist reconstruction of mathematics the theories of sets, of real numbers and of functions are particularly prominent.

About 1930 J. Łukasiewicz's (10) inspired and careful constructions definitely introduced the new many-valued logics. One of these was the *trivalent* logic, in which the logical values of the propositions are three, which can be indicated by the symbols 1, $\frac{1}{2}$, 0, which are interpreted respectively as: certainly true, doubtful, certainly false. Łukasiewicz was guided by the idea of the indetermination of future events, and by modal logic (see Chapter IV, §8).

In trivalent logic, as in the system of the intuitionists, the principle of the excluded third is not valid; instead we have the following postulates:

(a) Every meaningful proposition in logic with three values always has the value 1 or else $\frac{1}{2}$ or else 0.

(b) No proposition can have two or more values at once.

In trivalent logic we generally choose as the main propositional function, functions that are a generalization of those already considered in bivalent logic; that is, if in the matrices which define the functions considered in trivalent logic we remove the lines relating to the values $\frac{1}{2}$, we obtain the corresponding matrices of bivalent logic (see Chapter XVIII, §6).

These generalizations can be made in various ways, for instance by negation (Np), by disjunction (Apq), by logical conjunction (Kpq), by material implication (Cpq):

(10) Łukasiewicz: we quote from Belletti's paper on many-valued logics.

p	Np	Apq	$1\ \tfrac{1}{2}\ 0$	Kpq	$1\ \tfrac{1}{2}\ 0$	Cpq	$1\ \tfrac{1}{2}\ 0$
1	0	1	1 1 1	1	$1\ \tfrac{1}{2}\ 0$	1	$1\ \tfrac{1}{2}\ 0$
$\tfrac{1}{2}$	$\tfrac{1}{2}$	$\tfrac{1}{2}$	$1\ \tfrac{1}{2}\ \tfrac{1}{2}$	$\tfrac{1}{2}$	$\tfrac{1}{2}\ \tfrac{1}{2}\ 0$	$\tfrac{1}{2}$	$1\ 1\ \tfrac{1}{2}$
0	1	0	$1\ \tfrac{1}{2}\ 0$	0	0 0 0	0	1 1 1

Other logics can be constructed with more than three values.

The many-valued logics are particularly interesting when placed in relation to the calculus of probability and the logic of modalities, and they are applied in present-day physics and in the philosophy of law (11).

The philosophical meaning of non-Aristotelean logics has given rise to advanced and subtle discussion. F. Severi, referring to Gödel's results (which we shall treat in §5 of this chapter) observes (12):

'All that remains therefore, in verifying the compatibility, is to realize what Enriques calls the relative character of the ascertainment, that is in the last analysis its reduction to the compatibility of the real, as resulting from *our* senses or intuition.

'The same criterion, or rather the exquisitely super-logical postulate of the compatibility *a priori* of the real, is needed to understand and verify the compatibility of the axioms of the new logics thought of as symbolical structures (distributive or not) of propositions. If these are to be understood by a sane man with a mind which can follow and listen to expositions, the judgment of these cannot but be founded on common logic, which is classical logic, whose principles descend straight from intuitive reality.

'The new logics, if one wishes to call them by their real name, without wanting to *épater le bourgeois*, are nothing but chapters of modern algebra or abstract algebra; and the assertions that in them can be invalid the Principle of Contradiction or that of the excluded third, are expressions of algebraic theorems, which, leaving aside euphemisms, we should end by understanding through that same logic whose principles have been valid for thousands of years.'

Severi also points out an important result obtained by Moisil

(11) See Reichenbach, **1**, **2**; Bobbio, **2**; Vaccarino, **2**, pp. 245–9.
(12) Severi, **3**, p. 37.

on trivalent logic, a result which appears rich in philosophical meaning (13): 'as the non-geometries are interdependent among themselves (remember the images of non-Euclidean geometry in ordinary geometry), the same thing happens in the various logics, in relation to classical logic. Moisil himself finds in the field of logic a theorem analogous to that of Beltrami in the field of geometry; that is a mapping of the trivalent logic of Łukasiewicz on classical logic such that every proposition of the first finds its interpretation in two propositions of the second.'

Many, even today, are attracted by Severi's idea of a logic that allows us to understand the fundamentals of the various specialized logics and to judge their validity; but in fact no logical proof can establish the existence of a supreme logic. In fact, whatever proof one considered would presuppose the existence of the very logic it was trying to prove. The existence of the supreme logic can be admitted only through intuition.

§3. *Neo-positivism and neo-empiricism. Carnap's exact language* (14)

The positivism of A. Comte (1798-1857), J. Stuart Mill (1806–73), H. Spencer (1820–1903), and E. Mach (1838–1916) admitted only the material facts (physical, biological, naturalistic) in the world of science. Yet there was still room, within the framework of nineteenth-century positivism, for universal and necessary truths which have lost all meaning for our contemporary neo-positivists.

Old and new positivists consider the empirical method as the criterion of truth, but while the nineteenth-century positivists deified the fact in a certain sense (Ardigò used to say: 'The fact is divine, the explanation human'), the neo-positivists abandoned the metaphysics of facts, and concentrated on the logical connections of the interpretation of experience, relating the meaning and value of an enunciation to the method of verifying it; that is, what cannot first of all be observed or measured does not exist physically. Science becomes essentially methodological.

(13) Severi, 4, p. 238.
(14) On neo-positivism and neo-empiricism, see: Geymonat, 1, pp. 6–8; Waismann 2 (on neo-empiricism in reference to mathematics); Abbagnano; Severi, 4, pp. 225–44; Barone; Carruccio, 21.

Closely linked with neo-positivism is neo-empiricism. Neo-positivism deals above all with researches in the physical world, neo-empiricism looks for the basis and meaning of mathematics.

The present-day neo-empiricist's point of departure is formulated by Geymonat thus: 'Thought must not be spoken of when it is not clearly expressed in speech or in some other way.' This naturally leads us to consider the analysis of language, identified with thought, as essential in philosophy, and the construction of the new logics appears very important. According to Carnap, the term 'exact language' (15) means 'the set of figures formed by a finite number of elements (the so-called linguistic signs), the form, combinations and transformations of which are established exactly from certain basic conventions (or syntactic rules of the language considered)'.

A mathematical theory can be considered Carnap's 'exact language'. How does Pieri's hypothetical-deductive system differ from this?

In the conception of the hypothetical-deductive system we speak of arbitrarily chosen postulates at the basis of the system, but the rules for deducing the consequences are those of a logic that is considered single and universal. But in Carnap's conception every system has its own logical and syntactical rules. Just as the non-Euclidean geometries come between Aristotle's demonstrative science and Pieri's hypothetical-deductive system, so the non-Aristotelean or non-Crisippian logics come between the hypothetical-deductive system and the exact language.

According to Geymonat (16), 'the task of logic is not to trust in one convention or another, but to build them all up, to develop them one after the other as instruments of our freedom. There is no categorical imperative in logic, as Carnap put it so well: "*In der Logik gibt es keine Moral*".'

What propositions of an exact language *L* have any meaning for the neo-empiricists? Only the postulates and their consequences in so far as they are true in *L* and the negations of one or the other in so far as they are false in *L* (17).

(15) Geymonat, **1**, pp. 128–9.
(16) Geymonat, **3**, pp. 130–1.
(17) Waismann, **2**, pp. 178, 286, 287.

As far as antinomies are concerned the neo-positivist's attitude differs from that of the mathematical logicians we have so far considered. He interprets the antinomies as 'undecided propositions ... for which the logical structure of the language does not allow us to give either a positive or a negative reply' (18).

Yet a profound difference exists between undecided propositions and antinomies. In an exact language L a proposition P is undecided, when P and $\overline{P}$ do not belong to L. Whereas antinomies exist in L if we can deduce P and $\overline{P}$ in L.

Even a logic that does not share all the philosophical positions of neo-positivism or neo-empiricism cannot fail to appreciate the work that originated in the *Tractatus Logico-Philosophicus* (Anglo-German edition, with a preface by Bertrand Russell, London 1922), of L. Wittgenstein (19), who founded the Wiener Kreis (Vienna Circle) (20); and K. Gödel's work, undertaken in the Vienna Circle which we will deal with in §5, is very important too in modern logical thought.

§4. *Rational system* (21).

Carnap's conception of an 'exact language' is based essentially on two principles: (1) when new logics are to be built up, the logical rules for deducing theorems from postulates must be made precise for every exact language; (2) rational thought is identified with the group of signs through which it is expressed.

Whether we like it or not, we are bound to take notice of (1) since new logics have been constructed that differ from the traditional logics, and so the rules for deduction must be made precise for every theory one wants to develop logically. For instance, are demonstrations *per absurdum* allowed? (According to traditional logic, yes: according to the logic of the intuitionists, no.) Is it permissible to operate with sets containing an infinity of elements? (According to Aristotle, no; according to Cantor, yes.) And so on.

(18) Geymonat, **3**, pp. 130–1.
(19) Wittgenstein, **1** and **2**.
(20) The 'Centro metodologico di Torino' has fostered studies in the philosophy of science, starting from the 'Wiener Kreis': we recall Abbagnano ... Persico; Abbagnano ... De Finetti; Atti del Congresso di Studi metodologici, Torino, 1954; Casari; Rossi Landi.
(21) Carruccio, **20** and **21**.

But this does not touch on the question of whether a supreme logic does or does not exist.

Point (2), however, is not justified by all this, and, as §7 of this chapter will show, it cannot be accepted. However, leaving aside for the moment the question of the possibility of expressing a theory integrally in symbols, we will set down the following definition of a *rational system*, which we shall refer to later.

A rational system is built up of a finite number m of postulates $P_1, P_2 \ldots P_m$, and from theorems which are derived from them by applying n rules $r_1, r_2 \ldots r_n$ for deduction, these also finite in number.

§5. *Gödel's theorem on the non-contradictoriness of systems. Metamathematics*

To return to the problem of the non-contradictoriness or consistency of the rational systems: according to Enriques, a judgment on the compatibility of a system of mathematical postulates can be based on physical and psychological experience, on intuition, or on a logical demonstration, this latter being the only one which can be counted as rigorous proof; but the value of such a proof is only relative, because the compatibility of the system is deduced from the compatibility of another system (it is not demonstrated directly) (22).

Is it possible, now, to get over this relative character? In 1908 B. Levi (23) established that it was impossible to prove the non-contradictoriness of logical laws, and later a more general result of Gödel's, published in 1931, replied in the negative to this question (24). Waismann enunciated Gödel's result thus (25): 'It is impossible to demonstrate the non-contradictoriness of a logical mathematical system using only the means offered by this system.'

Gödel gives the demonstration of his theorem in two forms: the first of these is short and suggestive and can be referred to any rational system; as Gödel says himself, it has a certain

(22) Enriques, **6** (1926), ch. III, par. 18, pp. 113–15.

(23) 'Can the laws of thought cause contradiction? It is not easy to admit this, but it is impossible to *prove* the impossibility of such a contradiction; for a proof must rely upon the rules of logic, and must also accept their compatibility' (Levi, **1**, pp. 187–216).

(24) Gödel; Waismann, **2**, pp. 142–4; Carruccio, **9**; Ladrière.

(25) Waismann, **2**, p. 142.

similarity to the paradox of the Liar, and Richard's paradox; the second, the fundamental thought of which appeared in the first, is developed in detail and refers to a system P founded on the five axioms of Peano's arithmetic, and on those on which Whitehead and Russell's *Principia Mathematica* are based; it mentions applications to other systems, but leaves some gaps, which Gödel himself recognizes.

Other mathematical logicians (26) took up and perfected his reasoning, but there have been objections to it, too, for instance from Carnap (27).

But as far as the development of our own considerations is concerned, a simple piece of reasoning based on the theorem of the Pseudo-Scotus (see Chapter X, §4) will do; this theorem can be demonstrated at once on the basis of Hilbert's calculus of propositions (see Chapter XVIII, §5), observing that

$$(A \ \& \ \overline{A}) \to B$$

is an expression which is always true for any B, if we admit the antecedent of the material implication considered.

We can now demonstrate the following theorem (28). *It is not possible to demonstrate the non-contradictoriness of a rational system by using only the means offered by the system itself.*

In fact, suppose we have managed to prove, with the means offered only by the rational system S, with a chain of formally perfect deductions, that S itself is not contradictory.

But for a proof in S to be valid we must know that the system S is non-contradictory, because if it were, any proposition could be proved in S (in particular that S is non-contradictory).

But if what we want to prove is just that very non-contradictoriness of S, it is proved that a proof of the non-contradictoriness of S by the means offered by the system S always begs the question.

(26) See bibliography in Church, 2.

(27) Carnap does not approve Gödel's way of thinking: it presupposes that only plausible methods of proof are allowed, but excludes some plausible methods (see Lombardo Radice).

(28) This proof may be found at the end of Carruccio, **9**, §3. In Carruccio, **17**, pp. 99–101, there is a statement of the discussions concerning the proofs of Gödel's theorem.

It has been remarked that, while Gödel proves that the non-contradictoriness of S cannot be proved in S, here we prove only that such a proof, if it existed, would have no value. Such a result may suffice.

On Gödel's theorem, see also Casari, p. 88.

In fact, to prove the non-contradictoriness of S is absolutely impossible.

Let us suppose we have managed to prove the consistency of S with the means offered by another system T. This proof would effectively attain its object if we could prove the consistency of T; but this could not be done with the means offered by T; we should therefore need to have recourse to another system U, and so on. Either we keep meeting new systems all the time, or else in the succession S, T, U . . . we meet some system already met with on the succession itself. The consistency of S cannot be definitely proved: it can only be established through a postulate on the basis of physical or psychological experience, or on intuition.

Gödel's theorem is one of the most suggestive results of the new discipline called metamathematics, which deals with the theory of mathematical proof.

§6. *Some consequences of Gödel's work*

Two possible results of Gödel's work suggest themselves: the first is the following: there exists at least one rational consistent system Σ_1 such that, given any other system S, either this is contradictory, or else, if it is consistent, its consistency can be proved, having admitted the consistency of Σ_1. It is hardly necessary to illustrate the importance the system Σ_1 would have, whenever it existed, in the solution of many fundamental mathematical questions. (It might perhaps be thought that Σ_1 was identifiable with a system based on logical axioms, or with a very general theory of aggregates. . . .) If this possibility were verified, Gödel's result would have a limited importance, because as the consistency of a system S could not be demonstrated by means intrinsic to S, this same consistency could be deduced from a single privileged system Σ_1, which one would hope would be evidently consistent.

The second possibility is that a system Σ_1 with the properties mentioned does not exist.

Recent researches have brought us to the second alternative (29), and have shown the importance of Gödel's theorem.

To demonstrate another of its consequences in the field of real numbers, here are three definitions.

(29) Carruccio, **17**, §5.

Definition 1. We will call well-defined a real number of which it is possible to calculate values to within a prearranged degree of approximation.

Numbers are therefore well defined whose successive reduction to continued fractions can be calculated, the numbers expressed as developments in convergent series, or else as an infinite decimal whose successive figures can be calculated, etc.

Definition 2. We call *ultrarational* a real number for which we cannot demonstrate that it is rational or irrational.

We shall demonstrate that such numbers exist and that therefore the definition is legitimate.

Definition 3. A rational system S is called completely known when, from the postulates and rules for deduction which are at its base, we can obtain the theorems of the system, one after the other.

Through the systems which are completely known, even if there exist different orders in which it is possible to obtain the theorems, it should be possible to fix an order in which every theorem of the system is successively reached by a finite number of logical steps.

We can say that if a completely known rational system contains an infinite number of theorems it is possible to establish a one-to-one correspondence between the natural numbers and propositions of the system itself (30).

Theorem. There exist well-defined ultrarational numbers (31).

In fact: consider the set of the formulae that express the postulates and the theorems of a completely known rational system S, containing an infinite number of theorems, these formulae are placed in biunivocal correspondence with the series of natural numbers. To every proposition

$$f, f_1 \ldots f_n$$

which is not in contradiction with one of the preceding, we will make correspond by a determinate law the natural numbers that differ from zero:

$$a, a_1 \ldots a_n \ldots$$

(which can also, for the sake of simplicity, be taken as equal to one another).

(30) Carruccio, **16**.
(31) Carruccio, **9**.

Now form the following continued fraction:

$$a + \cfrac{1}{a_1 + \cfrac{1}{a_2 + \cfrac{\cdots}{\begin{matrix} a_n \\ \cdot \\ \cdot \\ \cdot \end{matrix}}}}$$

If in the succession of theorems of S we meet a theorem in contradiction with one of the preceding, the succession is interrupted and stops at the term a_n corresponding to the last theorem of the system S which does not give rise to explicit contradiction. The continued fraction will in every case be equal to a real well-defined number, rational if the succession is limited, irrational if it is not.

If the system S is contradictory this fact will have to be established with a finite number of logical steps. But if S is non-contradictory we cannot demonstrate its non-contradictoriness, nor foresee therefore that the succession $a, a_1 \ldots a_n$ will be non-terminating and that the number expressed in the continued fraction considered will turn out to be irrational. Nor can it be demonstrated that the number in question is rational, because that would be equivalent to saying that S is contradictory, contrary to what we have supposed. There exist, therefore, ultrarational well-defined numbers.

The interpretation of this has given rise to arguments which we shall not examine (32).

Observation. There are well-defined real numbers which have not so far been proved to be rational or irrational, in spite of mathematicians' efforts. One of these is Euler's constant γ, for instance. It may be that some of these numbers are ultrarational.

With processes similar to these it has been shown that there

(32) Carruccio, 17, §5.

exist pairs of well-defined numbers which cannot be proved to be equal or unequal, that there exist series which are certainly convergent or divergent, but for which it is not possible to prove which of the two cases is in fact verified; and many other like phenomena (33).

These all show that mathematics cannot be exhausted in a single rational system, since in any system S there always exist statements which are undecidable in S (that is, which cannot be proved and also cannot be disproved) for instance, the consistency of S with its results.

§7. *The problem of expressing a rational system in symbols* (34)

From Leibniz's efforts, which we have mentioned, to construct an ideography that expressed in symbols the logical relationships (*characteristica universalis*) to the systems of Peano, Whitehead and Russell, and Hilbert, modern mathematics has aimed to translate rational theories into symbols.

At this point a problem arises: is it possible to translate a rational system totally into symbols? Or is there always, in every rational system, an unexpressed 'something' which cannot be expressed through any system of symbols?

For reasons of space we will not go into the very interesting observations made by Beppo Levi and Geymonat on the subject, but simply deal with the question itself, on which the neo-empiricism of the Vienna Circle is based; this, identifying (in a broad sense) thought and word, seems to exclude the possibility of a 'something' in thought which does not find expression in language.

Consider any disposition D (with repetition) of m symbols n at a time and see if D can express the primitive propositions and the rules for deduction at the base of a given rational system S; if the propositions and rules are known we can gradually obtain the symbolic expressions of the theorems t_k, of S.

I say that the disposition D can, logically speaking, always be interpreted in various ways which lead to various successions of theorems, and these can be expressed through various successions of arrangements of symbols.

(33) See Waismann, **2**, ch. IX; Carruccio, **15** and **20**.
(34) For this matter, see Carruccio, **13**; with passages of Levi and Geymonat; Severi, **3**, p. 36.

In fact: let a rational being M express through the arrangement D of symbols, as perfectly as possible, the primitive propositions and rules for deduction of a rational system S, which is in his mind, to which system a non-empty succession of theorems t_k belongs. Let M show the disposition D to another rational being R for him to obtain, once he has understood the meaning of D, the theorems t_k. Note that no 'original language' (in Geymonat's terminology) is presumed to be common to M and R, and already established before M shows R the disposition D; but on the very basis of D must be founded the language that characterizes the system S.

In such a case all R can do is try to interpret the meaning of D from the various possibilities.

One of these interpretations is the one in M's own mind, which leads to the reconstruction of the system S; this is not, of course, the only possible interpretation—a more obvious one, for instance, sees the symbols that make up D simply as an aggregate of objects of thought, shown to R in a particular order, which he must simply contemplate, without working anything out. According to this interpretation, the symbolic expression of the theorems built up by R on the basis of D would be without elements.

A rational system S is therefore well defined in M's mind, because he knows how to find the successive theorems of S and their expressions, but D can always be interpreted in other ways, which give rise to other successions of symbols expressing theorems; and the expression D in symbols of S does not wholly express M's knowledge of the system S.

This proves that a rational system cannot be expressed through symbolism, even when symbolism is understood as any exact language in the sense defined by Carnap.

These views on the question of inexpressibility are substantially those of St. Augustine in *De Magistro* (35).

§8. *General remarks on recent developments in logic* (36)

It cannot be denied that many people are almost bewildered by their first impression of the findings of contemporary logic. How, they feel, can mathematics, which boasts the most perfect form of expression of rational thought, hide what cannot be

(35) See ch. IX, §1 of this book.
(36) Carruccio, **10**, pp. 53–4.

expressed? Do you find what cannot be demonstrated lurking in demonstrative science? Is all this trying to undermine the very basis of mathematics?

This attitude is similar to the Pythagoreans' feelings in the face of the discovery of incommensurables and the theory of irrationals. But some of this bewilderment, at least, may disappear with further thought on the subject.

The results obtained concerning the impossibility of expressing a rational system S totally in symbols, and of transmitting my thoughts to others by means of symbols, does not prevent a rational theory from existing in my mind, which cannot be placed in biunivocal correspondence with sensible facts which are the signs of a language; and this theory must develop individually in each man's mind, as a personal effort.

This is very important from the didactic point of view, which holds that mathematics should be taught according to Socrates' so-called amaieutic method, linked by Plato with the theory of remembrance and given new meaning and importance in St. Augustine's philosophy.

Thus a teacher of mathematics cannot transmit the knowledge of a rational system from his own mind to that of his pupil; he must make the pupil reconstruct the system itself in his own mind, through his own intellectual effort.

The fact that a rational system cannot be proved to be non-contradictory shows a deeper impossibility of proof than in the case of the duplication of the cube, the trisection of an angle, or the squaring of the circle. These problems once depended on the use of particular instruments; but the impossibility of proving the non-contradictoriness of a non-contradictory rational system, and of solving certain questions of analysis, depends on the use and the limitations of human reason.

These matters take us to the limits of the world ruled by our logic, and give us glimpses of things beyond it. But here a question arises: does anything, even if only in the world of thought, really exist beyond what our logic can deal with? This question is linked with the problems of what has meaning in mathematics, or more precisely, what has meaning in a determinate system S (37).

From the point of view of contemporary logic, we must give

(37) Carruccio, **17**, §3.

a 'syntax of meaning', that is, establish with what rules we mean to use the term 'meaning'. To the neo-empiricists only the postulates, their consequences, and the negations of one or the other have a meaning, and there is nothing to stop us accepting their way of understanding the term 'meaning'. But once we have accepted this 'syntax' we must accept all its consequences. In particular, being given the impossibility of effectively proving the affirmation: the rational system S is not contradictory, there is no meaning in speaking of non-contradictory rational systems, nor may we deal with a well-defined real number, for instance, Euler's constant γ, before proving whether it is rational or irrational, because in these numbers may hide a non-meaning, and so on. But may not the very size of the field to which neo-empiricist criticism could be extended be its weakness? Where can the mathematician be free of non-meanings, when he is not allowed to speak of consistent rational systems?

It seems suitable, therefore, to adopt a less restricted syntax of meaning, for instance this: a proposition has a meaning in S when it refers to objects of logical thought, such, that is, that they satisfy the conditions expressed in the three principles of logic: identity, non-contradiction, excluded third. This means that statements relating to objects of thought that appear in the postulates by means of which they are implicitly defined, or constructed from logical operations agreeing with the postulates themselves, have a meaning.

In fact many of us accept intuitively that a rational system is non-contradictory or contradictory, that a well-defined number is rational or irrational, even if it is not possible for the human mind to establish with a finite number of logical steps which of the two possibilities holds.

All this may lead to a new understanding of the concept of objectivity in mathematics, mathematics composed not merely of arithmetic and Euclidean geometry, as in ancient times, but of all consistent rational systems. The recognition that a part of these systems is not contradictory, from which one could obtain the non-contradictoriness of other parts, must be made by an act beyond logic—dare I say an act of faith? All this gives rise to the living, disturbing problems of contemporary thought, which open up ever wider horizons.

Bibliography*

ABBAGNANO, N.: *La metodologia delle Scienze nella filosofia contemporanea* (in ABBAGNANO ... DE FINETTI).

ABBAGNANO, BOBBIO, BUZANO, CODEGONE, FROLA, GEYMONAT, NUVOLI, DE FINETTI: *Saggi di critica delle Scienze*, Turin, 1950.

ABBAGNANO, BUZANO, BUZZATI-TRAVERSO, FROLA, GEYMONAT, PERSICO: *Fondamenti logici della Scienza*, Turin, 1947.

ACKERMANN, W.: *Grundgedanken einer typenfreien Logik* (in *Essays dedicated to A. H. Fraenkel*, Jerusalem, 1961).

ADAM, CH.: *Vie et Œuvres de Descartes* (in DESCARTES, t. XII).

ADAM, CH., ... WAHL, J.: *Descartes*, Paris, 1937.

AGOSTINI, A.:
  1 *La memoria di Evangelista Torricelli sopra la spirale logaritmica riordinata e completata* ('Pubblicazioni scientifiche a cura dell' Accademia Navale', Livorno, 1949).
  2 *Il metodo delle tangenti fondato sopra la dottrina dei moti nelle opere di Torricelli* ('Periodico di Matematiche', Nov. 1950).
  3 *L'opera matematica di P. Mengoli* ('Archives Internationales d'Histoire des Sciences', Paris, n. 13, 1950).

ALBERTUS MAGNUS (St.): *Opera Omnia*, Lyons, 1651.

AMALDI, U.: *Sui concetti di retta e di piano* (ENRIQUES, 7, part I, Vol. I).

AMODEO, F.: *Origine e sviluppo della geometria proiettiva*, Naples, 1939.

APOLLONIUS of PERGA:
  1 Apollonii Pergaei: *Quae graeca extant cum commentariis antiquis*, edited by J. L. Heiberg, Leipzig, 1891–93.
  2 *Les coniques d'Apollonius de Perge*. French translation by P. Ver Eecke, Bruges, 1924.
  3 *Treatise on conic sections*, edited ... by T. L. Heath, Cambridge, 1896. (New edition, Cambridge, 1961.)

---

* The bibliography refers chiefly to the works the author has considered in composing this volume.

I particularly recall the works of my late masters: Federigo Enriques, Ettore Bortolotti, Giovanni Vacca.

I remember with sorrow my late wife Emma, who helped me with the early editions. I am grateful to my friends Francesco Lerda, Pietro Lingua, and to my son Enea, who have helped me with this edition.

Mathematics and Logic in History and Contemporary Thought

ARCHIMEDES:

1 *Opera omnia cum commentariis Eutocii*, ed. Heiberg, Leipzig, 1910–15.

2 *The Works of Archimedes, with the Method* . . . edited by T. L. Heath, New York and London, 1955.

ARISTOTLE:

1 Aristotelis: *Opera omnia graece et latine*. ed. F. Didot, Paris, 1878.

2 *Categoriae.*

3 *De Interpretatione.*

4 *Analytica priora.*

5 *Analytica posteriora.*

6 *Topica.*

7 *Physica Auscultatio.*

8 *De Coelo.*

9 *De Anima.*

10 *Mechanica.*

11 *Metaphysica.*

AUGUSTINE (St.):

1 *De Magistro.*

2 *Il Maestro.* Italian translation, introduction, commentary and appendices by A. Guzzo, Florence, 1927.

3 *De Ordine.*

4 *De Genesi ad litteram.*

5 *De Civitate Dei.*

6 *De Trinitate.*

7 *De Diversis quaestionibus* LXXXIII, liber unus.

BACON, R.: *Opus Majus*, Venice, 1750.

BAILLET, A.: *La vie de Monsieur Des Cartes*, 1691.

BARONE, F.: *Il neo positivismo logico*, Turin, 1953.

BELLETTI, A.: *Logiche polivalenti* ('La nuova critica', n. IV, Florence, 1957).

BENEDICTY, M.: *La geometria algebrica astratta e il concetto di varietà algebrica* ('Archimede', fasc. 4–5, Florence, 1953).

BERZOLARI, L.: *Enciclopedia delle Matematiche elementari*, Milan, 1930–53.

BJÖRNBO, A. A.: *Studien über Menelaos Sphärik* ('Abhandlung zur Geschichte der mathematischen Wissenschaften mit Einschluss ihrer Anwendungen', Leipzig, 1902).

BOBBIO, N.:

1 *Scienza del diritto e analisi del linguaggio* (in ABBAGNANO . . . DE FINETTI).

2 *La logica giuridica di Eduardo Garcia Mainez* ('Rivista internazionale di filosofia del Diritto', 1954, fasc. V–VI).

BOCHEŃSKI, I. M.:

1 *Nove lezioni di logica simbolica*, Rome, 1938.

Bibliography

2 *Précis de logique mathématique.* F. G. Kroonder, Bussum, 1948.
3 *Ancient formal Logic*, Amsterdam, 1951.
4 *Formale Logik*, Freiburg–Munich, 1956.
5 *A history of formal logic*, Notre Dame, Indiana, 1961.

BOEHNER, Ph.: *See* OCCAM.

BOMBELLI, R.: *L'algebra pubblicata a cura di E. Bortolotti*, Bologna, 1929.

BOMPIANI, E.: *Che cosa contiene la 'Geométrie' di Cartesio* ('Periodico di Matematiche', Bologna, 1921).

BONOLA, R.:
1 *La geometria non-euclidea* (Historical and critical study of its development) Bologna, 1906.
2 *Sulla teoria delle parallele e sulle geometrie non-euclidee* (in ENRIQUES, 7, part I, vol. II).
3 *Non-Euclidean Geometry*, New York, 1912

BORTOLOTTI, E.:
1 *Influenza dell'opera matematica di Paolo Ruffini sullo svolgimento delle teorie algebriche*, Modena, 1903.
2 *Lo studio di Bologna e il rinnovamento delle scienze matematiche in Occidente* ('Annuario della R. Università di Bologna per l'anno accademico 1920–21').
3 *Gli inviluppi di linee curve e i primordi del metodo inverso delle tangenti* ('Periodico di matematiche', July 1921).
4 *Le prime applicazioni del calcolo integrale alla determinazione del centro di gravità delle figure geometriche* (Paper read to the Royal Academy of Sciences of Bologna, 28 May 1922).
5 *Lezioni di geometria analitica*, vol. I, Bologna, 1923. (Historical introduction.)
6 *La scoperta e le successive generalizzazioni di un teorema fondamentale di calcolo integrale* ('Archivio di storia della Scienza', Sept. 1924).
7 *La memoria 'de infinitis hyperbolis' di Torricelli* ('Archivio di storia della scienza' fasc. I e II, 1925).
8 *L'algebra nella scuola matematica bolognese del secolo XVI* ('Periodico di matematiche', May 1925).
9 *I progressi del metodo infinitesimale nell'opera geometrica di Evangelista Torricelli* ('Periodico di matematiche') Bologna, Jan. 1928.
10 *Le prime rettificazioni di un arco di curva nella memoria 'de infinitis spiralibus' di E. Torricelli* (Paper read to the Royal Academy of Sciences of Bologna, 21 April 1928).
11 *Studi e ricerche sulla storia della matematica in Italia nei secoli XVI e XVII*, Bologna, 1928.
12 *I cartelli di matematica disfida e la personalità psichica e morale di Girolamo Cardano* (Imola, 1933).
13 *La scienza algebrica degli Egizi e dei Babilonesi* ('Memorie dell' Accademia delle Scienze di Bologna' Ser. IX, t. II, 1934–35).

14 L'infinito e l'infinitesimo nella matematica antica ('Memorie dell'Accademia delle Scienze di Bologna', Ser. IX, t. V, 1937–38).

15 L'opera geometrica di E. Torricelli ('Monatsheften für Mathematik und Physik', 48 Band, Leipzig and Vienna, 1939).

16 Primi algoritmi infiniti nelle opere dei matematici italiani del sec. XVII ('Bollettino dell'Unione Matematica Italiana', Bologna, 30 June 1939).

17 Lemmi e postulati attinenti ai concetti di infinito e di infinitesimo attuali ('Bollettino dell'Unione Matematica Italiana' Bologna, Nov.–Dec. 1939).

18 Le fonti della matematica moderna, matematica sumerica e matematica babilonese. ('Memorie della Accademia delle Scienze di Bologna', Ser. IX, t. VII, a. 1939–40).

19 Storia della matematica nella Università di Bologna, Bologna, 1947.

20 Storia della Matematica elementare ('Enciclopedia delle Matematiche elementari', edited by L. Berzolari, Vol. III, part II, Milan, 1950).

BOUTROUX, P.: *L'idéal scientifique des Mathématiciens*, Paris, 1920.

BRUNET, D. and MIELI, A.: *Histoire des sciences. Antiquité*, Paris, 1935.

BRUNSCHVICG, L.: *Les étapes de la Philosophie Mathématique*, Paris, 1912 (and other editions).

BRUSOTTI, L.: *Questioni didattiche* (in BERZOLARI, Vol. III, part II).

BUYTAERT, E. M., O.F.M.: *Bibliography of Fr. Philotheus Boehner O.F.M.* ('Franciscan Studies', vol. 15 n. 4 Dec. 1955).

CALÒ, B.: *Problemi trascendenti e quadratura del cerchio* (in ENRIQUES, 7, part II).

CALOGERO, G.: *Aristotele* (Article in the *Enciclopedia Italiana Treccani*, with bibliography).

CANTOR, G.: *Gesammelte Abhandlungen* ... edited by E. Zermelo, Berlin, 1932.

CAPONE BRAGA, G.: *Galileo e il metodo degli indivisibili* ('Sophia', Padua, July–Dec. 1950).

CARDANO, G.:
  1 Opera omnia, Lyons, 1663.
  2 Ars magna.
  3 De Regula aurea.

CARLINI, A.: *Aristotele. Il principio logico*, Bari, 1924.

CARNAP, R.:
  1 Foundations of Logic and Mathematics, Chicago, 1939.
  2 Fondamenti di Logica e Matematica. (Italian translation, notes and introduction by G. PRETI) Turin, 1956.

CARRUCCIO, E.:
  1 Il 'Nuovo Metodo' di Leibniz. (With historical notes.) ('Periodico di Matematiche' Nov. 1927).

Bibliography

2 *Applicazioni della legge di dualità sulla sfera alla teoria degli isoperimetri* ('Periodico di Matematiche', May 1932).

3 *Note sul poema astronomico e astrologico di Manilio* ('Archeion', Rome, a. 1936, fasc. IV).

4 *Notizie storiche sulla geometria delle api* ('Periodico di Matematiche', Bologna, January 1936).

5 *La quadratura delle curve secondo Newton* ('Periodico di Matematiche', Bologna, Feb. 1938).

6 *Costruzione dell'ettagono regolare secondo Archimede e i matematici arabi* ('Periodico di Matematiche', Bologna, Dec. 1938).

7 *L'estrazione di radice cubica, mediante inserzione di due medie proporzionali fra due segmenti dati in Leonardo Pisano* ('Periodico di Matematiche', Bologna, fasc. 4, 1939).

8 *Galileo precursore della teoria degli insiemi* ('Bollettino dell'Unione Matematica Italiana', April–June 1942).

9 *Considerazioni sulla compatibilità di un sistema di postulati e sulla dimostrabilità delle formule matematiche* ('Pontificia Academia Scientiarum', Acta vol. X, n. 2, paper presented in 1945).

10 *Orizzonti e frontiere della Logica matematica* ('Studium', Rome, Jan.–Feb. 1946).

11 *Presentazione del carteggio matematico di Paolo Ruffini ordinato a cura di Ettore Bortolotti* ('Accademia di Scienze Lettere ed Arti di Modena', commemoration of 22 June 1946, Modena, 1947).

12 *I fini del 'calculus ratiocinator' di Leibniz, e la logica matematica del nostro tempo* ('Bollettino dell'Unione Matematica Italiana', Bologna, August 1948).

13 *Il problema dell'esprimibilità in simboli del sistema ipotetico-deduttivo* ('Sigma', n. 6–7, Rome, 1948).

14 *Recenti sviluppi della logica matematica . . .* ('Atti del convegno di Pisa, 23–27 settembre 1948', Città di Castello, 1949).

15 *Il problema della razionalità del reale* ('Archimede', Florence, Feb. 1949).

16 *Sulla potenza dell'insieme delle proposizioni di un dato sistema ipotetico-deduttivo* ('Bollettino dell'Unione Matematica Italiana', Oct. 1949).

17 *Sulle dimostrazioni di coerenza dei sistemi ipotetico-deduttivi*, ('Università e Politecnico di Torino-Rendiconti del Seminario Matematico', vol. 10; 1950–51).

18 *Torricelli precursore dell'analisi infinitesimale* (Published in the volume 'Evangelista Torricelli nel terzo centenario della morte', Florence, 1951).

19 *La matematica nel pensiero di Cartesio* ('Rivista di matematica dell'Università di Parma', 2, 133–152, 1951).

20 *Costruzione di un sistema razionale più ampio di un dato* ('Archimede', July–Oct. 1952).

21 *Riflessioni critiche sui fondamenti del neoempirismo* ('Atti del Congresso di studi metodologici', 17–20 Dec. 1952, Turin, 1954).

22 *La logica matematica nel passato e nel presente della scienza* ('Scientia', Oct. 1954).

23 *Sul significato filosofico della Logica matematica contemporanea* ('Humanitas', X Brescia, 1955, 1).

24 *I fondamenti dell'analisi matematica nel pensiero di Agostino Cauchy* ('Bollettino della Unione Matematica Italiana', Bologna, June 1957); also (with additions) in 'Rendiconti del Seminario Matematico' of the University of Turin (Vol. 16, 1956–57).

25 *Influenza agostiniana sulla concezione delle matematiche nella scuola di Galileo* ('Bollettino dell'Unione Matematica Italiana', Bologna, June 1957).

26 *Prodromi delle logiche non-aristoteliche nell'antichità e nel medioevo* ('Actes du VIIIe Congrès International d'Histoire des Sciences', Florence–Milan, 3–9 Sept. 1956).

CASARI, E.: *Logica dei predicati* ('Centro di studi metodologici', published by the Unione Industriale di Torino, Turin, April 1957).

CASSINA, U.:

1 *In occasione de septuagesimo anno de Giuseppe Peano* ('Schola et vita', 1928).

2 *Vita et opera de Giuseppe Peano* ('Schola et vita', Milan, 1932, n. 3, with list of publications of G. Peano).

3 *L'œuvre philosophique de G. Peano* ('Revue de Métaphysique et de morale', 1933).

4 *Sulla Logica matematica di G. Peano* ('Bollettino dell'Unione Matematica Italiana', April 1933).

5 *L'œuvre philosophique de G. Peano* ('Revue de Métaphisique et de morale', 1933).

6 *Storia del concetto di limite* ('Periodico di Matematiche', Bologna, 1936).

7 *Sul principio della scelta e alcuni problemi dell'infinito* ('Rend. sem. mat. e fis.', Milan, 10, 1936).

8 *Parallelo fra la logica teoretica di Hilbert e quella di Peano* ('Periodico di Matematiche', 1937).

9 *Ideografia e logica matematica* ('Periodico di Matematiche', April 1952).

10 *Sulla dimostrazione di Wallis del postulato quinto di Euclide* ('Periodico di Matematiche', Bologna, Oct. 1956).

CASTELNUOVO, G.:

1 *Lezioni di geometria analitica*, Milan, 1924.

2 *Calcolo delle probabilità*, Bologna, 1926; another edition, Bologna, 1947.

3 *Sulla risolubilità dei problemi geometrici con gli strumenti elementari; contributo della geometria analitica* (in ENRIQUES, 7, part II).

4 *Probabilità, calcolo delle* (Article in the *Enciclopedia Italiana Treccani*).

5 *Le origini del calcolo infinitesimale nell'era moderna*, Bologna, 1938; Milan, 1962.

Bibliography

CAVALIERI, B.:
1 *Geometria indivisibilibus continuorum nova quadam ratione promota*, Bologna, 1635; another edition, 1653.
2 *Exercitationes Geometricae sex*, Bologna, 1647.

CHISINI, O.:
1 *Aree, lunghezze e volumi nella geometria elementare* (in ENRIQUES, 7, part I, vol. II).
2 *Sulla teoria elementare degli isoperimetri* (in ENRIQUES, part III).
3 *Analysis situs* (Article in the *Enciclopedia Italiana Treccani*).
4 *Isoperimetri* (Article in the *Enciclopedia Italiana Treccani*).

CHURCH, A.:
1 *A bibliography of symbolic logic* ('The Journal of Symbolic Logic', Vol. I, 1936, pp. 121–218).
2 *Brief bibliography of symbolic logic* ('Proceedings of American Academy of Arts and Sciences'. vol. 80, n. 2, May 1952).

CICERO:
1 *De finibus bonorum et malorum*.
2 *De fato*.
3 *Academica Prima*.

CIPOLLA, M.: *Storia della Matematica dai primordi a Leibniz*, Mazara, 1949.

CLIFFORD, W. K.: *Lectures and Essays*. Vol. I, London, 1879.

COLERUS, E.: *Von Pythagoras bis Hilbert*, Vienna, 1937.

CONFORTO, F.: *L'opera scientifica di Bonaventura Cavalieri e di Evangelista Torricelli* ('Atti del Convegno di Pisa', 23–27 Sept. 1948, Città di Castello, 1949).

CONTI, A.: *Problemi di terzo grado: Duplicazione del cubo—trisezione dell'angolo* (in ENRIQUES, 7, part II).

COSSALI, P.: *Origine e trasporto in Italia dell'Algebra*. II, Parma, 1797.

COUTURAT, L.: *La logique de Leibniz*, Paris, 1901.

CROCE, B.:
1 *La logica come scienza del concetto puro*, Bari, 1909.
2 *La filosofia di Giambattista Vico*, Bari, 1911.
3 *Logic as the Science of Pure Concept*, London, 1917.
4 *Philosophy of Giambattista Vico*, London, 1913.

DEDEKIND, R.:
1 *Stetigkeit und irrationale Zahlen*, Braunschweig, 1872.
2 *Was sind und was sollen die Zahlen?* Braunschweig, 1887.
3 *Essenza e significato dei numeri. Continuità e numeri irrazionali.* (Translation from the German, with historical and critical notes by O. ZARISKI, Rome, 1926).

DE FINETTI, B.:
1 *Fondamenti logici del ragionamento probabilistico* ('Bollettino dell'Unione Matematica Italiana', Bologna, Dec. 1930).
2 *Sul significato soggettivo della probabilità* ('Fundamenta mathematicae', t. XVIII, pp. 298–329, 1931).

3 *Probabilismo, Saggio critico sulla teoria delle probabilità, e sul valore della scienza* ('Bibl. di Filosofia', Naples, 1931).

4 *Sull'impostazione assiomatica del calcolo delle probabilità*, Trieste, 1949.

5 *Macchine 'che pensano' e che fanno pensare* ('Tecnica e Organizzazione'. nn. 3–4, 1952).

DESCARTES, R.: *Œuvres publiées par Ch. Adam et P. Tannery*, Paris, 1894–1913.

DICKSON, L. E.: *History of the theory of numbers*, I. Washington, 1919.

DIELS, H.; *Die Fragmente der Vorsokratiker*, Berlin, 1906–1910, vol. I.

DIOGENES LAERTIUS: *De clarorum philosophorum vitis, dogmatibus et apophthegmatibus libri decem*, Paris, ed. Didot, 1878.

DIOPHANTUS of ALEXANDRIA:

1 *Opera omnia cum graecis commentariis*, ed. P. Tannery, Leipzig, 1893–95.

2 *Les six Livres arithmétiques et le livre des nombres polygones . . .* ed. P. Ver Eecke, Bruges, 1926.

DÜRR, K.: *The propositional Logic of Boethius*, Amsterdam 1951.

EINSTEIN, A.:

1 *Über die spezielle und die allgemeine Relativitätstheorie.* Braunschweig, 1917.

2 *Sulla teoria speciale e generale della relatività.* Translated by G. L. Calisse, Bologna, 1921.

3 *Relativity: the special and the general theory. A popular exposition.* Translated by R. W. Lawson, London, 1920: enlarged edition, New York, 1947; London, 1954.

ENRIQUES, A.: *Polemica anti matematica nell'antichità* ('Periodico di Matematiche', Bologna, 1921, pp. 63–66.)

ENRIQUES, F.:

1 *Conferenze di geometria non-euclidea*, Bologna, 1918.

2 *Per la storia della logica*, Bologna, 1922.

3 *Gli elementi di Euclide e la critica antica e moderna*, Bologna, 1925–30–32.

4 *La definizione come problema scientifico* ('Periodico di Matematiche', Bologna, 1927).

5 *Lezioni di geometria proiettiva*, Bologna, 1926.

6 *Problemi della scienza*, Bologna, 1920; another edition, Bologna, 1926.

7 *Questioni riguardanti le matematiche elementari, raccolte e coordinate da F. Enriques*, Bologna, 1924–1927.

8 *L'evoluzione delle idee geometriche nel pensiero greco* (in ENRIQUES, 7, part I, vol. I).

9 *I numeri reali* (in ENRIQUES, 7, part I, vol. I).

10 *Spazio e tempo davanti alla critica moderna* (in ENRIQUES, 7, part I, vol. II).

11 *Massimi e minimi nell'analisi moderna* (in ENRIQUES, 7, part III).

Bibliography

12 *Le matematiche nella storia e nella cultura*, Bologna, 1938.

13 *L'importanza della storia del pensiero scientifico nella cultura nazionale* (Lecture and discussion at a meeting of combined classes at the R. Accademia dei Lincei, 6 Feb. 1938).

14 *Il significato della storia del pensiero scientifico*, Bologna, 1938.

15 *Dimensioni* (Article in the *Enciclopedia Italiana Treccani*).

ENRIQUES, F. and CHISINI, O.: *Lezioni sulla teoria delle equazioni e delle funzioni algebriche*, Bologna, 1915–1934.

ENRIQUES, F. and MAZZIOTTI, M.: *Le dottrine di Democrito d'Abdera*, Bologna, 1948.

ENRIQUES, F. and DE SANTILLANA, G.:

1 *Storia del pensiero scientifico*, vol. I, Milan–Rome, 1932.

2 *Compendio di storia del pensiero scientifico*, Bologna, 1937.

EUCLID:

1 *Opera omnia*, ed. Heiberg, Leipzig, 1883–88 (see also ENRIQUES 3).

2 *The Thirteen Books of Euclid's Elements*, translated, with introduction and commentary, by T. L. HEATH, 3 vols., Cambridge, 1908.

EULER, L.: *Solutio problematis ad Geometriam situs pertinentis* ('Petropolitani Commentarii', VIII, 1736).

FAGGI, A.: *Parmenide di Elea e il concetto dell'essere* ('Atti Reale Acc. delle Scienze di Torino' 1931–32).

FANO, G.: *Geometria non-euclidea, introduzione geometrica alla teoria della relativitá*, Bologna, 1935.

FERMAT, P.: *Œuvres*, ed. P. Tannery and C. Henry, t. I, 1891.

FRAJESE, A.:

1 *La teoria dell'uguaglianza dei triedri nel suo sviluppo storico* ('Periodico di Matematiche', July 1934).

2 *Alle origini della geometria proiettiva* ('Bollettino dell'Unione Matematica Italiana', Bologna, June–July 1940).

3 *I dialoghi di Platone e la storia della matematica* ('Sophia', Padua, Jan.–March 1943).

4 *Attraverso la storia della matematica*, Rome, 1949.

5 *Storia della matematica e insegnamento medio* ('Bollettino dell' Unione Matematica Italiana', Dec. 1950).

6 *Osservazioni sulla teoria delle parallele in Euclide* ('Bollettino dell'Unione Matematica Italiana', Bologna, March, 1951).

7 *La geometria greca e la continuità* ('Archimede', Florence, May–Aug. 1951).

FREGE, G.:

1 *Die Grundlagen der Arithmetik*, Breslau, 1884.

2 *Aritmetica e logica*, edited by L. Geymonat, Turin, 1947.

GALILEI, G.:

1 *Le opere*, edizione nazionale.

2 *Pensieri, motti e sentenze, tratti dall'edizione nazionale per cura di A. Favaro*, Florence, 1935–36.

377

Mathematics and Logic in History and Contemporary Thought

GENOCCHI, A.: '*Annali di matematiche*', 1855.

GERGONNE, J. D.: *Considérations philosophiques sur les éléments de la science de l'étendue* ('Annales de Mathématiques', t. XVIII, Jan. 1826).

GEYMONAT, L.:
1 *Studi per un nuovo razionalismo*, Turin, 1945.
2 *Storia e filosofia dell'analisi infinitesimale*, Turin, 1947.
3 *La crisi della logica formale* (in ABBAGNANO ... PERSICO).
4 *I fondamenti dell'aritmetica secondo Peano e le obiezioni 'filosofiche' di B. Russell* (in LEVI, B. ... CARRUCCIO, E.).

GIGLI, D.:
1 *Numeri complessi a due e a più unità* (in ENRIQUES, 7, part I, vol. II).
2 *Combinatoria, Analisi* (Article in the *Enciclopedia Italiana Treccani*).

GLIOZZI, M.: *Galileo e la scienza dei suoi tempi* ('Sapere', 15 Dec. 1941).

GÖDEL, K.: *Über formal unentscheidbare Sätze der Principia Mathematica und verwandter Systeme I* ('Monatshefte für Mathematik und Physik'. Leipzig, 1931, pp. 173–98).

GOLDBECK: *Galilei Atomistik und ihre Quellen* ('Bibliotheca mathematica', edited by G. Enestrom, Leipzig, 1902, pp. 84–112).

GULDIN–P. GULDINI.: *Centrobaryca*, Vienna, 1641.

GUZZO, A.:
1 *Posizione e deduzione in Euclide* ('Università e Politecnico di Torino—Rendiconti del Seminario Matematico'. vol. 13, 1953–54).
2 *La Scienza*, Turin, 1955.

HEATH, T.:
1 *A History of Greek Mathematics*, London, 1921.
2 *Mathematics in Aristotle*, Oxford, 1949.

HERON of ALEXANDRIA: *Opera quae supersunt omnia*, ed. L. Nix and W. Schmidt, Leipzig, 1899–1904.

HILBERT, D. and ACKERMANN, W.:
1 *Grundzüge der theoretischen Logik*, Berlin, 1928, 1937, 1949, 1959.
2 *Principles of mathematical Logic*, New York–London, 1950.

KANT, I.:
1 *Kritik der reinen Vernunft*, Riga, I. ed. 1781, II. ed. 1787.
2 *Critique of Pure Reason*, tr. N. K. Smith, I. ed., London, 1929.

KLEIN, F.: *Vergleichende Betrachtungen über neuere geometrische Forschungen*, Erlangen, 1872.

KLÜGEL, G. S.: *Conatuum praecipuorum theoriam parallelarum demonstrandi recensio*, Göttingen, 1763.

KNEEBONE, G. T.: *Mathematical Logic and the Foundations of Mathematics*, London, 1963.

KÖTTER: *Die Entwickelung der synthetischen Geometrie* ('Jahresbericht ... Deutscher Mathematiker Vereinigung', Band V, Heft 2, Leipzig, 1901).

Bibliography

LADRIÈRE, J.: *Les limitations internes des formalismes*, Louvain–Paris, 1957.

LAGRANGE, G. L.: *Œuvres*, edited by M. A. Serret, Paris, 1867–92.

LAPORTE, J.:
1 *Le rationalisme de Descartes*, Paris, 1945.
2 *La connaisance de l'étendue chez Descartes* (in ADAM, CH. . . . WAHL, J.).

LASSVITZ, K.: *Die Geschichte der Atomistik vom Mittelalter bis Newton*, Hamburg and Leipzig, 1890.

LEIBNIZ, G. W.:
1 *Leibnizens mathematische Schriften*. Ed. C. J. Gerhardt, Berlin, Halle, 1849–63.
2 *Die Philosophischen Schriften*. Ed. C. J. Gerhardt, Berlin, 1875–90.
3 *Der Briefwechsel von G. W. Leibniz mit Mathematikern*. Ed. C. J. Gerhardt, Berlin, 1899.
4 *Opuscules et fragments inédits*. Ed. L. Couturat, Paris, 1903.
5 *Philosophical Writings*, tr. M. Morris, London, 1957.

LEONARDO PISANO: *Scritti*, ed. B. Boncompagni, Rome, 1857–62.

LERDA, F.: *Principi della Matematica intuizionistica* ('La nuova critica', n. IV, Florence, 1957).

LEVI, B.:
1 *Antinomie logiche?* ('Annali di Matematica' Ser. 3, vol. XV, 1908).
2 *Nota di logica matematica* ('Reale Istituto Lombardo di Scienze e lettere' Ser. II, vol. LXVI, Milan, 1933).
3 *La nozione di 'dominio deduttivo' e la sua importanza in taluni argomenti relativi ai fondamenti dell'analisi* ('Fundamenta Mathematicae', t. XXIII, Warsaw, 1934).
4 *Logica matematica* (Article in the *Enciclopedia Italiana Treccani*).
5 *Correria en la logica* ('Universidad nacional de Tucuman', rev., serie A, Matematicas y fisica teorica, vol. 3, June 1942, n. 1).
6 *A proposito de la nota del Dr. Pi Calleja sobre Paradojas lógicas y principio del Tertium non datur* ('Matematicae notae', a. IX fasc. 3–4).

LEVI, B., ASCOLI, G., SEGRE, B., BARONE, F., GEYMONAT, L., BOGGIO, T., CASSINA, U., CARRUCCIO, E.: *In memoria di Giuseppe Peano. Studi raccolti da A. Terracini*, Cuneo, 1955.

LEVI CIVITA, T. and AMALDI, U.: *Lezioni di Meccanica razionale*, vol. I, Bologna, 1923.

LISTING, J. B.:
1 *Vorstudien für Topologie*, Göttingen, 1848.
2 *Census räumlicher Complexe*, Göttingen, 1862.

LOBACHEVSKY, N. I.: *Geometrische Untersuchungen zür Theorie der Parallelinien*, Berlin, 1840.

Mathematics and Logic in History and Contemporary Thought

LOMBARDO RADICE, L.: *Ordinali transfiniti e principio del terzo escluso* ('Rendiconti di Matematica e delle sue applicazioni', Ser. V, vol. IX, fasc. 3–4, Rome, 1950).

LORIA, G.:
1 *Evangelista Torricelli e la prima rettificazione di una curva* ('Rendiconti Acc. Lincei', 1897).
2 *Le scienze esatte nell'antica Grecia*, Milan, 1914.

ŁUKASIEWICZ, J.: *Philosophische Bemerkungen zu mehrwertigen Systemen des Aussagenkalkuls* ('Comptes rendus des séances de la Societé des Sciences et Lettres de Varsovie' XXIII, a. 1930 fasc. 1–3).

MACH, E.: *La mécanique*, Paris, 1904.

MADDALENA, A.: *I Pitagorici*, Bari, 1954.

MANILIUS, M.: *Astronomica*, Leipzig, 1908.

MARCOLONGO, R.: *Lo sviluppo della meccanica sino ai discepoli di Galileo* ('Mem. R. Acc. Lincei', Ser. 5, vol. XIII, 1910).

MARITAIN, J.: *Eléments de philosophie*, II. *L'ordre des concepts*, I, *Petite logique*, Paris, 1946.

MENELAUS of ALEXANDRIA: (See BJÖRNBO).

MIELI, A.: *Il trecentenario dei 'Discorsi e dimostrazioni matematiche' di G. Galilei* ('Archeion' Sept. 1938).

MILHAUD, G.: *L'œuvre de Descartes pendant l'hiver 1619–1620* ('Scientia' 1918).

MONDOLFO, R.: *L'infinito nel pensiero dei Greci*, Florence, 1934.

MONTUCLA, J. F.: *Histoire des Mathématiques . . .*, Paris, 1789–1802.

MULLACH, F. W.: *Fragmenta philosophorum graecorum*, 3 vol., Paris, 1865–81.

NALLINO, C. A.:
1 *Avicenna* (Article in the *Enciclopedia Italiana Treccani*).
2 *Averroè* (Article in the *Enciclopedia Italiana Treccani*).

NATUCCI, A.: *Teorema di Descartes sui poliedri* ('Archimede', Nov.–Dec. 1949).

NEWTON, I.:
1 *Tractatus de quadratura curvarum*, London, 1704.
2 *Philosophiae naturalis Principia Mathematica*, London, 1687.

NOTARI, V.: *L'equazione di V° grado: teorema di Ruffini–Abel* (in ENRIQUES, 7, part II).

OCCAM, WILLIAM of: *The Tractatus de praedestinatione et de praescientia Dei et futuris contingentibus, of W. Ockam*, ed. with a study on the Medieval Problem of a Three-valued Logic, by Ph. Boehner O.F.M., New York, 1945.

OMAR KHAYYAM: *Algebra*, translated and edited by F. Woepke, Paris, 1851.

ORESME, N.: *Questiones super geometriam Euclidis*, ed. by H. L. L. Busard, Leiden, 1961.

Bibliography

PADOA, A.:
1 *La logique deductive*, Paris, 1912.
2 *Massimi e minimi delle funzioni algebriche* (in ENRIQUES, 7, part III).

PAPPUS of ALEXANDRIA:
1 *Pappi Alexandrini, Collectionis quae supersunt*, ed. F. Hultsch, Berlin, 1876.
2 *Pappus d'Alexandrie, La collection mathématique*, introduction, translation by P. VER EECKE (2 vols.), Paris-Bruges, 1933.

PASCAL, B.: *Pensées*, Paris, 1836.

PASCAL, E.: *Esercizi critici di calcolo differenziale*, Milan, 1921.

PASTORE, A.: *La logica del potenziamento*, Naples, 1936.

PEANO, G.:
1 *La numerazione binaria applicata alla stenografia* ('Atti della R. Acc. delle Scienze di Torino', Vol. XXXIV, disp. I, 1898–99, Turin, 1898, pp. 47–49).
2 *Introduction au tome II du 'Formulaire' de Mathématique*, t. V ('Revue de Mathématique', t. VI, Turin, 1896–99).
3 *Formulaire de Mathématique*, Turin, 1894, 1895, 1897, 1899, 1903.
4 *Aritmetica generale e algebra elementare*, Turin, 1902.
5 *Formulario mathematico*, Turin, 1908.
6 *Super Theorema de Cantor-Bernstein* ('Revista de Mathematica', t. VIII, Turin, 1902–06).
7 *Le definizioni in matematica* ('Arxivs de L'Institut de ciences', a. 1, n. 1, Barcelona, 1911; other edition, 'Periodico di Matematiche', Bologna, May 1921).
8 *Opere scelte*, a cura dell'Unione Matematica Italiana, Rome, 1957–59.
9 *Formulario mathematico*, Rome, 1960.

PEET, T. E.: *The Rhind Mathematical papyrus, British Museum 10057 and 10058*. Introduction, transcription, translation and commentary by T. Eric Peet, Liverpool, 1923.

PETRUS HISPANUS: *Summulae logicales*, ed. I. M. Bocheński O.P., Turin, 1947.

PICCOLI, G.: *Definizioni proprie e improprie nelle matematiche* ('Archimede', Florence, July–Aug. 1950).

PIERI, M.: *Uno sguardo al nuovo indirizzo logico-matematico delle scienze deduttive*, Catania, 1907.

PLATO:
1 *Theaetetus*.
2 *Phaedo*.
3 *Euthydemus*.
4 *Meno*.
5 *Sophist*.
6 *Timaeus*.
7 *Statesman*.
8 *Republic*.
9 *Laws*.
10 *Il Fedone con note e introduzione di* M. VALGIMIGLI, Palermo, 1921.

PLEBE, A.: *Retorica aristotelica e logica stoica* ('Filosofia' Turin, a. X fasc. III, July 1959).

PLINY: *Naturalis historiae libri XXXVII.*

PLUTARCH: *The banquet of seven Wise Men.*

POINCARÉ, H.:
  1 *La science et l'hypothèse*, Paris, 1908.
  2 *La valeur de la science*, Paris.
  3 *Science et Méthode*, Paris, 1909.
  4 *Dernières pensées*, Paris, 1913.

POLITANO, M. L.: *Sull' 'Analysis situs' di Leibniz* ('Archimede', July–Oct. 1957).

PROCLUS DIADOCHUS: *In primum Euclidis Elementorum librum commentarii* (ed. G. Friedlein), Leipzig, 1873.

PTOLEMY (CLAUDIUS PTOLEMAEUS): *Opera quae extant omnia*, ed. Heiberg, Leipzig, 1898–1907.

REICHENBACH, M.:
  1 *Über erkenntnistheoretischen Problemlage und den Gebrauch einer dreiwertigen Logik in der Quantenmechanik* ('Zeit Naturforschung', 1951, pp. 569–75).
  2 *I fondamenti filosofici della meccanica quantistica*, Turin, 1954.

RONCHI, V.: *Storia della luce*, Bologna, 1939.

ROSSI-LANDI, F.: *Il pensiero americano contemporaneo. Filosofia Epistemologia Logica.* Editor: F. ROSSI-LANDI; contributors: F. BARONE, L. BORGHI, G. DORFLES, P. ROSSI, F. ROSSI-LANDI, U. SCARPELLI, V. SOMENZI, G. VACCARINO, A. VISALBERGHI, Milan, 1958.

RUFFINI, P.: *Opere matematiche*, edited by E. Bortolotti, Rome, 1953–54.

RUFINI, E.:
  1 *La preistoria delle parallele e il postulato di Euclide* ('Periodico di Matematiche', Bologna, Jan. 1923).
  2 *Il 'Metodo' di Archimede e le origini dell'analisi infinitesimale nell'Antichità*, Bologna, 1926; Milan, 1961.

RUSSELL, B.:
  1 *Principles of Mathematics*, Cambridge, 1903.
  2 *An inquiry into Meaning and Truth*, London, 3rd ed., 1948.

RUSTOW, A.: *Der Lügner*, Leipzig, 1910.

SABBATINI, A.: *Sui metodi elementari per la risoluzione dei problemi geometrici* (in ENRIQUES, 7, part II).

SACCHERI, G. G.: *Euclides ab omni naevo vindicatus*, reprinted with an English translation by G. B. Halsted, Chicago and London, 1920.

SCARPIS, E.: *Sui numeri primi e sui problemi dell'analisi indeterminata* (in ENRIQUES, 7, part III).

SCHIAPARELLI, G.: *Scritti sulla storia dell'astronomia antica*, p. I, t. I, Bologna, 1925.

Bibliography

SCHOY, C.: *Graeco-arabische Studien* ('Isis', Vol. VIII, Brussels, 1926).

SCHRECKER, P.: *La méthode cartésienne* (in ADAM, CH. . . . WAHL. J.).

SEVERI, F.:
1 *Lezioni di analisi algebrica*, Bologna, 1933.
2 *I fondamenti remoti e prossimi della geometria algebrica* ('Università e Politecnico di Torino. Rendiconti del Seminario Matematico', 1950–51).
3 *Intuizionismo e astrattismo nella matematica contemporanea* ('Atti del terzo congresso dell'Unione Matematica Italiana, 1948', Rome, 1951).
4 *La crisi del pensiero moderno* ('Nuova Antologia', Rome, March 1951).
5 *Leonardo e la matematica* (Studies on Leonardo da Vinci, scientist and philosopher, taken from 'Scientia', 1952–53).

SIERPINSKI, W.: *Leçons sur les nombres transfinis*, Paris, 1928.

STEINER: *Sur le maximum et le minimum des figures dans le plan, sur la sphère et dans l'espace en général*, 'Journal de Crelle', 1842.

TERRACINI, A.: *Iperspazio* (Article in the *Enciclopedia Italiana Treccani*).

THEODOSIUS of TRIPOLI:
1 *Sphaericorum Elementorum libri tres*, ed. Heiberg, Leipzig, 1928.
2 *Les sphériques de Théodose de Tripoli. Œuvres traduites avec une introduction et des notes, par* P. VER EECKE, Bruges, 1927.

THOMAS AQUINAS, St.:
1 *Divi Thomae Aquinatis Doctoris Angelici, Quaestiones quodlibetales.*
2 *In Aristotelis libros perì Hermeneias et posteriorum analyticorum expositio*, Ed. Marietti, Turin, 1955.
3 *Summa theologica.* (English edition, London, 1957.)

TIMPANARO CARDINI, M.: *I sofisti, frammenti e testimonianze*, translation, preface and notes by M. Timpanaro Cardini, Bari, 1954.

TONELLI, L.:
1 *Fondamenti di calcolo delle variazioni*, vol. II, Bologna, 1921 and 1923.
2 *Integrale, Calcolo* (Article in the *Enciclopedia Italiana Treccani*).

TORRICELLI, E.:
1 *Opera geometrica*, Florence, 1644.
2 *Opere*, Faenza, 1919–44.
3 *De infinitis spiralibus*, introduction, rearrangement and revision of the original manuscript, translation and commentary by E. CARRUCCIO, Pisa, 'Domus Galilaeana', 1955.

TRENDELENBURG, F. A.: *Elementa Logices Aristoteleae in usum scholarum ex Aristotele excerpsit convertit illustravit F. A. Trendelenburg*, Berlin, 1868.

VACCA, G.:
1 *Sui precursori della Logica Matematica* ('Revue de Mathématiques', VI, Turin, 1896–99, pp. 121–5 and 183–6).

2 *Sui manoscritti inediti di Leibniz* ('Bollettino di bibliografia e storia delle scienze matematiche', 1899).
3 *Sulla logica simbolica* ('Leonardo', a. IV, s. III, pp. 366–8, Florence, 1906).
4 *La scienza nell'estremo oriente* ('Scientia', vol. XI, 1912).
5 *Nepero e l'opera sua* ('Seminario matematico della'Università di Roma', 12 Dec. 1914).
6 *Sul commento di Leonardo Pisano al libro X degli Elementi di Euclide e sulla risoluzione delle equazioni cubiche* ('Bollettino dell'Unione Matematica Italiana', April 1930).
7 *Sul concetto di probabilità presso i Greci* ('Giornale dell'Istituto Italiano degli Attuari', a. VII, n. 3, July 1936).
8. *L'opera matematica di Gerolamo Cardano nel quarto centenario del suo insegnamento in Milano* ('Rendiconti del Seminario Matematico e Fisico di Milano', vol. XI, 1937).
9 *Binomio* (Article in the *Enciclopedia Italiana Treccani*).
10 *Descartes (parte matematica)* (Article in the *Enciclopedia Italiana Treccani*).
11 *Torricelli Evangelista* (Article in the *Enciclopedia Italiana Treccani*).
12 *Origini della scienza*, Rome, 1946.
13 *La matematica dei Romani* ('Studi Romani', May-June 1954; lecture given in 1939, Rome).

VACCARINO, G.:
1 *La scuola polacca di logica* ('Sigma', 8–9, Rome, 1946).
2 *L'implicazione stretta e la logica delle modalità*, in ROSSI-LANDI.

VAILATI, G.:
1 *Scritti*, Leipzig–Florence, 1911.
2 *Sulla teoria delle proporzioni* (in ENRIQUES, 7, part I, vol. I).

VER EECKE, P.: (See APOLLONIUS, 2; DIOPHANTUS, 2; PAPPUS, 2).

VICO, G. B.:
1 *Opere*, edited by G. Gentile and F. Nicolini, Bari, 1914.
2 *Opere*, edited by F. Nicolini, Milan–Naples, 1953.

VIOLA, T.: *Sulle origini della prospettiva* ('Il Filomate', Milan, July–Aug. 1948).

VITALI, G.: *Sulle applicazioni del postulato della continuità nella geometria elementare* (in ENRIQUES, 7, part I, Vol. I).

VITALI, G. and SANSONE, G.: *Moderna teoria delle funzioni di variabile reale*, p. I, Bologna, 1935.

VITRUVIUS: *De Architectura libri X*.

VIVANTI, G.: *Il concetto d'infinitesimo*, Mantua, 1894.

VOLTERRA, V.: *Saggi scientifici*, Bologna, 1920.

WAISMANN:
1 *Einführung in das mathematische Denken*, Vienna, 1936.
2 *Introduzione al pensiero matematico*, Turin, 1939.

Bibliography

WANTZEL, P. L.: *Recherches sur les moyens de reconnaître si un problème de géométrie peut se résoudre par la règle et le compas* ('Journal de Mathématiques', t. II; 1837).

WESSEL, C.: *Essai sur la représentation de la direction*, Copenhagen, 1897.

WITTGENSTEIN, L.:

1 *Tractatus logico-philosophicus*, London, 1922.

2 *Tractatus logico-philosophicus*, testo originale, versione italiana a fronte, introduzione critica e note a cura di G. C. M. COLOMBO, S. J., Milan–Rome, 1954.

WHITEHEAD, A. N. and RUSSELL, B.: *Principia mathematica*, Cambridge, 1910.

WHITESITT, J. E.: *Boolean Algebra and its Applications*, Addison–Wesley Publ. Co., 1961.

WINDELBAND, G.: *Platone*, Palermo–Genoa.

ZAPPELLONI, M. T.: *Il Teorema di Tolomeo e le formule di addizione delle funzioni circolari* ('Periodico di Matematiche', Bologna. 1928).

ZARISKI, O.:

1 *Gli irrazionali nella matematica greca* (in DEDEKIND, *3*).

2 *Continuità e numeri irrazionali* (in DEDEKIND, *3*).

3 *Sul principio d'induzione completa* (in DEDEKIND, *3*).

ZEUTHEN, H. G.:

1 *Die geometrische Construction als Existenzbeweis in der antiken Geometrie* ('Math. Ann.', Band. 47, 1896).

2 *Théorème de Pythagore: origine de la géométrie rationnelle* ('Rendiconti del II Congresso internazionale di Filosofia', Geneva, 1904).

INDEX

Abacus, Babylonian, 14
Abbagnano, N., 357, 359
Abel, H. N., 176, 238
Abelard, 163
Abstract geometry, 249–50
Achilles and the Tortoise (paradox), 33–4
Ackermann, 350; see Hilbert and Ackermann
Acta eruditorum (Leipzig), 230
Adam, C., 173
Aganis, 252
Agatharcus, on stage set, 36
Agostini, A., 214, 216, 219
Ahmes, the scribe (Rhind papyrus), 16
Albert, artist and scientist, 168
Alberti, L. B., 239
Albertus Magnus, 93
Alcuin: founds Palatine Academy on Charlemagne's orders, 158; *Propositiones ad acuendos iuvenes*, 158–9
d'Alembert, 175, 237; on theory of parallels, 258
Alexander the Great, 76
Alexandrine age, last geometric studies of, 122–3
Algebra: Babylonian, 14; algebraists of Italian Renaissance, 168–72; later algebraists, 172–177; Descartes's contribution to algebraic symbolism, 172–4; algebra as Arabic development, 158; algebraic geometry, 248–50
Al Khuwarizmi, Arab mathematician, 158
Amadeo, F., 239
Amaieutic method (Socrates), 367
Amaldi, 82
Ameinia of Croton, 29
Ammonius Sacca, founder of Neo-Platonism, 127
Analysis: combinatorial, 286 seqq.; modern infinitesimal, 192–238
Analysis situs, 247

Analytic functions, Fantappiè's work on, 237
Analytic geometry, origins of, 178–91; forerunners of, 178; aspects of Descartes', 182–7; Descartes' contribution to, 179–89; Fermat's, 178–9
Anaritius, Arab mathematician, 252
Anaxagoras, 34, 35–6, 105, 200
Anaximandrus, 23
Angles: internal, of triangle, 91–2; sum of these, 259–63; of contingency (Democritus), 42, (Nemorario), 96; (Newton), 313; hypothesis of obtuse, 272
Anselm, St., of Aosta, 163
Antinomies, logicians' classes of, 344–50
Antiphon of Athens, 38
Apollodorus, 22
Apollonius (and his followers), 117, 119–23; demonstrates axioms, 85, 186; work of, generally, 117, 118, 131; in Arabic, 158
Apuleius, 125
Aquinas, St. Thomas, 163, 164, 202; and the mathematical infinite, 166; *In libros peri Hermeneias*, 164; defines logic, 161
Arabic mathematics, 156–8
Arcerian Codex, 144
Arcesilaus (Sceptic), 79
Archimedes, 20, 107–16, 131, 154, 311; on Democritus, 41; and the method of exhaustion, 107–11; mechanical method of, 111–15; studies in cyclometry, 59; attitude of, to 'non-Archimedean' magnitudes, 98–9; knowledge of area of ellipse, 205; his *Method*, 195; his *Sand-Reckoner*, 115, 144; his Treatise on Spirals, 115; infinitesimal methods of, 103 seqq.; continuers of work of, 192–3;

387

Index

389

Index

Infinitesimal: methods: classical, 103–16; modern analysis, 192–238; geometrical ideas, Anaxagoras', 35; influence of Torricelli, 218; later developments, 233–7; criticism of principles of, 237–8; infinitesimal calculus; see Leibniz; Newton's development of, 233, 234; joint work of Leibniz and Newton, 225–33; Descartes and, 220–5

Integral, calculation of, the, 206

Intuitional mathematics, 353–6; see Brouwer

Ionic School of Thales, 23

Irreducible case, imaginary numbers, 174–5

Isidore, Bishop of Seville, 154, 158

Isidore the Elder, of Miletus (architect), 154

Isidore the Younger; 154

Isoperimetrical: theory, ancient, 131; isoperimetric property of the circle, 139–40

Jacobi, 236

Jevons, W. S., 316

John XXI, Pope, see Petrus Hispanus

Jordan, C., 238, 306, 308

Jordan Nemorarius, *Arithmetica*, 166

Journal of Symbolic Logic (U.S.A.), 339

Judgments and rules of conversion, 65–7

Jürgens, 308

Justinian, closes Plato's Athenian School, 154, 158

Kant, E., 75, 248, 351; on three dimensions of space, 248; cf. Newton and Galileo, in physics, 284

Kepler, gives impulse to infinitesimal analysis, 196, 204; in projective geometry, 240

Khayyam, Omar, 158, 286; *Algebra*, 286

Klein, F., 148, 247, 250, 278, 298

Klügel, G. S., 257

Koch, H. von, 309

Koenig, 301

Koenigsberg's seven bridges (problem), 247–8

Kötter, 244

Ladrière, J., 325

Lagrange, G. L., 176, 236, 249, 242; interpretation of mechanics, 248; memoir on parallels, 258; spherical trigonometry, 242

Laguerre, E., 246–7

Lambert, J. H., 256–7, 259, 265, 269, 316

La Place, S., *Théorie Analytique des Probabilités*, 287; *Essai Philosophique des Probabilités*, 287

Laporte, J., 180

Lassvitz, K., 200

Lebesgue, 237

Legendre, A. M., 259; on parallels, 258. *See* Saccheri

Leibniz, G. W., 18, 64, 181, 225, 230, 231, 247, 286, 306, 324, 340; *characteristica* of, 315–16; 365; as heir of Plato, 47; on Cartesian geometry, 181; on imaginary numbers, 174–5; combining with Descartes, 187–8; on Torricelli, 218; his first work on differential calculus, 230–3; on the void, 283; on order in space, 301; infinitesimal analysis of, 225, 230–3; principle of indiscernibles, 347

Leodamas, 53

Lerda, F., 354

Leucippus, 34, 39

Levi, B., 302, 316, 339, 348–9, 350, 360, 365

Levi-Civita, T., 273, 310, 313

Levy-Bruhl, or non-Aristotelean logics, 353

L' Huilier, Simon, 237

Light rays, 129–31

Lindemann, on transcendence of π, 39

Lionville, *Comptes Rendus*, 298

Listing, J. B., 248

Lobachevsky, N., 153, 269–72, 273, 277, 280–3, 284; *Pangeometry*, 188–9, 252, 255; and Gauss,

393

Index

Viète, S., 174

da Vinci, Leonardo, as artist and scientist, 167, 168, 196, 239

Viola, T., 239

Virgil, qu., 124

Vitali, G., 246, 252, 310; and Sansone, 292

Vitruvius, 36, 41, 124; *Architecture*, 124

Vivanti, G., 231

Void: affirmation of, 35; with mobile atoms, 35, 39, 283; Democritus' belief in, 283; Eleatics deny, 283; in philosophy and science, 283 seqq.; Descartes' position, 283

Volterra, V., 235, 237

Wachter, L., 258

Waismann, 81, 149, 300, 301, 306, 325–6, 357, 358, 360, 365

Wallis, J., 174, 219; on history of infinitesimal calculus, 252–4; *Institutio logicae*, 254

Wantzel, P. L., 122

Weierstrass, K., 236, 238, 309

Wessel, C., 175

Western mathematics, late mediaeval, 158–9

Weyl, H., 237–8

Whitehead and Russell, *Principia Mathematica*, 327, 343, 344–5, 347, 365

Whitesitt, J. E., 340

Windelband, G., 44

Wittgenstein, L., *Tractatus Logico-Philosophicus*, 359

World: finite, but not limited, 285

Xenophanes of Colophon, 29

Xenophon: and Plato's work, 42

Zappelloni, 385

Zara (game of chance, calculation of probability basis), 287

Zariski, O., 192

Zeno of Citium (Stoic), 78; *Zenon* (Diels), 32–4, 93

Zenodorus, theorem of, 136–9; in problems of maxima and minima, 132; on curves, 236

Zermelo's axiom and postulate, 294, 302, 305

Zeuthen, H. G., 25, 83

Milton Keynes UK
Ingram Content Group UK Ltd.
UKHW020014071024
449327UK00031B/2787